MASTERING ARTISAN CHEESEMAKING

PRAISE FOR *MASTERING ARTISAN CHEESEMAKING*

"I am truly knocked out by this wonderful book. *Mastering Artisan Cheesemaking* is simply superb, and well-nigh flawless. Caldwell's voice comes through so clear, friendly, and free of clinicality. I have learned so much that I didn't realize I didn't know. In fact, the book is so good I'm a little embarrassed that I didn't even know how badly I needed it, and feel I am a vastly improved master cheesemonger for having been blessed with reading it."

—**STEVEN JENKINS**, author of *Cheese Primer*

"*Mastering Artisan Cheesemaking* is the one book that tells you everything you need to know to become an award-winning cheesemaker. Caldwell's practical and straightforward explanations make this entire book (along with its amusing anecdotes) a great pleasure to read. She makes it clear that artisan cheesemaking is a serious endeavor; but her light-hearted approach will remove the fear factor and inspire you to make a go of it anyway. And for everyone who just loves to eat fine artisan cheeses, this book will elevate your respect for all that goes into their creation. Among the recent bounty of books on cheese, this one is a must-have."

—**MAX MCCALMAN**, author of *Mastering Cheese*

"Easy to understand and fun to read, *Mastering Artisan Cheesemaking* provides valuable information on every aspect of cheesemaking—an excellent resource for both cheese makers and cheese lovers."

—**SALLY FALLON MORELL**, president, The Weston A. Price Foundation, and cheesemaker

"I am in awe! For anyone who has ever written a book or sweated over a cheese vat, know that both require artistry, focus and discipline, dedication, time and more time, and perhaps a bit of creative insanity. In her new book, Gianaclis Caldwell argues eloquently for the strong relationship between art and science, for deeper understanding and appreciation of milk, and the chemistry and alchemy of a cheese vat. As the landscape of American artisan cheese heads toward 1,000 small-scale producers, she offers new insights, essential knowledge, and encouragement to experiment and succeed."

—**JEFFREY P. ROBERTS**, author of *The Atlas of American Artisan Cheese*

"Gianaclis Caldwell has combined her extensive knowledge as an award-winning cheesemaker with complementary information from a variety of sources to create a practical guide that will delight both aspiring and seasoned home-scale cheesemakers. *Mastering Artisan Cheesemaking* will be a valuable addition to any home cheesemaker's library."

—**PAUL KINDSTEDT**, author of *Cheese and Culture* and *American Farmstead Cheese*

"*Mastering Artisan Cheesemaking* is a beautiful book, rich in theory *and* practice. Whether you want to deepen your appreciation and understanding for cheese, produce for your family, or bring artisanal cheeses to market, you won't be disappointed."

—**SHANNON HAYES**, author of *Radical Homemakers* and *Long Way on a Little*

"If only *Mastering Artisan Cheesemaking* had been available a decade or so ago, my cheesemaking adventures would have been a lot less stressful. Already I see this as the new go-to book for my workshop attendees. Gianaclis Caldwell presents her method of making cheese as an intuitive process, and broaches the truly formidable task of the science of cheesemaking in an understandable way that only someone who has learned in the classroom and worked in the cheese room can. The section on surface-ripened cheese is especially good since these cheeses are more difficult to ripen. Go out and get this book, if you haven't already."

—**JIM WALLACE**, New England Cheesemaking Supply Company

"Gianaclis Caldwell, a farmstead cheesemaker of famously delicious cheeses, has—amazingly—revealed all her hard-earned secrets in this wonderful new book. Reading *Mastering Artisan Cheesemaking* is like taking an advanced cheesemaking class—not only do we learn how to make cheese at home, we learn the science behind the process, from flocculation to affinage. A must-read for anyone who wants to make cheeses at home."

—**NOVELLA CARPENTER**, author of *Farm City: The Education of an Urban Farmer*

"I have been waiting for a book like *Mastering Artisan Cheesemaking* for years. Gianaclis Caldwell's detailed, thorough, and accurate guide is a godsend for both home cheesemakers just starting out and experienced professionals. Gianaclis has answered all of my tough questions in this remarkable tome; my only regret is that I needed this guide years ago when I was just starting out."

—**KURT TIMMERMEISTER**, author of *Growing a Farmer* and commercial cheesemaker

"*Mastering Artisan Cheesemaking* is a must-have book for the aspiring cheesemaker or cheesemonger. Not only is it amazingly easy to use as a reference book—laying out cheese science in as clear and jargon-free a way as possible—it's also a fun time for the cheese-obsessed reader. Why does that rind look like that? How does that cheese get that amazing flavor? I'll admit, this book helped clear up some misperceptions I have held for twenty years about the way certain cheeses are made. I've read this once, but I know I will be referring to it over and over again."

—**GORDON EDGAR**, author of *Cheesemonger: A Life on the Wedge*

"Gianaclis Caldwell has poured her generous mind and heart into this book and reveals the secret life of one of humanity's most delicious foods. *Mastering Artisan Cheesemaking* is a wealth of in-depth information won by first-hand experience, yet it's friendly and reassuring, and skillfully unpacks the science and craft of cheesemaking for the interested hobbyist and the artisan alike. After reading it, I'm more fascinated by cheese than ever! My next project may well be a few little bloomy rinds. . . ."

—**MARGO TRUE**, food editor, *Sunset* magazine

MASTERING ARTISAN CHEESEMAKING

The Ultimate Guide for the Home-Scale and Market Producer

GIANACLIS CALDWELL

Foreword by Ricki Carroll

Chelsea Green Publishing
White River Junction, Vermont

Project Manager: Patricia Stone
Editor: Makenna Goodman
Copy Editor: Eileen M. Clawson
Proofreader: Eric Raetz
Indexer: Linda Hallinger
Designer: Melissa Jacobson

Printed in the United States of America
First printing August, 2012
10 9 24

Our Commitment to Green Publishing

Chelsea Green sees publishing as a tool for cultural change and ecological stewardship. We strive to align our book manufacturing practices with our editorial mission and to reduce the impact of our business enterprise in the environment. We print our books and catalogs on chlorine-free recycled paper, using vegetable-based inks whenever possible. This book may cost slightly more because it was printed on paper from responsibly managed forests, and we hope you'll agree that it's worth it. *Mastering Artisan Cheesemaking* was printed on paper supplied by Versa Press that is certified by the Forest Stewardship Council®.

Library of Congress Cataloging-in-Publication Data
Caldwell, Gianaclis, 1961–
Mastering artisan cheesemaking : the ultimate guide for home-scale and market producers / Gianaclis Caldwell ; foreword by Ricki Carroll, author of Home cheese making.
pages cm
ISBN 978-1-60358-332-9 (pbk.)—ISBN 978-1-60358-333-6 (ebook)
1. Cheesemaking. I. Title.

SF271.C354 2012
637'.3—dc23

2012023148

Chelsea Green Publishing
85 North Main Street, Suite 120
White River Junction, VT 05001
(802) 295-6300
www.chelseagreen.com

CONTENTS

LIST OF TABLES

FOREWORD

Just like Gianaclis, Robert and I started with goats in 1975, after moving from the suburbs to the lush country hills of Western Massachusetts. It wasn't long before our beautiful does gave more milk than we could drink, and we were off and running, making butter, ice cream, yogurt, kefir, and, of course, many types of cheese. This was at a time very early in the game when there was little information or help here in the United States for aspiring home cheesemakers. We wrote letters around the world, looking for experienced mentors, and experimented with the meager ingredients we could scrounge. We made a lot of feta because the brine would keep the mold at bay.

Soon afterward, while in England, we learned consistency and some variety, and upon our return to the States, we started sharing our newly gained knowledge with others in the same predicament. But there was nowhere to get supplies, ingredients, or equipment, so it was up to us. In 1978 New England Cheesemaking Supply Company was born. We brought in and sold materials that were not available on a small scale and started teaching workshops in our home. At that time we had fifteen wonderful goats and a beautiful Jersey cow in our backyard, and we taught classes that were quite technical in nature. In 1980 we wrote *Cheese Making Made Easy* (now called *Home Cheese Making*) and took our classes on the road; thus arose a subculture of cheese artisans around the country. We helped birth the artisanal cheese movement and for the last thirty-three years have watched it evolve and grow into the tremendous revolution that it has become today.

I often find myself in tears, with awe, gratitude, and inspiration as I watch the multitude of artisanal cheesemakers working to create their unique masterpieces, knowing how much effort goes into it and what a labor of love it is for this ancient art.

With the onset of the Internet, our classes changed, and people living in the cities became more interested in the DIY movement, so I quickly realized that something simpler was needed. These people did not have access to backyard animals, so I started teaching a more basic class, high in energy and filled with simple cheeses, all with milk bought at the grocery store. (Of course, I had to do some side-by-sides with raw milk for comparison to show some glorious possibilities.)

With the explosion of this art going mainstream and the number of people who wanted to go to the next level, we have gone back to teaching both beginning and more advanced technical classes. I teach the beginners with Jamie's help, and Jim Wallace teaches our more technical classes. And now, Gianaclis Caldwell has arrived as another expert teacher not to be missed.

Mastering Artisan Cheesemaking is a sensational way to get to that next step in the evolution of your home or artisanal cheesemaking. Gianaclis seamlessly brings the science to the art and offers technical support and information on an advanced scientific level. It warms my heart that we can now reach these standards of practice and go from the simple to the sublime, from the quick and easy fresh cheeses to the more intricate aged and deliciously mold-ripened ones.

Gianaclis offers us a complete understanding of why and how the processes that we may have taken for granted happen, thus making it possible for our cheeses to reach a higher level of perfection on the dining room table. The knowledge Gianaclis offers comes at a time when cheesemaking has taken off in leaps and bounds and has taken its rightful place in the kitchen next to home bread, beer, and wine making. It has also come at a time when artisanal cheesemakers are growing in numbers and others are waiting in the wings for their chance at new careers.

When we started I had no idea that our little niche would give way to such an explosion. I am humbled by the breadth and scope of this book and the detailed way in which the knowledge is presented.

As Gianaclis says in the book, "The more I learn about cheesemaking, the more I realize that so much is really in the hands of the cheesemaker—both the interpretation of the results and the implementation of control measures." This is so true and is something we have tried to convey as long as we have been in business. As the reader and cheesemaker, you will be trying to make the same cheese consistently, but one day it will come out with a slight variation, and then: the "aha!" moment, and you love it. Just make sure to take scrupulous notes, and you will know what you did differently. Also, don't forget to refer to the phenomenally detailed information Gianaclis shares here to find answers and be able to go forward creating what will become your unique cheese.

This book is packed with information, detailed techniques, controls, and more. The testing, the aging, precautions to take, troubleshooting, flavoring, equipment usage, sanitization, and an infinite amount more all make this book a literal *must* for those who are making or thinking about making cheese to complete their current libraries.

The extraordinary recipe section in this book offers a starting place for you to have an exceptional cheesemaking experience, and through your adaptation of them and adjusting them to your own *terroir*, you will be pleased beyond your wildest dreams. As Gianaclis puts it, "Start trusting your instincts . . . be ready in your cheesemaking for some surprises . . . it might be the best cheese you have ever created!"

Frank Kosikowski's dream at the first meeting of the American Cheese Society was of cheesemakers helping cheesemakers. And as Steve Jenkins once said decades ago: education, education, education. I can truly say that I have now lived to see it all. As we take the next step in the United States to improve skills, increase knowledge, and open ourselves to endless possibilities, we will continue to grow and create unique masterpieces, gain a deeper knowledge of the process, and come out all the better for it.

Remember that cheese is a way to store milk, farming is a way to restore our land, and helping each other is a way to restore our souls.

To Gianaclis: Congratulations on a job well done; your dedication and valuable help in the field, your attention to details, your enthusiasm, and your phenomenal amount of knowledge serve the community well. Thank you for opening new doors so we can collectively delve deeper into this magical, rewarding, ancient art of cheesemaking.

To the reader: Enjoy, create, and evolve—while I go and make some cheese.

In peace,
Ricki Carroll
a.k.a. the Cheese Queen

ACKNOWLEDGMENTS

One of the main reasons I wanted to write this book was to learn. It is a selfish reason, but the best one to haul you through the long hours of drudgery that is really the core of what is entailed in writing a book. I could not have learned what I did and organized it for others who also want to learn about this same topic without an amazing group of supporters and resources.

It goes without saying that the first thanks will always go to Vern (husband) and Amelia (at-home daughter), who not only didn't try to dissuade me from writing a second book but took on much of the workload of the farm and creamery so that I could plant myself for long hours on my balance-ball chair in front of the computer, go on research trips, or make experimental batches of cheese, instead of the mainstay cheeses for which Pholia Farm is known. In fact, Vern and Amelia both started marketing my successful experiments at the farmers' market as "book cheese." I would also like to thank the multitude of WWOOFers (interns who came to our farm through World-Wide Opportunities on Organic Farms) for their amazing help with the animals and cheese; stimulating curiosity and questions; and support and friendship.

I am so grateful for the invaluable input of my corps of diverse readers—from people far beyond myself in knowledge to those just beginning to make cheese. These wonderful, honest people gave invaluable input that helped make this book accurate and complete. So thank you, people!

Technical Experts: Dr. Lisbeth Goddik, Oregon State University; Jim Wallace, New England Cheese Supply; Dr. Mark Wustenberg, Tillamook County Creamey; Dr. Steve Zeng, Langston University

Cheesemakers: Pablo Battro, David Bleckmann, Cary Bryant, Vern Caldwell, Rhonda Gothberg, David Gremmels, Ann Hansen, Christy Harris, Karin Harris, Sam Koster, Meghan McKenna, Carly Payne, Jos Voltu

Culture Expert: Dave Potter, Dairy Connection

Cheeses of the World Expert: Max McCalman, coauthor of *Mastering Cheese*

Cheese Queen and Pioneer: Ricki Carroll

Another motivation for writing a second book was the chance to work with the people at Chelsea Green Publishing again. What an honest, dedicated, and true-to-the-cause group of people—and company. Special thanks to my editor for her great eye, for her broad sweeping viewpoint, and for letting me take this book to its full conclusion.

Just a few months before this book was printed, the cheese world lost a major player and a person I admired and was lucky enough to have gotten to spend a little time with, Daphne Zepos. While my personal time with Daphne consisted of only a few passing greetings and a memorable dinner when she took me out in San Francisco, her impact on my life and my own cheese career came home to me upon her death at the untimely age of 52. From the moment I first saw her speak at an American Cheese Society Conference in 2005 to the time when I booked her as a cheese judge for the American Dairy Goat Association cheese competition, Daphne was a role model for sophistication, intelligence, disposition, and mostly her passion for cheese. Her encouragement to me after my first book without a doubt helped build the confidence I needed to attempt this book. I am sad to not be able to share it with her and to thank her for all that she was—to so many.

INTRODUCTION

If you are reading this book, there's a good chance that you really love cheese—so much so that you want to make your own. Maybe you have already created your first satisfying batches or perhaps you are one of the growing number of licensed artisan producers in the United States or elsewhere. Wherever you are on your journey as a cheesemaker, this book is meant to be your guide, resource, and even inspiration.

If you never want to understand the science behind the process that converts a fluid, rather bland, perishable liquid into a solid, flavor-intense, long-lasting food, then this book is probably not for you (at least not yet). But if you're the type of person for whom deep understanding of a subject brings enhanced enjoyment of the process, then *Mastering Artisan Cheesemaking* is the book for you.

If you are new to making cheese, this book aims to demystify what can sometimes feel like a lot of scientific language, while still retaining the depth of information every serious cheesemaker—including the hobbyists—should know. If you are already a fledgling or even an accomplished cheesemaker, then let this book be your complete resource, troubleshooter, and guide to taking your craft to the next level. Whoever you are, my goal with this book is to help you become an intuitive, enthusiastic, educated, and consummate maker of truly great cheeses.

Different things lead people to cheesemaking. For me it was a desire to return to my self-sufficiency roots, a love for dairy animals, and the desire to provide healthy, affordable milk and cheese for my family. I never thought making cheese would turn into a profession, much less lead to a book (or two). Cheesemaking has brought me incredible satisfaction; I went from an art career, making work that I believed in, but with which it was difficult for people to identify, to making cheeses that a vast number can enjoy, appreciate, and even admire.

When I was a little girl, I wanted to be a botanist. I would carry little field guides around our 200-acre forest and farm, gather specimens, and attempt to study them under an ancient 10× power microscope that we had. Later, at different times in my childhood, I wanted to be a nurse, cook, mother, and—always—an artist.

We had a Guernsey cow named Buttercup when I was very young, and later I had two much-adored Jersey cows of my own, Daffodil and Butterscotch. We made our own butter, buttermilk, and yogurt. My mother attempted cheesemaking using the only instructions available at the time, a USDA pamphlet that she got from the local Extension office. I still have this little booklet, and while its instructions aren't bad, the reality of the lack of available cultures and quality rennet destined my mom's first hard cheese to the chicken yard.

When my husband, Vern (who used to come to watch me show my cows in 4-H while we were both still teens), was close to retiring from his career (really the *family's* career) in the United States Marine Corps, we got our first dairy goats, Nigerian Dwarfs. I had no idea that I would eventually love them even more than dairy cows! I began making cheese using a less-than-satisfactory book and quickly switched to Ricki Carroll's pioneering work, *Home Cheesemaking*. What magic those first batches were! I'll never forget how it felt to see, for the first time, milk being transformed into solid, tasty, amazing cheese. I loved to make it, my family loved to eat it—we were all hooked.

Within a year of getting the goats, in 2003, I entered the American Dairy Goat Association's annual cheese competition in the Amateur division, the incubation ground for many soon-to-be-professional cheesemakers, and I won Best in Show. I entered again the next year, this time with a hard cheese, and won Best in Show again. As our friend Ken Miller, co-owner of Pastoral Artisan Cheese shops in Chicago, Illinois, said, there is nothing like "a win beneath your wings" to make you feel that you have found your path. At this time, in the early 2000s, we were just starting to see articles and features about small farmstead cheesemakers. It was still a relatively unknown career path. I'll never forget when my brother-in-law scoffed at the idea of making goat cheese for a living. It's amazing to what degree and how quickly things have changed. Now most people, when told of our profession, quickly assume a dreamy look and comment, "You're living the dream!"

A couple of years into my learning, I was fortunate enough to take a cheesemaking workshop from the amazing Peter Dixon, from Dairy Foods Consulting in Vermont. The class was at Black Sheep Creamery in Washington State. Not only was the class great, but being able to work in a functioning creamery was an education in and of itself. Later that same year, I flew to New York to take a workshop on the biology and chemistry of cheesemaking taught by Patrick Anglade, another well-known and respected cheese instructor from France. That class was mind-boggling and well over my head at the time. Fortunately, not long after that, Paul Kindstedt wrote *American Farmstead Cheese* (Chelsea Green, 2005). Between the notes from the previous class and Paul's book, the fog began to lift from the science of cheesemaking.

My first book, *The Farmstead Creamery Advisor,* was published in 2010 and was focused on the more number-crunching and infrastructural aspects of starting a cheesemaking business. I loved the process of writing and began toying with the idea of writing the kind of cheesemaking book, with recipes, that I would want to read but had yet to find—one that would contain a broad depth of information but be easy to read and understand. While I have certainly learned many lessons about cheesemaking the hard way, that doesn't mean you should, too! I am hoping this book will fill in the blanks missing in most cheesemakers' educations.

I began the formidable task of studying the academic cheese technology books; rereading the many notes and handouts from cheesemaking seminars and classes that I have attended over the years; consulting with cheesemakers far more versed in specific specialty cheeses such as blues; and making every recipe that I have included in this book, with the exception of several of those provided by the profiled cheesemakers. It was so nice to have their help and contributions! (Unless otherwise noted, every photograph of cheeses and processes was taken here at our creamery.) I hope I have accomplished what I set out to do—to digest, interpret, and translate cheesemaking science and apply it so that all cheesemakers, great and small, will have access to the beautiful knowledge that surrounds our shared passion and making the best cheese possible.

Mastering Artisan Cheesemaking begins with a section called "The Art and Science of Making Cheese." In this section I will gradually introduce you to each facet of cheesemaking and explain the beautiful interplay of science and art that goes into creating truly great cheeses. The first chapter, "Ingredients for All Cheeses," will thoroughly introduce everything that goes into cheese as well as its properties and how they interact with each other. Next, in "Concepts and Processes for Successful Cheesemaking," you will learn how you as a cheesemaker can control and influence these interactions through understanding each step of the cheesemaking process.

Next is "The Fundamentals of Acid Development and Monitoring during Cheesemaking," in which we cover the details of understanding and measuring acid development during cheesemaking. Some readers—the beginner cheesemakers—might not feel quite ready for this topic, so those people could skip it for now. But eventually, the consummate cheesemaker will have to master this subject, so why not start pondering it at the very least? Next comes a chapter about the art of aging cheeses called "Aging Cheese Gracefully—The Art of Affinage," which includes several options for setting up a successful small aging unit, the many options for rind treatments, and a troubleshooting guide.

In "Spicing It Up: Adding Flavors to Cheese," I help you learn how to choose, prepare, and safely add herbs, spices, alcohols, and even cold smoke to cheese using any cheese recipe. The last chapter in part I is "Designing, Equipping, and Maintaining Your Home Cheesemaking Space." This chapter will give you many options for choosing small-scale equipment and properly cleaning and sanitizing both the equipment and the cheesemaking space. (The resource guide in the appendix will give you some ideas for where to find equipment and supplies.)

And then the fun part—the cheese recipes themselves. First I'll introduce you to the eight major processes by which I have divided the family tree of cheese. Each recipe chapter will cover the must-know information and the deeper science specific to the type of cheese being made.

The recipes themselves might look a bit different from most cheesemaking books. The first thing you will probably notice is that the cheeses are not called by traditional (often protected) names; instead they are called what they really are, which is a process, or a style. There are many cheeses made throughout the world that are very similar to each other, but they have different names. So instead of giving you multiple recipes for basically the same cheese (I could make a few changes in the recipes to make you *think* they were different cheeses . . .), I am offering detailed instructions for a process that will create a distinctive cheese, *and* I'll give you some examples of cheeses similar to traditional and modern cheeses. It is my belief that this method will not only help you become a better informed cheesemaker, but it also acknowledges and respects the protected nature of many of the world's most admired and known cheeses, without simply trying to copy them. Next, each recipe includes measurements for small home batches, as well as guidelines for increasing the batch size for larger, artisan commercial production. The recipes are all streamlined to avoid redundant instruction and processes (but these can readily be referenced in each chapter).

In addition to the recipes, I have included a few inspiring profiles. These are not cheesemaker celebrities but rather everyday artisans with a passion for exploring and making great cheese. They are *you*. Several of these awesome folks have also contributed recipes and tips so you can learn from their experiences.

This book will help you become an intuitive artist and a scientist—able to create beautiful, flavorful cheeses and deduce answers and troubleshoot problems when things don't go as they should. You will be able to design your own recipes based on the results you want. Finally, you will have an even greater appreciation for some of the great cheeses of the world and, even more importantly, for the great cheesemakers of the world—including yourself!

A Note About the Photographs

Unless otherwise noted, each cheese and process image in this book was taken of a cheese I made, either during our normal cheesemaking process or as a part of my research into the recipes in this book. (This includes most of the images of flaws, too!) I became quite good at using my camera's automatic timer and ability to take multiple exposures, as I was most often not only the cheesemaker but also the photographer.

A Word of Caution for the Commercial Producer

Many of the techniques and methods included in this book are practiced widely throughout the world, often under the guidance of "Best Practices" guidelines (followed by commercial producers of all sizes). In many countries, perhaps nowhere more so than in the United States, these best practices are evolving and changing almost daily. Even when a guideline seems firm under one jurisdiction, it is often up for interpretation in a different area and by a different regulator. Therefore it is advisable to verify your practices, whether that be something such as the use of wood shelves for aging cheeses, prematuration of milk (in some areas this is not allowed, while others may interpret this as the "start of the cheesemaking process" and therefore not falling under the rules for fluid milk), or adding vegetable charcoal to cheese (currently under scrutiny) with your local regulators before implementing in the production of commercial products.

I recommend joining the American Cheese Society and a regional, active cheese guild (if you have one) to keep abreast of changes and even play a role in our country's evolving, artisan cheesemaking community. By knowing your craft and continuing to educate yourself, you can be a part of helping develop and maintain the rights to produce one of the most amazing foods civilization has developed.

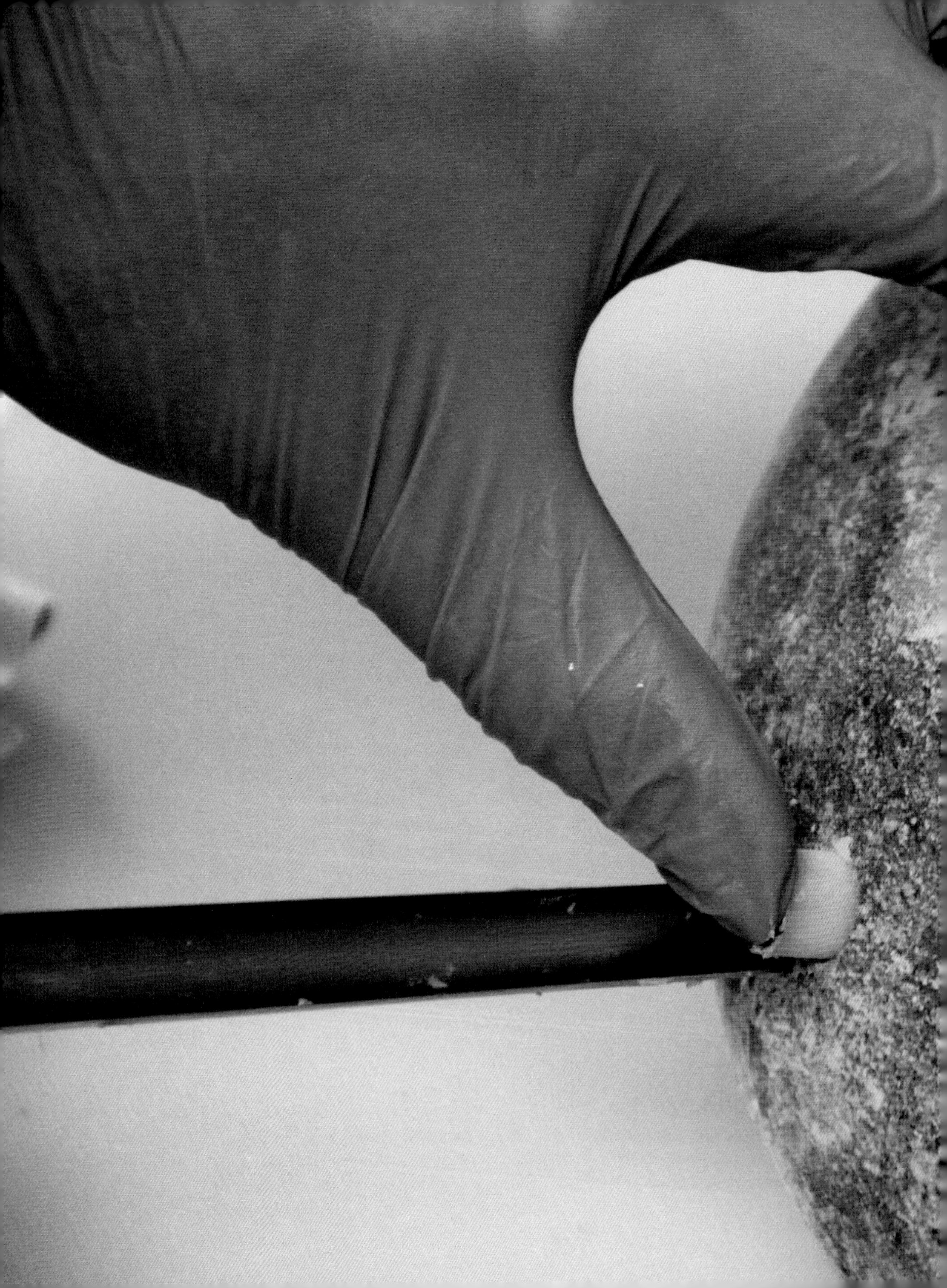

PART I:

THE ART AND SCIENCE OF MAKING CHEESE

1: INGREDIENTS FOR ALL CHEESES

An image that often comes to my mind when I'm making cheese is that of a staged theater piece, with a group of willing and able actors and Milk as the leading lady. The other ingredients, including culture, coagulant, and salt, are the supporting cast and the crew, some of whom are perhaps a bit difficult to work with. It's helpful to begin thinking a bit about the incredible orchestration of events, of all of which you, as the cheesemaker and director, are in charge. And indeed, you have a cast of characters that are all important, even though we may think of Milk as the most critical actor. Of course, without milk there would be no cheese, but without the other ingredients there would be only milk.

This chapter is filled with more than just information about the ingredients that go into cheese, as well as how those ingredients interact with each other to *form* cheese. In other words—lots of science! don't worry—this is not a dry, academic textbook, and you are not in a structured course. Feel free to skim through the headings and move past what you don't feel ready to absorb. Remember, even if you don't understand the science behind the processes, during cheesemaking the science is still taking place. I am pretty sure that by the time you are done with this book, you will be excited and stimulated to see how understanding and mastering the science will turn you into a true artisan.

As you read part I, keep in mind that there is a complex interplay between all of the ingredients involved in cheesemaking, and that it is impossible to completely segregate them as individual topics. So as you read you may think, "Hey, now she's talking about enzymes, and we haven't even covered them yet." Believe me, I understand! But this is perhaps the first important lesson: **Cheesemaking is not a simple series of events**, and that's what makes it such a stimulating challenge. In the beginning you have to accept a certain level of befuddlement and mystery, but I'll be doing my best to guide you safely through to a mastery of cheesemaking.

UNDERSTANDING THE BEAUTY AND COMPLEXITY OF MILK

To begin understanding the characteristics of milk, it is important first to acknowledge that the milk of every mammal, every species, and even every breed within a species can be unique. I am not going to throw a bunch of charts at you comparing the fat and protein content of whale, guinea pig, and human milk; while this type of information can be very interesting, it won't likely tell you about the milk you will be using for cheesemaking. Even beyond breed and species differences, milk will also vary based on what the animal eats, what season it is, how long it has been since the animal gave birth, and how healthy she is. Even the color of the milk (it isn't all pure white, you know) will change based on the above factors. Milk is very much a product of both nature and nurture. The following table is based on averages throughout the year for a few of the breeds and species that might be the source of your cheesemaking milk.

All milk is composed of sugars, fat, protein, minerals, vitamins, and enzymes. It is the unique combination in

Two tasty surface-ripened cow's milk cheeses: Dinah's Cheese (Kurtwood Farms, Washington State) on left, showing a natural yellow paste, and Seastack (Mt. Townsend Creamery, Washington State), which is naturally whiter

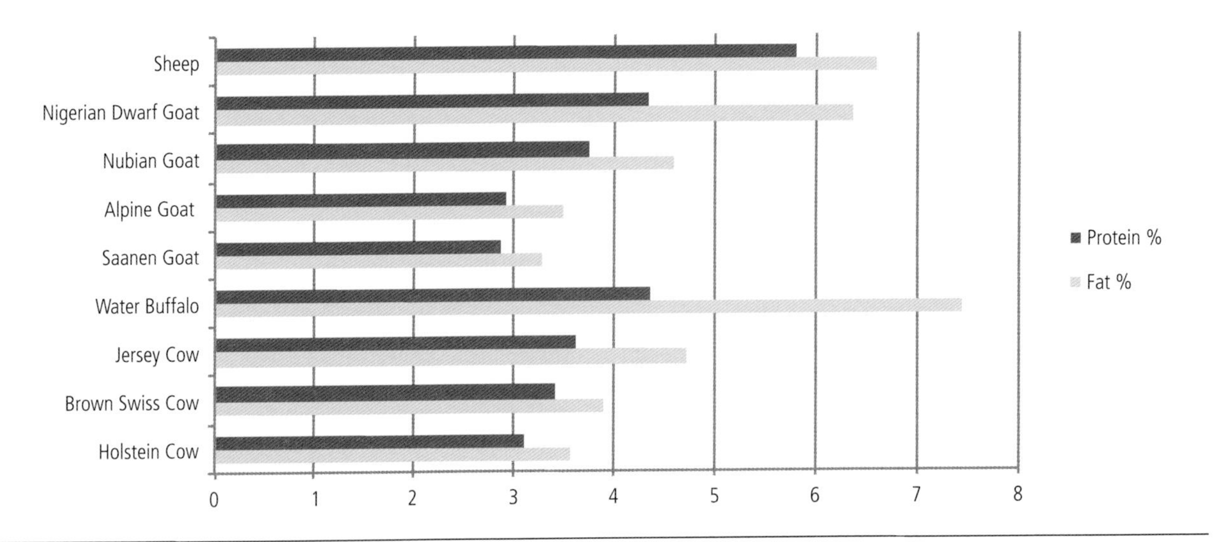

Chart showing some breed and species milk component differences. Keep in mind that usually the higher the components, the lower the milk volume compared to other animal breeds in the same species. Sources: Data from "2010 California DHIA Report," The University of Illinois Lactation Biology website, and Yves M. Berger, "Breeds of Sheep for Commercial Milk Production" (Spooner Agricultural Research Station, Univeristy of Wisconsin–Madison)

CHEESEMAKER MOTIVATIONS: THE PROVIDER

When we first started looking for a milk source, it was to provide healthy, unprocessed milk for our family. I started making cheeses immediately as a way both to use the surplus milk and to provide more "packaging-free" quality products for my family. I had no idea how much gratification I would get from making cheeses! It is difficult to describe the thrill of seeing that first pot of milk solidify into curds, then transform itself into delicious cheese. The fact that my family is made up of some pretty passionate cheese lovers only added to the gratification!

Some professional homemakers (and I am including anyone who chooses to remain at home and provide the valuable support that makes a family complete) turn to cheesemaking to save money and still others to provide products of better flavor and healthfulness than those available in the grocery store.

Cheesemaking at home can be a family affair and even a part of children's education, whether homeschooled or away-from-home-schooled. The cheesemaking process not only brings in all the elements of fairly advanced chemistry and microbiology classes but also an appreciation of food production. I believe that an understanding of what it takes to produce foods eventually leads to better eating and purchasing habits—not a minor fringe benefit.

If you are making cheese primarily to provide sustenance, you will most likely be more concerned with staple-type cheeses rather than with those with fancy rinds and short shelf lives (although you might fantasize about making a few of these types!). If you are providing cheese as a basic food, don't feel as though you need to apologize or provide excuses for not making the fancy stuff. Your reasons are the reason cheese exists: to provide nourishment.

the milk you choose, then what you as the cheesemaker do to these components that will create a superior (or not) cheese. So let's take an in-depth look at the constituents of milk.

MILK SUGAR—SWEET SECRETS

Milk sugar is the first component of milk to play a part in the cheesemaking process. The sugar provides food for the bacteria in the starter culture which is added to the milk; the starter culture digests the sugar and produces acid. This acid production begins the process that will eventually turn the milk into cheese. Truly, it is the control of this acid production that the cheesemaker will need to learn to regulate to become consistently successful. While you don't necessarily need to have a thorough understanding of the complexity of milk sugar to understand acid control, learning more will provide the basis for the proper selection of cultures and will help answer other questions that will arise during cheesemaking. So let's dig deeper into understanding the sweet secrets of milk sugar.

The sugar in milk is called *lactose*. You might know that when an animal gives milk it is called "lactation" or "to lactate." When going by their scientific name, sugars end with the suffix "ose," so you can probably see where milk sugar got its name. Lactose is made up of two simple sugars, *glucose* and *galactose*. During cheesemaking it is the breakdown and consumption of lactose (and its resulting simple sugars) that produces acid—*lactic acid*. Without the presence of acid, the production of cheese does not occur.

To break the bond that holds the two simple sugars together, a process must occur. When people drink milk, the enzyme lactase (we'll learn more about enzymes later, but for now you should know that enzymes act upon other substances to break a bond) is required to break the lactose into digestable sugars. Some people don't naturally produce this enzyme, making them what is known as "lactose intolerant." During cheesemaking, lactose is broken down through the actions of certain bacteria, which are usually added to the cheesemaking milk in the form of starter culture but can also be present in raw milk or low-temperature heat-treated milk.

As the bacteria consumes the milk sugars through the process of fermentation—first lactose, then glucose and galactose—lactic acid is produced. You will often see the term *lactate* (not to be confused with the verb

FERMENTATION DEFINED

While most of us are familiar with the term *fermentation*, and associate it with beer and wine making, sauerkraut, and sourdough bread, most people rarely think of cheese as a fermented food. But it is! Fermentation is the process through which carbohydrates and sugars (carbohydrates are combined sugars, rather than simple sugars) are broken down into by-products such as alcohol (there is your beer and wine), the gas carbon dioxide (hence the bubbles in your champagne and beer and the eyes in your Swiss cheese), and organic acids (thus the tang in our kraut and the acidity of cheese). Fermentation is a complex process that follows various pathways based on what is being fermented, which is known as the *substrate*. Fermentation can occur anytime organic matter is exposed to microorganisms such as bacteria and yeasts. This includes not only food but also compost and animal and human waste. As a means of preserving food, fermentation is one of the oldest—right alongside drying and salting. In fermentation guru Sandor Ellix Katz's new book, *The Art of Fermentation: An In-Depth Exploration of Essential Concepts and Processes from Around the World* (Chelsea Green Publishing, 2012), this amazing process is explored in great depth and passion.

lactate) used almost interchangeably with lactic acid. While it is molecularly slightly different, for the cheesemaker the terms can almost be synonymous. There are two variations of lactate. When you first get started making cheese, or even when at the intermediate level, you don't need to understand that there are two variations, but later in your cheesemaking education, you will see references to starter bacteria that produce one or both of these two types of lactic acid. So when that topic arises, I want you to be prepared! The two varieties of lactic acid are called D-lactate and L-lactate. The production of D-lactate can also create two other by-products that influence cheesemaking: the alcohol ethanol and the gas carbon dioxide. This is useful to know when the development of eyes (such as in Swiss alpine types of cheese) by gas production is a desired end result, as some bacteria are incapable of producing D-lactate. (We'll learn more about cheesemaking bacteria later in this chapter.)

In addition to fermenting lactose into two types of lactic acid, starter bacteria also vary in their ability to ferment the two simple sugars, galactose and glucose. If only one of the two is completely fermented, residual sugar will remain in the cheese curd. This remaining sugar can help feed other bacteria during the first stages of aging, which is desirable in some cheese types. For most pressed cheeses all sugar will have been metabolized and utilized after about two weeks of aging.

The amount of lactose in milk changes depending on what point the animal is at during her lactation.

LACTOSE INTOLERANCE—MYTHS AND MISUNDERSTANDINGS

Fairly frequently in articles in small dairy type publications, I read of people's belief that raw milk, especially goat's milk, helps them cope with their lactose intolerance. Words such as "I am severely lactose intolerant and will get sick if I don't drink goat's milk every day" are not uncommon. Being a passionate supporter of raw milk myself, and somewhat biased toward goat milk, it is wonderful to hear of these folk's passion, but difficult to digest the inaccuracy of their science. No matter how you feel about raw milk, from any beast, let's clear up a few facts about lactose and lactose intolerance.

It is important to remember that goat's milk (along with all of the most common milks consumed by humans) has almost the same amount of lactose as cow's milk. If you are truly lactose intolerant, meaning your body no longer produces—as it did when you were an infant—the enzyme lactase, these milks should cause as many digestive problems as does any other milk. Goat's milk is typically easier to digest, thanks to its smaller, more delicate fat globules and different protein composition, and some data suggests that raw milk may have lactase present because of adventitious bacteria that introduce the enzyme, but the evidence is not conclusive. So for the severely lactose-intolerant individual, fluid milk is likely to cause digestive problems unless the enzyme lactase is consumed at the same time or is added to the milk after harvesting from the animal. In fact, the issue of lactose intolerance is of such importance to the fluid milk industry that some milk is now being marketed in which the lactose has been broken down to simple, digestable sugars before it reaches the market.

Where cheese consumption is concerned, a lactose-intolerant person can consume aged cheeses, as lactose is virtually nonexistent in such products. The fresher and more moist the cheese, however, the more likely that some lactose will remain, so these should be avoided.

CRYSTALS—THE "CRUNCHIES" IN CHEESE

If you are a lover of extra-aged hard cheeses, you probably also adore the sensation of biting into an aromatic, granular chunk and feeling your teeth crunch through tiny, grainy crystals that cheese lovers often refer to as "crunchies." There are a couple of types of crystals that can occur in cheese. The first type is formed from the amino acid tyrosine and is desirable in certain cheeses. Tyrosine crystals typically form in high-protein, long-aged cheeses such as Parmesan and aged Gouda. The second type of crystal is formed when calcium combines with residual lactate to form calcium lactate crystals—which is considered a defect. Calcium lactate crystals are far more common on the surface of certain cheeses, cheddar being the most likely victim.

While milk components such as fat and protein (you'll meet them in more detail in a bit) usually increase later in the lactation, after the animal has been milking for several months, the percentage of lactose usually decreases. This can influence the rate at which acid is developed by the starter culture, as there is simply less fuel for the process. As the rate of acid production changes, the final product will be affected as well, due both to the level of acid and the rate at which it develops, as well as the subsequent by-products produced by the starter bacteria that the milk sugar feeds.

MILK FAT—FILLING OUT THE FLAVOR

Milk fat's role, much like the purpose of fat in cooking, is to provide an appealing texture, a well-rounded flavor, and—through the changes that it will experience during aging—wonderful aromas. The amount, composition, and quality of milk fat in cheese milk are all important factors to understand during cheesemaking. These three facets can be somewhat manipulated by the cheesemaker to alter the outcome of a recipe. For example, Parmesan is traditionally made with partly skimmed milk. If you follow the same recipe but use milk higher in fat, the resulting cheese will be quite different from the traditional Italian original. Other cheeses, such as cream cheese and triple crème Brie, have cream added to increase the total fat content and therefore change the texture and taste of the final cheese. In the case of some high-production blue cheeses, the milk is skimmed and the cream homogenized, then added back into the milk. From this you can probably get a good idea of the importance of different levels of fat in the creation of different types of cheeses. Before we talk more about the amount, composition, and quality of milk fat, let's take a closer look at the structure of the fat globules.

ANATOMY OF THE MILK FAT GLOBULE

Milk fat globules are quite large in comparison to other components in the milk solution. The globules are surrounded by a membrane known as the *milk fat globule membrane* or MFGM. This membrane contains most of the enzymes (more on enzymes later) that occur naturally in milk. The main exception to this is the enzyme *lipase* (which breaks down fats/lipids), which does not exist in the fat globule membrane. This makes sense, for if the enzyme resided in the membrane, it would destroy the globule.

The inside of the membrane contains the fats, mostly (about 98 percent) as triglycerides, also known as triaglycerols. A glyceride is a glycerol molecule, which you can think of as the basic molecule for all fats, bonded to one, two, or three fatty acids. Triglycerides (the kind with three fatty acids) are the fats that make up most of the food oils and fats that we consume. The milk of cows, goats, sheep, and water buffalo (all called *ruminants*, which are animals that "chew a cud") contain a greater variety of fatty acids than any other milk types. This is quite nice for cheesemakers, as some

HOMOGENIZATION FACTS AND MYTHS

Homogenization is a mechanical process designed to alter the natural behavior of fat globules and keep them from creaming. The process typically involves sending the milk under high pressure through a screen with openings sized smaller than the fat globules. The process reduces the size of the fat globules and denatures (changes the natural behavior of) cryoglobulins, the proteins that encourage creaming. (Isn't it funny the amount of industrial energy that goes into saving us the labor of shaking the milk jug before pouring?)

You may have heard or read that goat's milk is "naturally homogenized." While it's true that goat, sheep, and water buffalo milk all lack cryoglobulins and will cream very slowly by comparison, they are *not* naturally homogenized.

For the cheesemaker this alteration of the membrane's normal structural composition means that cheese curd made from homogenized milk tends to not drain and lose whey properly during the draining process (called *synerisis*), so it forms a curd that may be higher in moisture than desired. Remember the blue cheese I mentioned a bit ago that uses homogenized cream that is added back into the skimmed milk? Are you wondering why they would bother with this extra step? Well, now that you know what homogenization does to the fat globule, you might be able to imagine that the denatured fat is more susceptible to attack and breakdown by enzymes. This more rapid fat breakdown helps these cheeses develop the piquant, spicy flavor that may be desired by the cheesemaker (and the cheese eaters!).

fatty acids are particularly good at producing volatile compounds (which means that they can easily be carried on the air; in other words, you will be able to smell them) that provide aroma and flavor. These *short-chain* fatty acids—*butyric, caproic, caprylic,* and *capric*—are present in varying amounts depending on the species.

Milk fat globules are less dense than the surrounding milk fluid, with the result that, given the right amount of time and conditions, they will gradually float to the top of the milk and form a layer of cream. In fact, this process is called *creaming*. Conditions are best for creaming when the milk is cool and not moving. Cream rises faster in cow's milk, thanks in part to the presence of *cryoglobulin*, a protein that is attracted to and absorbed onto the globule when temperatures are cool (hence the prefix *cryo-* for cold and the root *globulin*) which encourages the massing together of the fat globules. Once clumping, or *agglutinating*, begins, more globules are attracted at an increasing rate, and the process speeds up. The absence of cryoglobulins in sheep, goat, and water buffalo milk means that the fat globules rise much more slowly.

The breakdown of the fats is called *lipolysis* (*lipo* for fat and *lysis*, meaning "to break"). It's the process of lipolysis that releases the volatile fatty acids that are responsible for major flavor and aroma development—sometimes good, sometimes not so good. We'll talk more about that in a bit.

THE AMOUNT OF FAT IN MILK

When trying to determine the amount of fat (usually measured in a percentage) in milk, you may be able to get this information from the milk producer. If not, there is data available online and in publications based on the average fat production according to species and breed. Within species, fat content can vary as much as almost 100 percent. Take, for example, the milk of the average Saanen goat at 3.3 percent and the average Nigerian Dwarf goat at 6.4 percent. Some charts and other sources also list seasonal variances in milk fat, which is generally at its highest percentage late in the animal's lactation and during the cold winter months.

Remember that during cheesemaking you are attempting to capture the fat globules by trapping them in a protein network. Therefore, the correct amount and type of protein (more on that later) is required to capture the butterfat. A very high level of butterfat without adequate protein content will lead to a loss of the excess fat. Different cheese types each have an ideal percentage of fat and protein. This is referred to as the *protein-to-fat ratio*. In the large commercial-scale production of cheese, this ratio is controlled by a process you will learn more about in the next chapter, called *standardization*. The artisan cheesemaker will probably not try to "standardize" the milk but rather will learn to adapt her process *to* the milk. In fact, this may be the biggest difference between an artisan and an industrial cheesemaker!

> The artisan cheesemaker will probably not try to "standardize" the milk but rather will learn to adapt their process to the milk—that's the biggest difference between an artisan and an industrial cheesemaker!

HOW THE COMPOSITION OF FAT CAN VARY

The composition of fat in milk—mainly fatty acid content and globule size—varies depending on species, breed, and diet. Goat's milk typically has smaller globules than does cow's milk. In the bovine world Ayrshire and Holstein cow's milk have a somewhat smaller globule than does the high-fat milk from Jersey cows. Studies in Denmark and Sweden have shown that as total fat content in an animal's milk increases, so does the size of the individual globules, which is not necessarily desirable. This matters to the cheesemaker because it is believed that it is easier to capture smaller fat molecules in the cheese curd.

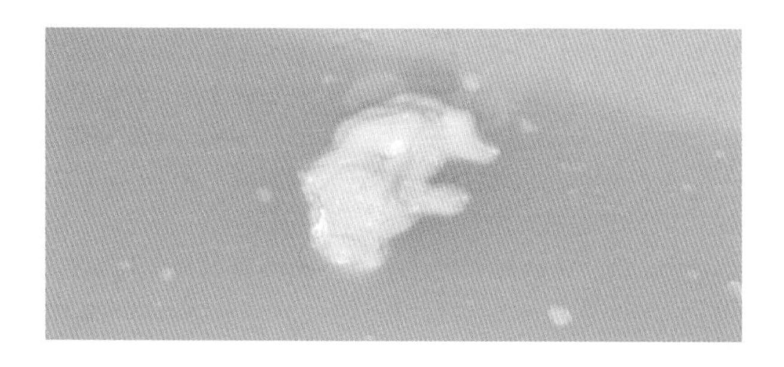

The extra butterfat from rich Jersey cow's milk formed little clumps of butter on the whey during the making of a hard cheese.

If your curd, at the time of draining and placing in forms (more on this later), is consistently greasy, large fat molecules might be to blame. To find out if you are losing fat in the whey, a sample of whey can be sent to a lab for analysis, or more simply, observe the surface of the milk during cheesemaking—especially toward the end—and you might see little fatty puddles floating on the surface.

Goat's milk typically has the highest concentration of certain short-chain fatty acids—caproic, caprylic, and capric, which are the main fatty acids responsible for aroma and flavor. Sheep's milk, as well, is higher in these three but also contains more butyric acid than does cow's milk. This gives sheep's milk an overall advantage in potential flavor and aroma development.

There is much discussion and controversy over the inclusion of animal fats in the human diet and their relationship to general health. It is erroneous to conclude that all animal fat is equal; the diet of the animal can alter the balance of fatty acids in both milk and meat. Take, for example, the documented benefits of one particular fatty acid, conjugated linoleic acid (CLA), which is produced by ruminants and is present to a greater degree when animals are fed a more natural diet of pasture. CLA is proven to have anticarcinogenic properties and also positively influences weight control.

ASSESSING THE QUALITY OF MILK FAT

The fat in milk can be easily damaged before and during the cheesemaking process, leading to rancid and unpalatable flavors. Bacteria that enter the milk during milking and processing contain lipases, enzymes that given the right conditions will break down fats. This can be prevented, or limited, through proper sanitation during milking and processing and proper, rapid chilling to limit the growth of these bacteria. Too much mechanical agitation, through rough milk pumps or excessive stirring, is another leading cause of premature fat breakdown.

You can perform a simple check for changes due to fat breakdown by comparing a fresh sample with one that has been stored. Place the samples in glass jars with about 2 inches of airspace, and let both come to room

> **FAT SECRETS FOR THE DAIRY PRODUCER**
>
> Fat production in milk does *not* come from feeding more fat to the animal—except in some disturbing experiments (which I do not recommend) involving the feeding of unnaturally encapsulated nutritionally desirable fats to milking animals. Instead, it is the presence of long fiber in feeds that is involved in a complex breakdown in the animal's digestive system, which increases fat production in milk. But this is only possible if it is already within the animal's genetic potential, so it's not just feeding but breeding that makes the difference. Even the interval between milkings can influence the fat content of milk, with the shorter interval being associated with higher levels. Milk that is mastitic (from an udder with some level of infection) will have a lower fat content than normal. All of these are just good reminders of why thoughtful herd management and milk quality, and therefore superior cheese, are so important.

temperature. Agitate the jar, then open it right below your nose. Breathe deeply. You can taste the samples, as well, as long as you are comfortable with the milk's safety. Compare the two. Ideally, they will be virtually indistinguishable from each other. If there are any odors or flavors that are barnyard, musty, rancid, or otherwise unfresh, then the milk has suffered some damage.

MILK PROTEIN—THE BUILDING BLOCKS FOR CHEESE

Just like a body builder, cheese needs protein to create its structure. It is a network of protein that will trap the milk fat and transform the liquid milk into curd. Not only does milk protein provide the basis for forming the cheese but it supplies much of the valuable nutrition that cheeses provide. And perhaps equally important for some, proteins will eventually turn into the compounds and substances that make cheeses so tasty to eat. As you just learned about milk fat, the amount and type of protein in the milk and your ability to maximize its capture in the cheese will make a vast difference in the type and quality of cheese you can make. So if you are ready to learn more about the building blocks of cheese, let's continue.

For the cheesemaker it is important to understand that milk protein is classified into two main types: cheese proteins, otherwise known as *caseins* (kay-seens), and whey proteins. The caseins are the proteins that can be captured during the coagulation process, while the whey proteins remain in the fluid by-product of cheesemaking, whey. (The industrial-scale dairy industry has developed processes for retaining more of these proteins in the cheese, increasing overall yield and lowering the loss of the valuable nutritional aspects of protein.) Cheese normally contains all the essential amino acids (the building blocks of protein) necessary for human nutrition. Aged cheeses tend to be more easily digested than fresh cheeses and milk, thanks to both the lack of lactose (which many adults cannot process well) and the breakdown of protein into its smaller components. It is easy to see how cheese can be not only a pleasure to eat but can also play an important role in providing high-quality protein in a balanced diet. Let's take a closer look at the structure and behavior of proteins during the cheesemaking process, factors affecting protein content and quality, and how proteins are concentrated by using both rennet and acid to coagulate milk during the making of cheese.

THE ANATOMY AND NATURE OF CHEESE PROTEINS

There is ongoing research into the exact nature of how caseins exist in milk. While the model of a clump or cluster of proteins, known as the *casein micelle*, is still widely accepted, the new research hints at a slightly different structure. For the cheesemaker, though, the

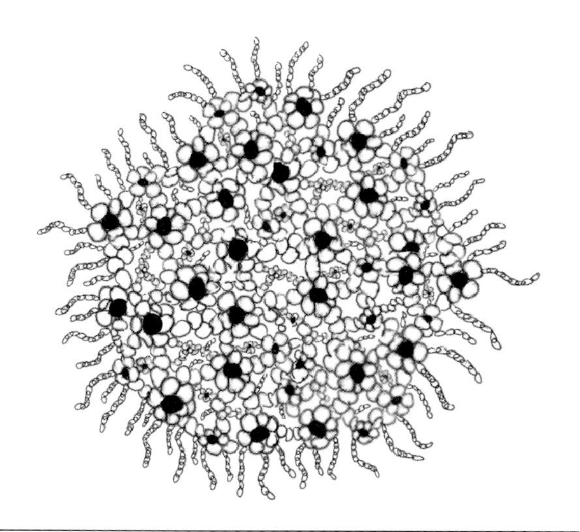

Conceptual drawing of casein micelle showing proteins (white circles) and calcium phosphate (dark spheres) clumping with protruding kappa-casein chains

actual structure is not as critical to understand as how the proteins react to rennet and to changes in pH (for acid coagulated cheeses). So let's talk about the behavior of protein clusters and how acid and rennet allow us to concentrate the milk into curd.

Casein micelles are small clusters that float suspended in the milk. They are made up of primarily individual casein "units" from two main groups of proteins: α-casein and β-casein (named using the Greek symbols for alpha and beta). These units are thought to be held together or bonded by the mineral calcium phosphate. Toward the outside of the clusters, another milk protein, kappa (κ) casein, is imbedded. In theoretical illustrations the κ-caseins are usually depicted as hairlike protrusions projecting out from the main structure. The protruding end has a negative electrical charge. This charge means that it repels other micelles (remember the chemistry basic of opposites attracting and "likes" repelling). The projecting segment of the κ-casein is water loving (the scientific term is *hydrophilic*) and easily stays dispersed and suspended in the milk, since milk is made up mostly of water. (See the illustrations on page 137.)

Each α- and β-casein unit within the micelle is made up of chains of smaller units called amino acids. During the aging of cheese, the amino acid chains are broken down to create flavors and texture. The specific characteristics of these changes depends on the amino acid at the end of these broken chains, known as the *terminal amino acid*. It is amazing, really, to think that something as small as an amino acid can have an odor or a flavor.

MILK ALLERGIES

If people suffer from a true allergy, as opposed to lactose intolerance or an other digestive difficulty with milk, the component in milk that they are reacting to is a protein. The most likely suspect for most individuals with a milk allergy is an α-casein called alpha s1. This protein, as you will learn in this section, is usually more prevalent in cow's milk than in goat's milk, although some goat breeds genetically test equally strong for its production. In addition, allergy sufferers may react to more than one protein type in milk. Allergies vary in severity, but people with a true milk allergy should only attempt to drink other milks, such as goat's, under the guidance of a physician.

FACTORS AFFECTING THE AMOUNT AND QUALITY OF MILK PROTEIN

One of the main influences on the type and percentage of protein content in milk is the species and breed of animal. Cows typically produce 80 percent cheese proteins sand 20 percent whey proteins, while goats average 70 percent cheese proteins to 30 percent whey proteins, leading to a slightly lower yield when the curd is coagulated using rennet. In addition, goats, in general, have more beta (β) casein than alpha (α) casein. As α-casein is associated with the formation of a stronger protein network, a higher percentage of α-casein will increase the strength of the curd mass and the overall yield of cheese.

While most breeds within each species maintain a fairly regular proportionate ratio of protein to fat with fat normally being higher than protein. So if the fat content goes up seasonally, the protein content usually goes up as well. Milk from breeds with a higher fat content, such as Jersey cows, Nigerian and Nubian goats, and sheep, therefore typically also have a higher protein content.

Cold storage of milk, as occurs on most farms and in many cheesemaking facilities, causes a loss of protein from inside the clusters (micelles) into the liquid portion of the milk (where it is referred to as "soluble protein"). Mastitic milk from infected udders and colostrum, which is the first milk excreted by the mother immediately after giving birth, both have higher levels of soluble protein (the type that stays in the liquid portion of milk instead of inside the micelles), making them a poor choice for cheesemaking milk for the same reason.

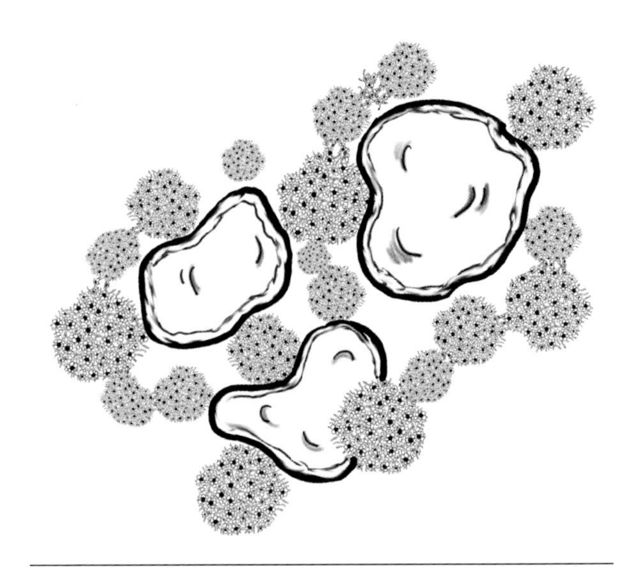

Conceptual drawing showing casein micelles beginning to form a network trapping large fat globules

COAGULATING MILK PROTEINS USING RENNET

Kappa casein exists in the smallest proportion in milk (compared to α-caseins and β-caseins), but it is the first protein to be acted on by the cheesemaking process. Rennet, or any other coagulant that contains an enzyme, such as chymosin, has the ability to split, or cleave, the chain of amino acids in κ-casein at a specific point. When you add such a coagulant to the milk, it immediately begins "cleaving" the κ-casein from the outside surface of the micelle—you can think of it as giving the micelle a "haircut." In addition to removing the negatively charged end of the κ-casein, the cleaving also exposes a water-hating, or hydrophobic, behavior, meaning that the clusters will repel water and become, if you want to think of it this way, more affectionate with their own kind. Once the κ-casein has been cleaved, the recently shorn casein micelles start bonding together, and with the help of calcium phosphate in the liquid portion of the milk, a thickened curd mass known as the *coagulum* can form, and cheese becomes possible.

COAGULATING MILK USING ACID

Coagulation can also be brought on by acid, either added in the form of a food acid such as vinegar or lemon juice or by the production of lactic acid by a starter culture. As the pH of milk drops close to 4.6, caseins reach their *isoelectric point*; in other words, they lose their negative electrical charge and therefore their ability to repel each other. In addition, calcium phosphate, which was formerly bonded within the micelle, becomes soluble in the acidic environment. All this produces a condition in which a coagulum is formed—but one that is quite different in texture and structure from the rennet-produced mass.

PROTEOLYSIS, PEPTIDES, AND FLAVOR—THE MAGIC OF CASEIN BREAKDOWN

Earlier I described the process of breaking amino acid chains for the development of flavor and texture in cheese. Now let's take a look at how this breakdown of protein, *proteolysis*, occurs. Enzymes naturally present in milk, provided by the addition of rennet and contained within the cell walls of beneficial bacteria already present in the milk or in the added cultures, will contribute greatly to the cheesemaking process. These proteolytic enzymes are able to break down proteins by cleaving (cutting) the chains of amino acids. They each target specific sites on the chain and break them at that point. This cleaving creates shorter chains of amino acids, called *peptides*.

Peptides have their own flavors depending on the amino acid located at their end (the terminal amino acid). Common flavors for peptides are bitter, sweet,

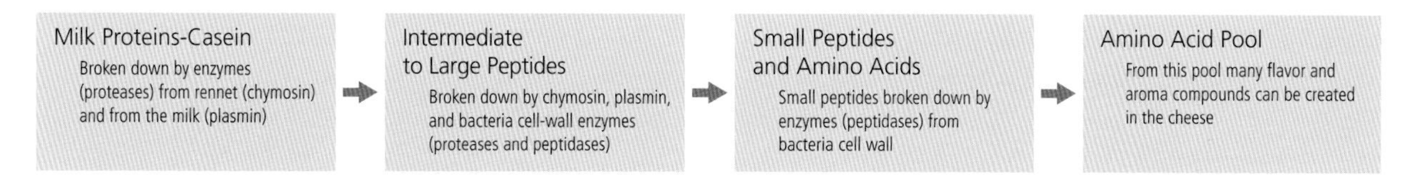

A simplified diagram of protein breakdown

brothlike, and rubberlike. A cheese tasted during the "peptide stage" of aging will commonly have bitter flavors that will likely disappear after longer aging. If the aging environment is not favorable, a cheese can get stuck in this bitter stage. But if maturation continues peptides are broken down to individual amino acids (called free amino acids) that can be metabolized further into tasty little things such as aldehydes, amines, and keto acids (organic and chemical compounds). One amino acid, tyrosine, is well known for forming crunchy, pleasant crystals in hard cheeses that are aged for an extended period of time (see the previous sidebar on crystals in cheese).

While the breakdown of protein during cheesemaking, specifically while it is aging, is a desirable thing, premature breakdown of protein in fluid milk is a problem for the cheesemaker. What can cause this early damage? Any proteases, which are enzymes capable of breaking down protein, will negatively affect the milk if introduced at the wrong time during the cheesemaking process. Common sources include psychrotrophic (cold-loving) bacteria, which may contaminate the milk during collection and processing; plasmin (a protease naturally present in milk); mastitic and late-lactation milk (which is higher in plasmin and somatic cells [white blood cells]); and other bacterial contaminants.

I mentioned earlier that protein content may change, depending on a variety of influences. With higher protein levels comes higher calcium content. In just a minute you will learn more about how these higher levels affect cheesemaking.

MINERALS—KEEPING IT ALL TOGETHER

We all know that minerals are essential for life, but they are also essential for making cheese. Milk contains quite a few of these elements, but one in particular, calcium, is the standout winner of "most important mineral" when it comes to converting fluid milk into cheese. The calcium content of the milk and how it changes during cheesemaking is one of the most fascinating aspects of the science behind the art of the process. So let's get better acquainted with the minerals in milk and the critical role they play during cheesemaking.

Minerals in milk are often called *salts* or *ash*. "Salts" is chemistry's term for minerals, and "ash" is used because the salts can be separated from the fluid milk by burning off the liquid portion, leaving an ash. To keep it simple, I will only refer to them as minerals.

Cow's milk has been identified as containing twenty-five minerals, and while only present in small quantities, they play an important role in cheesemaking by facilitating reactions and changes in the process of converting milk to cheese. *Calcium* and *phosphate* are the two minerals in milk of primary importance to the cheesemaker. Keep in mind that many of these minerals often occur in more than one form—sometimes they are bonded with another mineral or substance; such is the case with calcium phosphate and calcium citrate.

About two-thirds of calcium in milk resides inside the protein micelles. This type of calcium is called *colloidal calcium*. The remainder exists in the fluid portion of the milk and is called *soluble calcium*. Anything that limits the amount of colloidal calcium in the milk will limit the potential for a good curd set during cheesemaking. For example, when milk is stored at cold temperatures, some of the colloidal calcium leaves the micelles (becomes soluble) and is unavailable to assist with coagulation.

MORE ABOUT CALCIUM

Calcium's role in cheesemaking is to assist in forming a firm curd that traps fat and will hold up well to stirring, pressing, stretching of the curd, and so on. I like to think of calcium as a mortar that cements the protein blocks together. Without the proper levels of calcium available in the milk, you cannot achieve a good curd texture. It is easy to disturb the balance of calcium in milk; cold storage, heat treatment, acid development, and other things can negatively affect the balance. For this reason calcium chloride (a calcium salt) is often added to cheesemaking milk (prior to adding rennet) to encourage maximum coagulation potential (in chapter 2 we'll go over the rates used).

The cheesemaker can also purposefully manipulate the calcium content to make cheeses such as washed curd varieties (that retain more calcium) and bloomy rinds (which ultimately will contain less calcium). By the same token the calcium content can be negatively affected during the cheesemaking process. So just how do you as the cheesemaker change the calcium content of the cheese? It all comes down to acid. As with mineral deposits in your coffeepot or milk pail, calcium and other minerals will dissolve in the presence of acid of the right strength. Once your cheese has coagulated, you can still lose or retain minerals depending on the rate of acid development in the vat. For example, a washed curd cheese usually retains more calcium and is often called a "mineralized" cheese, thanks to the removal of whey (and therefore lactose), which results in slower acid development. By the same token cheeses, such as fresh goat cheese, that develop a higher level of acid over a long coagulation and ripening phase lose more calcium in the whey. This mineral balance in the final cheese is directly linked to the cheese's texture—a mineralized cheese is pliable and long in texture while a demineralized cheese has a short, crumbly texture, otherwise known as *friable*.

When calcium levels are high in the milk (often when protein levels are high as well), starter culture amounts might need to be a bit higher.

Calcium phosphate acts as a buffer in the milk, curd, and cheese. Buffering refers to the ability to resist pH change; you'll read more about that in chapter 3. To simplify, think of milk minerals as having the ability to absorb hydrogen ions—present in increasing volume the more acidic the solution—and therefore limiting acid increases. This is why a glass of milk (or Rolaids or Tums, whose active ingredient is calcium carbonate) can help relieve acid indigestion and heartburn. The cheesemaker has the potential to control the final calcium content of the cheese, which, as noted before, is critical for texture. When calcium phosphate levels are high in the milk, which is often when protein levels are high as well, starter culture amounts might need to be a bit higher, if you find that acid production is too slow thanks to the extra buffering capacity of the high mineral milk.

CITRATE AND ITS ROLE IN CHEESEMAKING

Citrate, otherwise known as citric acid, occurs naturally in milk. It exists for the most part bound to calcium and magnesium and plays an important role in cheesemaking. Citrate can be utilized by very specific bacteria to produce diacetyl, which is responsible for a pleasant buttery flavor and aroma. An important by-product of this metabolization is the gas carbon dioxide. It is the metabolization of citrate that is responsible for the formation of small eyes, especially in Dutch-type cheeses such as Gouda. When we learn more about cultures later in this chapter, you will learn which bacteria are able to metabolize citrate.

VITAMINS—FOR THAT HEALTHY GLOW

For the cheesemaker vitamins have a minimal role in cheesemaking. The most apparent contribution is made by *carotene*, a provitamin or previtamin substance that the body can convert to vitamin A. It's carotene that gives cow's milk cheese a yellow color. Goat, sheep, and buffalo milk, on the other hand, don't contain carotene, so products made from these types of milk are whiter by comparison. Carotene levels are higher during the summer months, but only if fresh grazing is available. In fact, it's that golden summer milk that inspired the addition of yellow/orange coloring to milk at other times of the year. One could argue that this added coloring has been taken to quite an extreme with our artificially orange cheddar cheeses and even in the very vibrant, carroty color of the traditional French cheese Mimolette. (Both orange cheddar and Mimolette are colored with a natural pigment called *annatto.*) A few other vitamins are retained in cheese, where they do contribute to the microbial activity during cheesemaking and aging.

Vitamins can be divided into fat soluble, or *lipophilic,* and water soluble, or *hydrophilic,* categories. Vitamin A and the provitamin carotene are fat soluble, so they can remain in the curd along with the milk fat globules to provide color to the cheese. Vitamin D, another fat soluble, is present in varying amounts, depending on the animal's diet and exposure to sunlight. As you can probably imagine, most of the water-soluble vitamins—such as the many B vitamins and vitamin C—are not retained in large proportion in the cheese but are instead flushed out with the whey. It is believed, though, that the B vitamins are used by some of the microorganisms in cheese, so even though not present at significant levels, they have a part to play in the development of flavor in cheese. Milk subject to ultrahigh pasteurization temperatures is deficient in B vitamins, especially thiamine. Another B vitamin, riboflavin, gives a greenish tint to whey, in case you ever wondered where that color came from!

This somewhat disturbingly orange cheese is labeled "Dutch Mimolette." Legend has it that the French cheese Mimolette originated from an attempt to copy the Dutch cheese Edam. This is an interesting turnabout of the copy being copied.

ENZYMES—THE ACTION HEROES OF CHEESEMAKING

The last important component of milk that the cheesemaker needs to understand is the action heros of cheesemaking—the enzymes. (These guys are so essential, in fact, that you will meet them again in association with the other two main ingredients, *culture* and *coagulant.*) The word "enzyme" is used a lot these days in advertising, boasting about a product's "enzymatic activity" or claiming that it "contains powerful enzymes." It's true, enzymes are potent ingredients; they have the power to cause and speed up chemical reactions that alter molecules and compounds. Pretty much every process in life involves enzymes, from the digestion of food to the decay of a bouquet of long-stemmed roses. Enzymes are not characters to be taken lightly, so let's get better acquainted with these busy little components.

There are as many as sixty enzymes found in milk—naturally occurring, from bacteria that enter the teat canal, and from "contamination" during and after the milking process. Enzymes are named after the substance they act on and usually end with the suffix "ase." *Lipases* break down lipids, or fats, and proteases (also called *proteinases*) break down proteins. These are the two enzyme groups that the cheesemaker should become the most familiar with.

Lipase is naturally present in milk, where it is called *native* or *milk lipase.* This type of lipase is generally weaker than the types produced by other microorganisms or harvested from the stomach of ruminants (this type is called *pregastric lipase*). Lipases produced by microorganisms may also be present in the milk from adventitious bacteria that entered the milk during collection. These adventitious organisms might be undesirable psychrotrophic bacteria, those microorganisms that can grow during the cold storage of milk. Pregastric lipases are often added to the milk to increase flavor in the cheese (we'll cover this type later in this chapter).

Native milk lipase is mostly destroyed by pasteurization, while what is produced by microorganisms in the milk usually survives heat treatment. Why does this matter? The breakdown of the fat molecules in milk is something that the cheesemaker wants to happen at the appropriate time, which is usually during aging. Anything that causes their breakdown earlier will lead to undesirable flavors. Homogenization, rough agitation, rewarming of cool milk (such as can occur when warm milk from one milking is combined with cool milk from another), and cold storage with contaminating bacterial growth will all degrade the fat globules and allow native or bacterial lipase to begin to prematurely break the fatty acids into short sections, otherwise known as short-chain fatty acids. It is these types of fatty acids that can be associated with off flavors.

While all milk has *native* lipase, goat's milk has more than cow's milk, and sheep's milk has hardly any. Since lipase breaks down fat, and milk fat contains many fatty acids that can be broken down to produce flavors (including rancid and animal/barnyard flavors), the content of lipase in the milk and the things that happen that make fat globules vulnerable to lipase are critical for the creation of good-quality cheese.

Proteases are the next milk enzyme that's of prime interest for the cheesemaker. As described previously, the job of these guys is to break down proteins. As with lipase there are natural proteases in milk. These are mostly destroyed by pasteurization, but additional proteases are provided in adequate quantity by the use of rennet and starter culture bacteria (more on that in the following section). Proteases break the proteins down in stages. The end goal is the development of flavor and the goal texture of the cheese type being made. These processes can only occur in the right environment, both in the vat and on the aging shelves.

The last milk enzymes I will mention here are *alkaline phosphatase* and *lactase*. Alkaline phosphatase is of interest because pasteurization eliminates this enzyme from milk. When regulatory officials need to check milk to verify that it is indeed pasteurized, they run a test that measures alkaline phosphatase. *Lactase*, an enzyme that breaks down lactose, is not naturally present in milk, although some data suggests that raw milk may contain lactase that is produced by adventitious bacteria.

NUANCES AND KNOWLEDGE ABOUT CULTURE

Milk is a nutritious environment attractive to many different types of microorganisms, some that we want and others that we don't. It is this somewhat paradoxical reality that you as the cheesemaker must learn to acknowledge, embrace, and orchestrate. When raw milk is used for cheesemaking, for example, there will already be some bacteria in the milk, but added culture will still be important. When pasteurized milk is used, choosing and using cultures becomes even more critical.

Cheese cultures and their proper use are a somewhat mysterious and tricky aspect of cheesemaking. This has not been helped by the fact that the scientific names of cultures seem to be forever changing. For someone (such as me) who has issues even spelling names such as *Leuconostoc mesenteroides,* or its subspecies *cremoris*, it doesn't help that the same bacterium has had other names in recent history. So in this section I hope to help you become a bit more comfortable with these incredibly important cheesemaking cast members, so you can better harness their creative talents. In the recipe chapters I'll cover specific culture choices and options for each cheese type.

It is the role of culture to help transform milk into a tasty and well-preserved food. Cultures are usually bacteria but also can be yeasts and molds. When they first enter the scene, they rapidly begin to consume the milk sugar (lactose) and turn it into lactic acid. It is this acid that will preserve the cheese during aging or storage, as well as create a pleasing flavor. The second role of cultures is to *die*. This may sound strange, but it is a crucial part of the process. When cultures die, their cell walls leave behind enzymes that will further break down proteins and fats into tasty and texturally pleasing elements. Finally, cultures help create surfaces and textures that make cheese unique in look and flavor, as with washed rind cheeses (stinky red cheese) and bloomy rind cheeses (white furry cheese).

The term *cheese culture* covers three main types of organisms—bacteria, yeasts, and molds—used to turn milk into cheese. Some of these are added to the milk (*starter culture*), some exist in the milk or enter it accidentally (*native culture*), and others are added at varying stages of the cheesemaking process (*ripening cultures*). To better understand how these three types of culture interact and influence your cheesemaking, let's discuss each one in more detail.

ADVENTITIOUS VERSUS ADVANTAGEOUS

The words "adventitious" and "advantageous" might sound like synonyms, but they have distinctive meanings that sometimes overlap, which doesn't always help keep things clear. "Advantageous" is used when something is beneficial or gives the object an advantage. You can remember it by using the word *advantage* as a reminder. "Adventitious," on the other hand, means that the subject appears on the scene without being invited—meaning spontaneously or sporadically. I can remember this word by thinking of the word *adventure* as the base word.

Bacteria, yeasts, and molds that enter the milk during milking or at any point once the milk leaves the animal's udder can be called either *contaminating* or *adventitious*. Some of these might be useful or beneficial to the cheesemaker so they are also advantageous. Since the word "contamination" sounds negative, I will be using the word "adventitious" when bacteria that are beneficial are being referred to and "contaminants" when the nonbeneficial type are the subject.

STARTER CULTURE

Starter culture is the type of culture most cheesemakers, even beginners, will likely already be familiar with. There are four main ways to add starter culture to milk: freeze-dried *direct vat set (DVS)*, also known as *direct vat*

KEEPING CULTURES AFFORDABLE

Many aspiring home cheesemakers encounter quite the sticker shock when purchasing prepacked, single-dose cheese cultures. Often, these convenient packets, also known as sachets, are sold in kits, through online home cheesemaking supply companies or at local retail shops featuring cheesemaking and home preservation supplies. While easy to use, these packets are much more expensive than the same culture bought in multiple-dose packets; you are paying for convenience. If you find the prices limit your options for making cheese, consider moving up to multiple-dose packets purchased from small commercial suppliers (found in the resource section of this book). In chapter 2 I will teach you how to properly measure and repackage larger amounts of culture. So don't let the prices you might encounter keep you from becoming a cheesemaker—and a thrifty one at that!

MILK QUALITY

Milk in a healthy udder is believed to be sterile; in other words, it is free of bacteria. The teat, however, has bacteria growing on the outside and living inside the teat canal. When milk—a near-neutral pH solution filled with nutrients for both mammals and bacteria—leaves the udder, it is immediately an attractive host to all sorts of adventitious organisms. Some of these are beneficial bacteria, but there is an equal potential for the hosting of bad bacteria. Some of these bad guys might only damage the quality of your cheese, or they might make someone sick. Very simply, the healthier the animal and the cleaner the milk collection process, the lower the bacteria count in the milk and the higher the proportion of good guys to bad. Once cleanly collected raw milk is rapidly chilled and stored, it is still not safe from the growth of contaminants and other adventitious organisms. Certain psychrotrophic bacteria, yeasts, and mold types love cold temperatures and can crowd your otherwise clean milk with bacterial growth that could cause issues for your cheese.

Pathogenic and spoilage contaminants (microorganisms that can make you sick) include coliforms, *Clostridia*, *Campylobacter*, *Staphylococcus aureus*, *Listeria monocytogenes*, and *Salmonella*. Coliforms usually make themselves known during the cheesemaking process but are not evident in fluid milk. But scary bugs like *Listeria* and *Salmonella*, both responsible for fatalities, are completely invisible and without flavor during production.

inoculation (DVI) or simply *direct set (DS)* (which I wish they would just call "ready to use"); *mother culture*, the oldest, most time-consuming, and most unpredictable method; *bulk starter*, which is basically mother culture that someone else has prepared for you and often sold as frozen pellets; and *whey culture*, which is whey from the previous day's cheesemaking used to start the next batch.

The hands-down most popular form in which to purchase and use starter culture is as freeze-dried, powdered *direct vat set culture*. While a bit more expensive than using a mother starter or frozen bulk starter, direct set cultures have several advantages, including ease of storage (they can be kept in a regular freezer) and reliability (the calculations of their acid-producing capacity have been done for you); they are also less labor intensive (when you want to make cheese, the culture is ready to use).

To begin to understand and harness the potential of starter bacteria, you must remember that they each have an ideal temperature at which they will thrive, and also at which they will die. Depending on the type of cheese you are making, you must choose the right type of starter culture to do the job. There are warm-loving bacteria, or *mesophiles*, and heat-loving bacteria, or *thermophiles*. There are also cold-loving bacteria, otherwise known as *psychrotrophes*, which are not desired and are never intentionally added to cheese milk.

Knowing the way temperature affects the bacteria will help you control the development of acid in the vat—the key to preparing curd for draining and aging. Fortunately, most culture companies help us out by providing advice and labeling their culture choices according to the style of cheese you want to make. But since there is **no single perfect bacteria or blend of bacteria for any single cheese**, you will see lots of overlap and what appears to be redundancy when you are shopping for culture.

NATIVE CULTURE

Many home and small commercial cheesemakers, and increasing numbers of moderately sized artisan producers, make raw-milk cheeses. But remember, no matter how clean the milk source is, there will be a population of bacteria, mold spores, and yeasts present. While I refer to these organisms as "native," it is important to remember that they do not live in the milk naturally while in the udder of the animal but instead enter it

CHEESE HISTORY:
NATURALLY COAGULATED CHEESE

There are a few traditional cheeses still made without the use of any added culture bacteria. These cheeses rely upon the assumption that the milk will contain a significant portion of natural/native beneficial bacteria that when given the right conditions (temperature and time) will grow, produce acid, and (with or without the help of coagulant) thicken the milk into curd. This is referred to as soured, clabbered, or clotted milk. You can imagine that in times past starter culture was not available and mother cultures were difficult to maintain for most home cheesemakers (by home cheesemakers I mean almost every family who lived in a rural setting—a family cow or goats were prized possessions, and cheesemaking in some form was an essential domestic activity). In addition to the lack of an outside source for starter culture, milking and processing equipment was made from materials and in a fashion that allowed them to harbor a plethora of bacteria. This was not such a bad thing, as the colonies of bacteria nestled in cracks and surfaces acted as little culture factories, reliably inoculating the milk. The lack of detergents and sanitizers insured the perpetuation of this situation. Given time and consistent use, a fairly safe process actually existed. To duplicate this process today might be thought of by some as potentially inconsistent and even a bit risky, but there was a time when it was the best working option for making cheese.

A selection of naturally coagulated cheeses from well-known Virginia cheesemaker Rona Sullivan. Photo courtesy of Dickie Haskell

Two historical examples of clabbered milk cheese are Devonshire clotted cream and the German cheese Schmierkase, which many compare to American cottage cheese. Each relies on natural bacteria to first populate the milk or cream. Both also rely on heat to help complete the cheese (and in the process kill off most, if not all, of the bacteria, helping make a safer product). For more on how to replicate these historical cheeses, see chapter 8.

Modern cheesemaker Rona Sullivan, of Sullivan's Pond Farm, Virginia, makes all her cheeses using clabbered milk, which she calls "bonnyclabber" after the historical cheeses made from a similar process by Scotch-Irish settlers in the region. Rona's process involves allowing the raw goat's milk to clabber without the addition of any starter culture, rennet, or heat, after which it is drained, pressed, and aged before sale.

during and after the milk collection. It is also important to remember that while most of these microorganisms are usually not harmful, many are. If you are a believer in whole, pure food, as I am, it is very easy to get caught up in the "Raw Rules!" mantra, but to be a believer without really understanding what you are revering might eventually get you into trouble. See the sidebar on milk quality for more on common contaminants in milk.

Many native bacteria in milk are known as *nonstarter lactic acid bacteria* (NSLAB) because they aren't major producers of acid so are not especially helpful as starter cultures. NSLAB are considered highly beneficial and help raw-milk cheeses develop unique flavors and ripen faster. They are so helpful, in fact, that strains are being developed commercially to be added back into pasteurized milk to help emulate the ripening qualities and flavor profiles of raw-milk cheeses.

RIPENING CULTURE

While it is easy to think of the terms "culture" and "bacteria" as synonymous, this is not necessarily true, especially when it comes to ripening (also known as secondary or adjunct) cultures. Think of starter culture as the folks that get the cheesemaking process started and ripening cultures as the bacteria, molds, and yeasts that add to the process during aging and ripening. Some good examples of ripening cultures are propionic acid bacteria that create the large eyes in Emmental cheese (also known in the United States as Swiss cheese); white molds for producing cheeses such as Camembert and Brie; blue molds for ripening blue cheeses; coryneform (*Brevibacterium linens*) bacteria for use in creating red, smeary, smelly cheeses such as the infamous Limburger; and yeasts that are used in combination with both *B. linens* washed cheeses and some white mold (bloomy rind) cheeses. The recipe chapters contain charts showing specific ripening cultures and their uses. Native milk bacteria (the NSLAB mentioned above) also serve as secondary ripening cultures, especially in raw-milk cheeses. Ripening cultures are usually measured and labeled in "doses" rather than units. I'll cover these cultures in more depth in the recipe chapters where they are used.

CLASSIFYING AND DEFINING CULTURES

Starter bacteria are grouped by several methods. The most common is based on their ability to grow at certain temperatures—either mesophilic or thermophilic. Another is grouped by their ability to break lactose down to one or more varieties of lactic acid. Those that can break lactose into more than one by-product are further divided by their ability to metabolize another component of milk—citrate—at the same time as lactose into desirable by-products. Understanding these differences can help you select very specific cultures to achieve a unique flavor profile for your cheese.

DEFINING A CULTURE BY GROWTH TEMPERATURE

The primary way most cheesemakers divide culture is by the temperature range at which they will best produce acid. Both mesophilic and thermophilic bacteria have a wide range of temperatures at which they will survive, but a shorter range at which they will thrive. This is important to know so that you will understand how, even if the temperature isn't ideal, the bacteria can still offer its services in the cause of creating a superior cheese. Even if the temperature exceeds the range at which the bacteria will survive, they still can contribute their cell wall enzymes to the cause. The table that follows gives optimal and survival temperature ranges for common cheesemaking cultures.

DEFINING A CULTURE BY LACTIC ACID PRODUCED

Another, slightly more scientific way to group bacteria is by the type of lactic acid that they produce and the by-products of one of these two lactate types. Starter bacteria ferment lactose into two varieties of lactic acid, L-lactate and D-lactate. These two varieties are structurally mirror images of each other (which in chemistry terms is an *isomer*). While the pathway that results in the production of L-lactate has no other by-products, the process that produces D-lactate *usually* also produces carbon dioxide (CO_2), ethanol, and small amounts of acetate. Some starter bacteria also produce both L- and D-lactate. Bacteria are called *homofermentors* if there is only one resulting product (L-lactate or D-lactate with no other by-products) and *heterofermentors* if they ferment sugar to more than one end product (L- and D-lactate, CO_2, ethanol, and acetate). The production of CO_2 will, quite obviously, result in some small eye formation, and the production of ethanol and acetate will factor in to other by-products for flavor and aroma later.

DEFINING A CULTURE BY ITS CITRATE METABOLIZING POTENTIAL

Finally, the heterofermentors are divided by their ability to metabolize citrate. Citrate is present in milk in very small quantities, but its breakdown by bacteria is quite important for the development of flavor and aroma in many varieties of cheese. Citrate is not metabolized separately but is instead used at the same time as lactose. When citrate is involved in the fermentation process, the by-products include acetoin, acetate, and diacetyl (in addition to lactate and carbon dioxide). A letter system exists to label a culture as to whether it contains a citrate-metabolizing bacteria (also called citrate + or cit+ for short) or not; unfortunately, it uses the same letters as above: L and D. I find this confusing. "L" cultures are citrate + cultures containing *Leuconostoc mesenteroides* ssp. *cremoris.* "D" cultures contain citrate + *Lactococcus lactis* ssp. *lactis* biovar. *diacetylactis*, known also by its old quite similar name of *Lactococcus* ssp. biovar. *diacetylactis,* "diacetylactis" being the origin of the "D" designation. If a culture contains both cit + *Lactococcus* and *Leuconostoc*, it's called a "DL" culture. If it contains neither, it is an "O" culture. D cultures typically produce flavor and aroma faster than do L cultures (or DL cultures). This is something to keep in mind when making fresh cheese versus aged.

Once the bacteria run out of food (lactose and the two simple sugars it is composed of), they die off, leaving behind enzymes that were contained within, mostly in their cell walls. These enzymes contribute greatly to the further breakdown of butterfat and protein. If you are making a cheese that is meant to age for any length of time, choosing the right culture can make the difference between an okay cheese and a really great one.

Let's take a closer look at some specific starter cultures and their strengths and talents. Remember, many of these cultures are paired with others to complement and increase both acid production and flavor and aroma development. In the recipe chapters you will find charts of suggested cultures for use in specific types of cheese. Table 1–1 lists starter bacteria used primarily in the production of cheeses whose temperature in the vat does not exceed 102°F (39°C). Table 1–2 lists bacteria whose cooking temperature is much higher and can survive up to about 131°F (55°C).

PHAGE AND CULTURE ROTATION

Multidose packages of starter culture are produced in a "series"; for example, as MA 4001, 4002. To understand the reason for this and the importance of alternating which number in the series you use, you need to first learn a little bit about the cheesemaker's bane—phage (pronounced by most as phah-jsh).

Phage, more correctly referred to as *bacteriophage*, are virus particles that infect starter lactic acid bacteria. They use the bacteria as a host to replicate themselves, attach to receptor sites on the surface of the soon-to-be host cell, then inject their own DNA into the cell. The cell immediately begins replicating the invader's DNA and grows a bunch of new "baby" phage. Once the cell can no longer hold its new crop, the phage release an enzyme that blows the cell open and sets the now greatly increased number of attackers loose on more unsuspecting bacteria. In the best case this will lead to slow acid development in the vat, but in the worst case it causes what is often called "dead vats," in which no acid production occurs and the entire batch is lost. And not only the entire batch—the whole cheesemaking process must cease until the phage can be destroyed or a new, resistant starter culture can be found.

There is another form of attack that occurs that is not quite as aggressive and can actually "immunize" the host cell. In this case phage joins up just as in the above scenario, but instead of replicating new phage, it joins its DNA with the chromosomes of the host and remains a part of the host. The bacterium, while sharing its DNA duplicating powers, is now immune from attack by the former, cell-rupturing type of invader. At times it sounds like a great premise for a blockbuster movie.

Phage, being viruses, are very hardy; they can even survive for many years frozen. For these reasons they are quite difficult to get rid of. They also prefer raw milk, according to most data, and definitely are present more in whey than in any other milk-related medium.

TABLE 1-1. MESOPHILIC BACTERIA CULTURES (FOR CHEESES COOKED NO HIGHER THAN 102°F/39°C)

Name, Abbreviation (abbreviation used varies between suppliers)	Previous Name (often still used)	Optimal Growth Temp Range	Max Temp	Roles (in order of priority)	Cheese Types	Notes
Lactococcus lactis ssp. *lactis* LL	*Streptococcus lactis*	77–86°F (25–30°C)	108°F (42°C)	Acid production	All types (except high heat)	LL is the main culture used in all cheeses not cooked above 102°F (39°C)
Lactococcus lactis ssp. *cremoris* LLC or LC		77–86°F (25–30°C)	108°F (42°C)	Acid production	All types (except high heat)	A bit more salt- and temp-sensitive than LL
(Cit+) *Lactococcus lactis* ssp. *lactis* biovar. *diacetylactis* LLD or LD	*Lactococcus* ssp. biovar. *diacetylactis*	64–77°F (18–25°C)	102–104°F (39–40°C)	Acid production, CO_2 (gas), diacetyl (buttery flavor)	All types, (except high heat) where small eyes are acceptable	Produces gas faster than *Leuconostoc*
Leuconostoc mesenteroides ssp. *cremoris* LMC or LM	*Leuconostoc cremoris*	70–72°F (21–22°C)	100°F (38°C)	CO_2 (gas), diacetyl (buttery flavor)	All types, (except high heat) where small eyes are acceptable	Consider adding this while milk is cooler to get max gas and diacetyl production

Consequently, whey culture and bulk culture that is grown at the cheesemaking facility is more susceptible to phage contamination than starters prepared at another facility. Companies that manufacture starter culture must maintain strict protocols and testing to ensure that phage is not a problem. For this reason phage-resistant strains of the same culture are produced and labeled with a different number. By using rotations of the same bacteria in resistant forms, you can help limit the likelihood of any phage present in your environment zeroing in on and infecting the starter culture.

> Phage are often blamed for all sorts of problems with the cheesemaking process, problems that could easily be caused by other issues. Don't be a phagophobe!

Besides rotating culture strains, there are steps you can take to limit the likelihood of an all-out takeover. First, removing whey from the cheesemaking area as soon as possible is probably the first and best step toward prevention. Some artisan cheese plants won't even allow the whey to be returned to the farm from which they get their milk for fear that the phage in the whey will return through environmental contamination of the milk at the farm, with each round trip acquiring more strength and ability to infect the lactic acid bacteria. If you are using whey as a starter or for brine, problems could arise over time.

The final preventive measure you can easily take to prevent phage colonization is by using chlorine, a highly effective phagicide, to sanitize equipment and floors. (This should be done only *after* making the cheese to avoid contamination.) Using a chlorine solution of 100 parts per million (see chapter 6 for more on mixing sanitizer), rinse all equipment and surfaces that have had contact with whey. If you are a commercial producer, pay close attention to cracks in the floor and

TABLE 1-2. THERMOPHILIC BACTERIA CULTURES (FOR CHEESES COOKED OVER 102°F/39°C)

Name, Abbreviation (abbreviation used varies between suppliers)	Previous Name (often still used)	Optimal Growth Temp Range	Max Temp	Roles (in order of priority)	Cheese Types	Notes
Streptococcus thermophilus **ST**	*Streptococcus salivarius*, ssp. *thermophilus*	104–113°F (40–45°C)	131°F (55°C)	Comes in two varieties, one that produces acid quickly and one slowly	Fast acid producer used in all types of high temp. cheeses; slow producer used primarily in bloomy rind cheeses	Often used in lower temperature cooked cheeses to contribute qualities during ripening
Lactobacillus delbrueckii **ssp.** ***bulgaricus*** **LB or LDB**		108°F (42°C)	127°F (53°C)	Acid	Stretched curd and semihard to hard cheeses	Does not grow at high temperature of ST but survives and grows as temp drops
Lactobacillus delbrueckii **ssp.** ***lactis*** **LBL**		108°F (42°C)	127°F (53°C)	Acid	Stretched curd and semihard to hard cheeses	Does not grow at high temperature of ST but survives and grows as temp drops
Lactobacillus helveticus **LH**		108°F (42°C)	126°F (52°C)	Acid	Stretched curd and semihard to hard cheeses	Does not grow at high temperature of ST but survives and grows as temp drops
Lactobacillus casei **ssp.** ***casei*** **LBC**		108°F (42°C)	113°F (45°C)	Acid and aroma	Can be added to pasteurized milk	Often found in raw-milk cheeses as NSLAB

other nooks and crevices where residual whey could establish a phage population.

With all this said, it's important not to become a "phage-o-phobe." Over and over phage are blamed for all sorts of problems with the cheesemaking process, problems that could easily be caused by other issues, such as soft coagulation, slow coagulation, slow acid development, and many other possibilities. It is understandable that we are all a bit wary about phage infestation, but don't jump to the phage conclusion without first learning about the other possible explanations.

LIPASE—FOR EXTRA FLAVOR

You first met the enzyme lipase when we talked about the native lipase in raw milk. Lipase is an enzyme that breaks fats into short-chain fatty acids and helps create flavor and aroma in cheese. Here we'll talk about the varieties of lipase that you can purchase and add to the milk to help produce a particular flavor profile in the cheese.

In more traditional times cheese was coagulated using not a purchased extract of rennet from a young ruminant's stomach, nor a laboratory-produced coagulant, but an actual piece of the stomach lining from an animal on the cheesemaker's farm (or nearby) or a paste made from the same lining. In fact, in some areas of the world and for some traditionally produced cheeses, rennet paste is still used. In addition to coagulant enzymes (chymosin and pepsin), the paste also contains lipase. The lipase in the stomach lining is much stronger than native milk lipase. And depending on the species from which the enzymes are harvested, the strength also varies. Calf lipase is the mildest, kid the intermediate, and lamb the strongest. You can also purchase a mixed variety of kid and lamb. You can also vary the amount used from that recommended in a recipe or make your own blend. Remember, the addition of lipase to a recipe is *always* optional, but you cannot come close to duplicating some traditional cheese flavors without using lipase.

In my research I was unable to locate any quality vegetarian-based lipase. It does exist, but from what the manufacturers told me, it is meant for large, industrial processed cheeses, not (in their words) "real cheese." It does not break down fats in a reliable, predictable fashion over aging.

CALCIUM CHLORIDE—KEEPING THINGS IN BALANCE

I think no cheesemaker, home or commerical, should be without a supply of calcium chloride. This natural additive can save your process at certain times of the season when the milk needs its assistance. Some cheesemaking books simply call for its addition in each recipe, as a way to kind of hedge their bets that the recipe will work for the beginner. The more advanced cheesemaker might be able to anticipate its need and know when to add calcium chloride. Let's learn some more about what it is and when to add it. You'll learn how much to use in chapter 2. Appendix C contains a reproducible chart on suggested dosage rates for calcium chloride.

Calcium chloride ($CaCl_2$) is a clear, bitter, weakly acidic mineral salt usually sold in a properly diluted solution that is ready for the cheesemaker to use (you can purchase it as a granule and dilute it yourself, however). It is a natural mineral salt and is legal to add, in the right concentration, to commercially produced cheese as well as that made for home use.

Calcium chloride is added to the cheesemaking milk before adding rennet, never after! It will help reduce the coagulation time and create a firmer curd. Some cheesemakers use it whenever working with goat's milk, and definitely, it should be used when using store-bought milk of any kind. It can also assist with a better coagulum being formed when working with milk from animals late into their milking cycle, such as in the fall and winter. Be sure to measure it carefully, as too much will cause a reverse effect on coagulation and can lead to a hard curd and a bitter taste in the cheese. (If you ever decide to taste the $CaCl_2$ solution, be sure to have some orange juice or some mouthwash on hand, as it's pretty awful.)

If you remember, back in the milk mineral section, you learned that calcium exists in several forms in the milk; as positively charged calcium ions (Ca^{2+}); bonded to phosphate ($Ca_3[PO_4]_2$) and existing within the casein micelles (colloidal); and bound to citrate. Because calcium plays such a major role in the formation of a coagulated mass as well as in the final texture of the cheese, anything that upsets the balance and volume of calcium will have its effect on every stage of the cheesemaking process. Let's go over a few of the most common things that will alter the availability of calcium and therefore conditions under which adding calcium chloride might be helpful.

COLD STORAGE

During cold storage of milk at refrigeration temperatures and below, the casein micelle loses some of its colloidal calcium. This loss also reduces the buffering ability of the milk, and a slight rise in pH might be observed. The loss of colloidal calcium can cause a weak curd during coagulation.

OVER-PASTEURIZATION

While not directly damaging to calcium, milk that has been pasteurized beyond the recommended temperatures and/or times can suffer from slow coagulation because the whey proteins have been altered (denatured) so that they, in essence, coat the surface of the casein micelle.

LOW PROTEIN LEVEL

Some milk is naturally lower in total protein; consequently, the casein content, along with the calcium content, will be lower. Goat's milk typically has a lower ratio of casein to whey protein and the least of the best cheesemaking protein alpha s1. When casein levels are lower, there is also less colloidal calcium phosphate.

HIGH SOMATIC CELL COUNT

High somatic cell counts that correlate with mastitis (udder infection) can lead to a higher starting pH of the milk, which can in itself cause a longer coagulation time. In addition somatic cells also introduce enzymes that will damage the protein micelles, causing a loss of colloidal calcium. Somatic cells are blood components, specifically white blood cells (leukocytes) that migrate into the udder to attempt to fight off an infection. Since blood and its components have a pH of over 7.0 (making them alkaline), they have the resulting effect of raising the pH of the milk.

HIGH ENZYMATIC ACTIVITY

Enzymes that break down proteins, proteases, specifically plasmin are naturally present in the milk but can increase in volume toward the end of lactation and during winter. Goats seem more prone to this effect than cows. As I said regarding somatic cells, the damage caused to the protein micelle by the enzymes causes a loss of calcium and a resulting prolonged coagulation time and softer curd.

CHOOSING YOUR COAGULANT: ANIMAL, VEGETABLE, OR MICROBIAL

For years I thought of rennet as simply the ingredient that made the milk coagulate, when in reality coagulation is just the most *visible* result of adding rennet to milk. The enzymes in rennet (coagulant) play an equally important role in the development of flavor and texture in cheese. So let's spend some time getting to know the true nature of rennet.

While some cheeses are coagulated using acid—either through adding an acid such as vinegar, lemon juice, or citric acid, or by letting the milk develop acid produced by starter culture—most cheeses call for the addition of a coagulant called rennet (the terms "rennet" and "coagulant" are used in this book interchangeably). Rennet, in its oldest and still most recognized form, is made from the stomach lining of a calf, kid, or lamb that has not yet eaten solid food. The stomach lining of the baby ruminant contains two enzymes, *chymosin* and *pepsin,* that both coagulate milk in the baby's stomach. This keeps the milk from passing too quickly through the baby's system, allowing it time to absorb the nutrients. Pepsin is the main ingredient in junket rennet tablets that you can find at some grocery stores. Junket rennet, however, is a poor choice for making cheese as it will coagulate milk, but will add bitterness as the cheese ages, and often contains other chemical ingredients that you wouldn't want in your cheese!

The gold standard of coagulants is *calf rennet*; although kid and lamb are of superior quality as

WHERE "RENNET" GOT ITS NAME

If you have an inquiring mind, you may wonder where the term "rennet" came from. The enzyme in the young ruminant's stomach that coagulates milk is now called chymosin, but not so long ago it was known as *rennin.* Another enzyme, renin (ree-nunh), is produced in the kidneys. In an attempt to limit confusion, the gastric enzyme was renamed chymosin. The kidney glands, by the way, are more correctly called the renal glands—hence the name of the enzyme renin.

DAVE BLECKMANN'S GUIDE TO MAKING THISTLE COAGULANT

For me the allure of growing and harvesting my own rennet and using it to make cheese in my own kitchen was irresistible. It was a natural follow-up to the hobby of home cheesemaking itself, allowing me to take one more step down the path of "doing it myself" when making cheese.

In my climate (USDA Zone 8b), cardoons that I started from seed grew to 4-foot-high plants in one season. They survived the winter to grow to 7 feet the second year. I used cardoon seeds (*Cynara cardunculus*) from Territorial Seed Company. The flowers bloomed in late summer in my climate and had harvestable purple stamens for 3 to 4 weeks.

The flowers, which initially resemble artichokes, should be harvested when the purple stamens are in full bloom, before they start to wither. Cut the flowers in half, and remove the purple section of each stamen by cutting just where it turns brown. Only the purple portion of the stamen is used, as this is where the milk-coagulating enzymes reside. If not using right away, spread them out on clean paper towels or a similar surface and dry for 3 weeks (I like to use a clean household furnace filter as a drying surface as it allows air to move freely through the drying stamens). After drying, seal the stamens in an airtight jar, where they will keep for 2 years.

To prepare, grind the stamens to a powder using a mortar and pestle or pulverize them in a blender. Use about 1 gram of powder for each liter of milk, or ⅛ ounce per gallon. If using fresh stamens, you can reduce the amount by about 15 percent, as the enzyme is more active. Steep the stamens in warm (86°F [30°C]), clean water for 30 minutes. Use 10 milliliters for each gram of powder, or 1.2 ounces for each ⅛ ounce of powder. Filter the liquid through a coffee filter, and discard the solids. Add the liquid to ripened cheesemilk at 86°F (30°C). Coagulation should occur in 30 to 45 minutes.

The active enzyme in thistle-based rennet, cynarase, cleaves proteins much more indiscriminately than do traditional rennet enzymes and tends to produce a softer cheese. When used with cow's milk, however, an unpleasant bitterness develops in the cheese a week or so after production. This does not occur with goat and sheep milk, so these are appropriate for creating aged cheese. There are a wide variety of cheeses that can be produced using thistle-based rennet, and nearly any recipe could be adapted to use it.

Some of the tools and steps in producing thistle rennet
Photo courtesy of David Bleckmann

David Bleckmann is an avid home cheesemaker, blogger, father, and coauthor with Sasha Davies of The Cheesemaker's Apprentice *(Quarry Books, 2012).*

well, they are not as readily available. (All of these animal-sourced coagulants are currently often called "traditional" rennet.) Some cheesemakers make their own rennet or rennet paste from their own kids, calves, and lambs. Calf rennet contains 88 to 94 percent chymosin and 6 to 12 percent pepsin. As the baby ruminant switches from milk to a solid diet, the ratio of chymosin to pepsin slowly flip-flops, until it is basically a mirror of 88 to 94 percent pepsin to 6 to 12 percent chymosin. Pepsin has less clotting strength but greater proteolytic strength (this ratio, called the *clotting to proteolytic ratio* [CPR], depends a lot on the species from which the pepsin was harvested). When you are purchasing animal rennets, the supplier should be able to tell you the ratio of chymosin to pepsin (it should be no less than 90 percent chymosin), the dilution ratio, and, of course, the species source.

My personal favorite coagulant is *microbial rennet*, which has gained in popularity and is now widely used at the artisan and industrial level. For those seeking to produce vegetarian or kosher cheese, or who are not comfortable with the unknowns of using calf-harvested products—such as humane raising issues and animal health, this coagulant is the perfect option. Most microbial rennet available today is produced by the microbe *Rhizomucor miehei*, known also as *Mucor miehei*. **Microbial rennet has an undeserved reputation for causing bitterness in cheese.** This type of rennet has a proteolytic enzyme that coagulates milk in the same manner as animal chymosin but has greater proteolytic potential than properly prepared calf rennet. That being said, however, it is also not retained in the pressed cheese at the same level that animal rennet is retained. The *Mucor miehei* enzyme behaves similarly to chymosin in most other respects.

Another widely available coagulant is *fermented chymosin*. This coagulant is produced using genetically modified (GM) microbes into which the animal gene for producing chymosin has been inserted. Fermented chymosin is considered vegetarian and behaves very similarly to the animal source rennet in its ability to coagulate milk. Because it is 100 percent chymosin, though, it differs slightly in its ability to break down proteins during aging. The enzyme produced by these genetically engineered microbes does not have to be labeled as GM itself, even though it is the result of genetic engineering.

Finally, there are true *vegetable rennets* (as opposed to the microbial *vegetarian coagulant*), made from plants. The most well known is made from members of the thistle family. Coagulant made from the cardoon thistle is used to produce several well-recognized cheeses, such as Serra da Estrela from Portugal and Torta de la Serena from Spain. This type of coagulant works best with high-fat, high-protein milk. It has a very strong ability to break down proteins, so it can be guilty of producing bitterness. For more on making thistle rennet, see the sidebar. Other plants that have the ability to coagulate milk include fig trees, nettles, burdock, hogweed, and Lady's Bedstraw. Plant rennets do not have quite the coagulating power of their animal and microbial cousins, but they can add some interesting variety to your cheese.

THE PROCESS OF COAGULATION

As I mentioned before, the first thing that rennet does is to coagulate the milk. This is a complex process brought about by the enzymes in the rennet interacting with the protein clusters in the milk. Successful coagulation depends on several other factors, including the pH and temperature of the milk, the amount of calcium in the milk, and the type and quality of casein (milk protein). We'll address some of these issues in chapter 2.

During aging the same enzymes that caused the milk to clump continue to break down protein, a process called *proteolysis*. When this breakdown proceeds in an optimal fashion, complex flavors, aromas, and textures are more likely to develop—so using the right amount of rennet is as important for coagulation as it is for aging.

We have already talked a bit about what rennet enzymes, specifically chymosin, do to the protein in milk. If you recall, the casein micelles are little clusters of milk protein (α-casein, β-casein, and κ-casein), clumped together and suspended in the milk. The micelles contain calcium phosphate, referred to as *colloidal calcium*, which acts as the mortar or glue that bonds the cluster together. The κ-casein coats the surface of the micelle and prevents further clumping, which keeps them suspended in the milk, or "in solution." When added to milk at the right temperature, chymosin (or another similar protease) immediately starts attacking κ-casein and breaks, or "cleaves," the portion of the amino acid chain that protrudes from the micelle. Once this process has been completed, the first stage, known as the *enzymatic phase*, of coagulation is complete. The second phase of coagulation, when you can actually see the milk thickening, is not dependent on rennet, but on the physical properties of the altered casein micelles. We'll cover that later in chapter 2, when we talk about coagulation in more depth.

BEYOND COAGULATION—THE ROLE OF RENNET DURING AGING

After coagulation is complete, there is still some residual rennet in the curd. Some of this will be lost in the whey or be denatured by the heat from the cooking process (this is especially true for Italian-style and pasta filata–type cheeses, which are cooked at much higher temperatures than most other cheese types). The rennet enzymes that are retained play a key role in the breakdown of proteins during cheese aging, contributing greatly to flavor and texture development, as mentioned earlier.

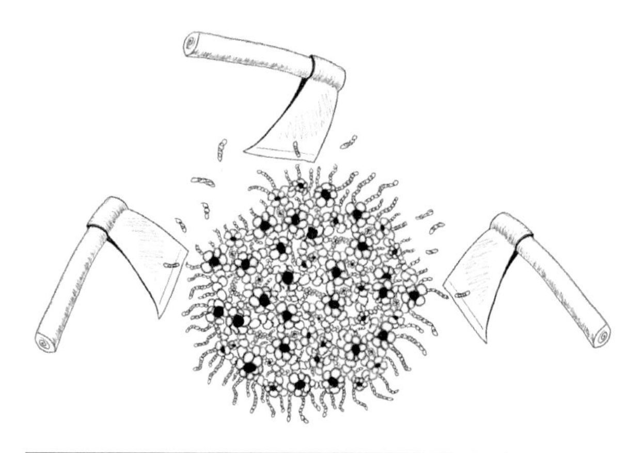

Author's illustration depicting enzymes (proteases) as hatchets as they cleave the kappa-casein protrusions from the casein micelle during enzymatic phase of coagulation

The use of too much of any coagulant can result in bitterness in the cheese and poor texture, thanks to an accelerated progression of protein breakdown. It's therefore very important that you know the right amount of rennet to use for the cheese type you are making. Put down that teaspoon, and get a graduated syringe for measuring! (We'll discuss proper tools in more detail in chapter 6.)

SALT—THE FINAL KEY INGREDIENT

Our last, but definitely not least, ingredient is salt, or sodium chloride. Salt serves several very important roles in the creation of a superior cheese. We'll go over those roles here, and then in chapter 2 we'll cover the details of how to best salt your cheeses—whether it be through adding salt to the curds, rubbing or sprinkling it on the outside of the cheese, or soaking the cheese in a saltwater solution called brine.

The first job of salt is to slow down, then bring to a halt, the production of acid in the cheese. It does this by stopping the starter culture bacteria, which are sensitive to salt, from growing. When they stop growing, they also stop producing lactic acid. By stopping the pH at the right spot, you help ensure the right finished flavor and texture in the cheese.

Salt also helps draw whey out of the cheese. This aids in finishing the draining of the curds. When properly drained, so that just the right amount of moisture remains, you help limit the growth of microorganisms, which will also help create the right finished flavor and texture.

Salt also helps protect the outside of the cheese (in the case of surface salted or brined cheeses) by preventing mold and bacterial growth on the surface of the cheese. It also provides some physical protection by helping dehydrate the rind and so creates a firmer surface that helps the cheese hold its postpressing shape, so it is less delicate during turning. It can also encourage salt-tolerant molds, such as white molds, to grow where their presence is desired.

The final role salt plays is to positively influence flavor, by both the added flavor of the salt itself and the flavors

Cheese being dry salted

it will help bring out. It also limits the growth of bacteria that would otherwise change the flavor of the cheese.

The temperature of the cheese during salting; the amount of fat in the curd; the size and shape of the cheese; and of course the concentration and components of the brine will all affect the rate of absorption and the loss of whey during salting.

SALT FOR TEXTURE, FLAVOR, AND AROMA

Over the course of this chapter, you have read about the casein micelle and its relationship with calcium phosphate, salt also plays a key role in this relationship. We learned how coagulant cleaves the hairlike protrusions of the outer layer of the casein clusters, which removes the repulsing charge and allows the clusters to clump and form a network in which is trapped butterfat. We also talked about how acid coagulation occurs (with or without the addition of rennet) when the pH is under 5.0 at room temperature and the casein micelles reach their isoelectric point and lose their electrical charge. This allows them to begin to form a network (but one less structured than when produced by rennet).

One of the other effects of the micelles reaching their isoelectric point is that they switch from being *hydrophilic* (water loving) to *hydrophobic* (water repelling). So if they can no longer bind water, the protein structure in the cheese can change from one of a springy, pleasant texture to a dry, harsh, brittle texture. What does salt have to do with this? The proper application of salt will help halt acid development and consequently help keep the cheese above that magic point at which the proteins reach their isoelectric point. It does this in a couple of ways: first, by impeding bacterial growth and the subsequent production of lactic acid; and second, by assisting *syneresis*, a process that removes more lactose, the food for the bacteria, preventing the pH from dropping too low and reaching the isoelectric point.

The first way salt increases flavor in cheese is the same way the right amount of salt will help the flavor of any food. Without getting into the chemistry of that

SALT-IN-MOISTURE AND WATER ACTIVITY

You may encounter these two concepts if you make cheese for long enough. Water activity (abbreviated a-w, which you can remember by thinking of as "available water") is the amount of water in a food that is available for use by other microorganisms. Without water, bacteria cannot function. Where cheese is concerned, you need just the right amount of water activity. But too much leads to spoilage by other bacteria and microorganisms.

Water in cheese can be "distracted" by adding salt. When salt is absorbed into a cheese and dissolves in the water (otherwise known as "salt-in-moisture") the water molecules are attracted to the salt and become unavailable for use by microorganisms. This same principle works with other ingredients that can be dissolved in water, such as sugar. Sugar has preservative qualities similar to salt in high-moisture foods—think about jams and jellies.

So a moist, wet cheese does not necessarily have a lot of water activity (or available water) if it has a high salt content. Feta is a great example of a cheese with a high level of salt-in-moisture.

phenomenon, let's just say our bodies need salt, but not too much. Salt all by itself is not tasty, but it has the amazing ability to magnify and enhance flavors when used sparingly. So the right amount of sodium chloride added to cheese will enhance the flavor by its mere presence. But it also assists with flavor development by helping prevent the growth of spoilage bacteria. Salt has no effect on aroma, however, as you can't smell sodium chloride. But you can certainly smell the results of undesirable bacterial growth in your cheeses!

FACTORS INFLUENCING THE RATE OF SALTING

The temperature of the curd, especially in milled curds such as cheddar, or of the wheel will affect how salt is absorbed for two reasons. First, if cheeses and curd are warmer than about 89°F (32°C), a surface layer of fat is likely to have coated the cheese or curds, which in turn blocks the absorption of salt and the drainage of whey. Second, if the curd is warm (but not so warm as to be leaking fat), whey drainage will be increased, and some salt may be washed away, or repelled, by the whey coming out of the cheese (this is more of an issue in dry-salted curd and cheese than it is in brine-salted cheese).

The length of time over which salt is applied to curd or to wheels can help with the amount of salt retained in the cheese simply because the first application of salt will bring a flush of whey out of the curd and cheese that can rinse away the salt (as mentioned above). So when dry-salting curds and wheels, recipes often indicate two or more applications instead of using the total amount of salt all at one time. Some cheesemakers compensate for this by estimating how much salt will be lost, then applying more.

A cheese that has a higher fat content will impede salt uptake, not just at the surface of the cheese, but throughout the interior. The large fat globules acts as a physical barrier to the movement of the sodium chloride into the cheese. High-fat cheeses, such as Romano, are often salted many times during the initial stages of aging or may need to be brined for longer than a similarly sized lower-fat cheese. See Table 2–4 for more information on the salt content of cheese varieties.

SALT-TOLERANT AND SALT-SENSITIVE MICROORGANISMS

The growth of microorganisms can be affected by salt in several ways. The growth of salt-sensitive microorganisms is slowed or halted while those that are salt-tolerant (also called *halotolerant*) might continue to grow well or even be stimulated by a certain amount of salt. The cheesemaker needs to understand that both unwanted and desired bacteria, yeasts, and molds will react differently to varying levels of salt.

Some "bad" bacteria are not inhibited by the addition of the 2 to 5 percent salt that is normal for most cheese types. Chief among these bacteria are the *coliforms*, a large family of bacteria, some of which are responsible for "early blowing" in cheese (more on that in chapter 4). While many coliforms are inhibited by salt, some are actually stimulated by a salt solution of 3 to 4 percent and not totally inhibited until the level reaches about 12 percent (which is not a normal salt level for any cheese).

Salt is used to promote the growth of some ripening cultures. Let's use blue cheese as an example. *Penicillium roqueforti*, the blue mold that makes blue cheese a kind of greenish-blue, really thrives when a small amount of salt is present—about 1 percent (in other words, for 100 pounds of cheese, 1 pound of salt). To take advantage of this, blue cheese curd can be partially salted in the vat. A bit more than 1 percent is added, since the moist, still draining curd will wash away a good deal of the salt with the whey. This early addition to the curd jump-starts the blue mold growth. After the cheeses are formed and in the aging room, they are usually rubbed with dry salt several different times. This higher concentration of salt on the surface of the cheese usually means that the outside layer of a blue cheese will be white and free of blue mold.

Another ripening culture whose performance can be affected by salt is the eye-forming propionic acid bacteria used to create eyes in cheeses such as Emmental (often called Swiss cheese). These bacteria are very sensitive to salt levels over about 2 percent, so Emmental-type cheese has a lower salt level than almost any other cheese (see Table 2–4 for more examples). You should know that because of its high cooking temperature there is less need for high levels of salt, as much of the lactose (what would feed the starter bacteria and allow them to keep creating acid in the cheese) has already left the curd. This is good to remember with any curd that has been cooked to high temperatures—anything over 112°F (44°C) and washed-curd cheeses.

2: CONCEPTS AND PROCESSES FOR SUCCESSFUL CHEESEMAKING

Now that you are a little more familiar with the properties and characteristics of the ingredients that go into the cheese vat, let's take a close look at how each step of the cheesemaking process is meant to guide these properties toward an end goal—a specific type of cheese.

We'll start by talking about the different things that can be done to prepare the milk for cheesemaking—pasteurization and thermization, homogenization, and standardization. From there we'll move to the more familiar steps of cheesemaking—culturing and ripening; adding things such as coloring and calcium chloride; coagulation; cutting the curd; cooking, stirring, and washing curd; draining and pressing; and finally salting (I am saving aging and ripening for their own chapter). Remember that you can refer back to chapter 1 for explanations on the underlying science behind each of the processes in this chapter.

CHOOSING AND PREPARING MILK

If you are a really lucky home cheesemaker, you will have your own cow, does (female goats), or ewes (female sheep) and a steady, or at least predictable, source of milk. If you are nearly as lucky, you will have a local source for high-quality, farm-fresh milk. But if you are like many home cheesemakers, you will be using store-purchased milk to make cheese. I was in this latter category for my first few months of cheesemaking, and I am here to tell you that you can make some darned good cheese using milk from the grocery store. When choosing store-purchased milk, try to select the least processed variety available—this means no ultrapasteurized milk, as it will not form a good cheese curd, thanks to the extremely high heat treatment that denatures a good portion of the milk proteins. Ideally, nonhomogenized milk, often labeled "cream top," should be selected. You will learn more about just how these two processes alter milk in a bit. In a perfect world certified organic milk is a superior choice, but read the label closely; in most parts of the United States, organic milk is often ultrapasteurized to extend its shelf life.

If you are using unprocessed farm-fresh milk, be sure to educate yourself about preserving the quality of this product through good hygiene from the time it leaves the animal to the time you process it into cheese. You should also educate yourself about the hazards that can be introduced through disease-bearing animals (see the sidebar about raw milk on page 35). The pasteurizing section in this chapter will tell you more about how to gently process milk that is better worked with when not raw. For commercial purposes there are many regulations involved, in all countries, when choosing how to process milk once it leaves the animal.

In chapter 1 we covered the components of milk, and I talked a little bit about some of the breed and species differences. As you proceed through this chapter, keep those differences in mind. In general, you can make almost every version of every type of cheese from all of the major milk sources. The different milks will need slight nudging and adjustments to the process and will produce different end products. Don't be afraid to try them all, and don't fall into the trap of thinking that there is cheese and then there is goat cheese (or sheep's cheese). In my mind there is good cheese and then cheese that could be improved upon.

When working with store-purchased milk, some guidelines are in order. First, never choose ultrapasteurized milk or cream. Ultrapasteurized milk will be labeled as such, while regularly pasteurized milk will be labeled as Grade A pasteurized. When buying the milk, ask the dairy case manager when milk is delivered, then buy it when it's as fresh as possible. If

WHAT'S IN A LABEL:
A CLOSER LOOK AT CERTIFIED ORGANIC

There is no single regulatory body that certifies organic producers, but in general rules are similar, thanks to recent changes in the US federal regulations that were meant to close some loopholes that held large producers to different standards.

Organic certification requires some wonderful things, such as an animal's access to pasture; natural fertilizers and land management free of herbicides and pesticides; no tail docking of cows; and no use of antibiotics and growth hormones. But here is the reason that our farm will *never* seek organic certification: If an animal requires the emergency use of antibiotics to save her life, she must be subsequently removed from the producing herd. While the regulations do not require that she leave the property, it is the rare farmer that can afford to keep a separate herd of retired animals that needed antibiotics one time to prevent death. When explaining this to customers at the farmers' market, we give people the example of their own children that they are hoping to feed certified organic cheese—rare is the person that wouldn't allow antibiotics to be administered to a seriously ill loved one. We feel that we owe our animals a similar amount of care and loyalty. So if you are working with a small farmer, understand that they may still follow all the principles of organic management, and perhaps exceed the humane requirements, but will still not seek official certification. Again, all the more reason to know your farmer and know their practices. Trusting a person is usually better than relying on a label.

you cannot talk to the manager, look for milk with an expiration date that is the farthest from the purchase date. Some cheese instructors recommend buying skim milk, then adding cream to duplicate nonhomogenized milk. This works really well if you have access to cream that is not ultrapasteurized or doesn't have any other alterations. Unfortunately, many of them have added dextrose (sugar) and thickeners. So give plain, store-purchased whole milk a try and also try working with skim milk and cream. Of course, if you can buy cream top (nonhomogenized) whole milk or goat milk from your local grocery store, you will be a few steps ahead!

For most of us the preparation of milk for cheesemaking is a very simple process: We take it out of the fridge (or right from the animal) and warm it up. But the real preparation starts when you choose what kind of animal will be providing the milk, how that animal is fed and cared for, and how the milk is harvested with regard to cleanliness and chilling. So let's go over a few things that you can look for when sourcing farm-fresh milk or when harvesting your own. This is not by any means a complete guide to producing milk but I hope will provide some foundation information that will help you understand the importance of these factors.

FINDING MILK FOR HOME CHEESEMAKING

If you don't have your own on-farm source for milk and you don't want to use the grocery store variety, finding a reliable, high-quality local source of milk might be your biggest challenge. Most states vary in how they regulate the sale of milk from an unlicensed source, such as a home dairy. If you are lucky enough to live in a state where limited sales of raw milk are legal, then you might be able to find a source by simply putting up a flyer at a local livestock feed store or posting a want ad on an online listing, such as Craigslist.

In some states so-called "cow shares" or "herd shares" are fairly common, whether legal or not. The concept of a cow share is that you actually own or share ownership of a cow or goat that is not kept on your property. You then pay the farmer to feed and milk "your" cow or goat, and you get a portion of the milk. There are many farmers operating a cow share who

CHEESEMAKER MOTIVATIONS: THE SURVIVOR

For some folks preserving food takes on new meaning when your life is focused on either preparing for a hardship where food supplies would be scarce, living a remote lifestyle that precludes the purchase of outside supplies, or living a self-reliant lifestyle in a nonremote location. The survivor (sometimes called survivalist) will probably focus on cheeses that are easy to preserve, such as brined cheeses, hard aged cheeses, and even dried cheeses (I have a good friend that dries her feta-type cheese into little cheese sticks she can take on trips and store for long periods of time).

The survivor is unlikely to worry about perceptions of others and will feel the full satisfaction of seeing a cheese pantry filled with aging wheels and tubs of brined cheeses (I feel that way when the hay barn is filled with fragrant stacks of feed and the woodshed is stocked with neatly stacked cords of split and aged wood). Cheesemakers that fit this category may rely on many of the original, more ancient methods for preserving and aging cheese than many of their modern peers. After all, cheese was originally a survival food!

carefully abide by the legalities that, fortunately, still give each of us the right to eat products produced by and from livestock we own. But there are others for whom it is obviously a technicality. I am suggesting that you be aware of your involvement and responsibilities should you enter into such an agreement. And if you do, become involved enough in the operation to feel confident about the animal's health, its care, and the manner in which the milk is handled. Truly own the opportunity! For more on state regulations and some sources for raw milk, visit www.realmilk.com.

ANIMAL HEALTH

No matter how well the milk is cared for once it leaves the animal, it can't make up for the poor or less than superior quality milk that will be produced by animals whose health is compromised. This includes both the animal's physical *and* mental health!

Where mental health is concerned, factors such as overcrowding; poor housing's lack of access to natural feeds, such as grass for cows and sheep and browse (shrubs and trees) for goats; and impatient handling all lead to stress in the animal. As with humans, a stressful environment leads to compromised health. For milking animals this is usually reflected by an increased number of udder infections, known as *mastitis*. Even before a visible case of mastitis, called *acute* mastitis, occurs, many animals will have cases, sometimes chronic, of what is called *subclinical* mastitis, meaning there are no visible signs in the milk or upon inspection of the udder. Subclinical mastitis is tracked by the counting of white blood cells in the milk. This count is called a *somatic cell count* (SCC). The SCC also counts a few other cells, such as those that are shed from the interior of the udder during milking. Overall, though, high SCC is indicative of poor udder health. On farms where the animals are well observed and are under less stress, you typically see lower SCCs.

As with the mental health of the animal, physical health greatly affects the quality of the milk. Beyond the obvious concern for diseases that can be transmitted from the animal to those who consume unprocessed milk, animals with physical ailments will be more prone to mastitis and will not be able to consume feeds, even when the best possible feed is available, as well as a healthy animal will.

PRESERVING MILK QUALITY DURING HARVEST

There is a wonderful axiom that I frequently quote in both cheesemaking classes and dairy classes: "Milk was never meant to see the light of day." It was meant to go directly from the mama into the baby's stomach. Every process we expose it to once it leaves the animal's udder brings a risk of reduced quality and perhaps increased risk where health of the consumer is involved. With that in mind every step once the milk is harvested should be meant to limit its exposure to anything, even the air in the milking parlor, that could introduce elements that could limit quality. The possibility of these factors being introduced is quite great, thus the "official" fear of raw milk.

As with the intricacies of animal health, the proper collection and handling of milk is truly a subject for its own book. But it all boils down to a few important points:

1. Anticipating and preventing the contamination of the milk during collection
2. Proper cleaning and sanitation of all equipment that will make contact with the milk
3. Rapid chilling of the milk after collection down to refrigeration temperature within two hours after the beginning of collection
4. Prevention of contamination during holding and the shortest holding time possible

With this information I hope you will gain a greater respect for the importance of humane, natural care of the animal and obsessive attention to the proper collection of the milk. There is no doubt that superior raw milk offers the greatest potential for making uniquely distinctive and unadulterated cheeses, but even if you are going to be making raw-milk cheese, please don't skip the next three topics: heat treatment (pasteurization and thermization), homogenization, and standardization. To become an accomplished artisan cheesemaker, you should know about this stuff.

Heat Treatment

Some cheesemakers (unfortunately, here I even mean many small farmstead cheesemakers) use pasteurization as a Band-Aid for dirty or careless milking hygiene. This is unfortunate, as you might be able to make "dirty" milk safe by pasteurizing, but you certainly cannot make it taste better. Cheese made from poor-quality milk that has been pasteurized to make it safe will always be inferior to cheese made from superior quality milk—pasteurized or raw!

There will be times when even the smallest-scale cheesemaker might want to heat-treat or pasteurize raw milk. Many home cheesemakers may have access to only pasteurized milk. So let's define pasteurization and heat treatment and then talk about what these treatments do to milk.

Pasteurization

Pasteurization involves heating milk to a specific temperature, then holding it at that temperature for a set time (see the milk heat-treatment temperature table). The goal of pasteurization is the elimination of disease-causing bacteria, or pathogens. A secondary result is extended shelf life (thanks to less bacterial activity in the milk). Of course, it's not only the harmful bacteria that are destroyed during high heat treatments but also the advantageous types that otherwise can help develop some unique characteristics during aging. At small cheesemaking facilities and at home, pasteurization is usually done at the lowest acceptable temperature. This is believed by many to be the gentlest method of pasteurization, with less change to the protein structures of the milk and less of a cooked flavor. Low-temperature pasteurization can be accomplished by using a hot-water bath or home pasteurizer (often home canners are sold as home pasteurizers).

On a bit larger scale, a "batch" or "vat" pasteurizer is used. In larger facilities milk is most often pasteurized at a higher temperature and for a shorter time. In this kind of pasteurizer, milk flows through a series of pipes, or plates, during which it is heated—then quickly cooled. (Some cheesemakers believe that this method is actually better for the milk, since the milk is kept hot for such a short time in comparison to the lower-temperature method.)

There are some unwanted effects of pasteurization, especially at the higher heat levels, that should be of concern to the cheesemaker. The two most important side effects are the destruction of the enzyme lipase and the possible denaturing of whey proteins. While regular pasteurization temperatures have little effect on the whey proteins, higher temperatures alter these proteins in a manner that causes them to interact with the surface of the casein micelle (which you might remember is a cluster of proteins that will form a network to create the cheese curd). Once on the surface of the micelle, they can interfere with coagulation and later with syneresis (drainage) during pressing, leading to poor body and cohesion of the curd mass. It is believed that pasteurization temperatures not above 161 to 163°F (72–73°C) do not significantly denature whey proteins. This is one reason that "store-bought" ultrapasteurized milk is not a good choice for cheesemaking. Pasteurization destroys most of the milk enzyme lipase (which plays a role in developing the flavor and aroma of an aging cheese). For this reason lipase is sometimes added back into the milk during the cheesemaking process.

While pasteurized cheesemaking milk is inoculated with starter and ripening cultures, raw milk has a population of adventitious bacteria that enter the milk through contact with the teat surface and from inside the teat canal, the so-called "nonstarter lactic acid bacteria" or NSLAB. You might remember that these bacteria are seen as beneficial for the cheesemaking process. They help ferment residual lactose (or one of lactose's simple sugars) and contain enzymes that greatly influence the rate of flavor development during aging. For this reason the industry is researching and developing culture strains that will attempt to help imitate the "raw-milk cheese" flavor when used in pasteurized milk.

With any method of pasteurization, if cheesemaking is to take place immediately following, the milk is only cooled to the called-for ripening temperature. In commercial scenarios the milk temperatures and times are recorded to meet licensing requirements. The chart below includes three alternate types of heat treatment, in addition to the common home and small scale methods: high temperature, short time; ultrapasteurization; and ultra-high temperature. These treatments are not recommended for cheese milk (more on that in a moment). But I wanted to include them so you can see the range of treatments done to increase shelf life of fluid milk, and you know what you're sourcing when buying processed milk.

TABLE 2-1. MILK HEAT-TREATMENT TEMPERATURES

Heat Treatment	Temperature	Time
Thermization	131–158°F (55–70°C)	2–16 seconds
Batch/vat pasteurization	145°F (63°C)	30 minutes
High temperature, short time (HTST)	161°F (72°C)	15 seconds
Higher heat, shorter time (HHST)	191°F (88°C)	1 second
Ultrapasteurization	280°F (138°C)	2 seconds
Ultrahigh temperature (UHT)	240°F (116°C)	4–15 seconds

Thermization

The process of thermization also involves warming milk and holding it at a specific temperature, but the temperature is lower than that of pasteurization. In some countries cheese that has undergone thermization can still be labeled as "made from raw milk." Thermization will kill many pathogenic bacteria and reduce coliforms (*E. coli* is one type of coliform) and cold-tolerant bacteria (psychrotrophes) that can damage milk during cold storage. So for milk that will not be used for cheesemaking within a short period of time, thermization prior to cooling and cold storage can help maintain high-quality cheesemaking milk.

Homogenization

You don't need to know a lot about homogenization to be a good cheesemaker, but a little knowledge about the process and how it changes the milk might come in handy. Homogenization is a process that sends whole milk, or cream, under high pressure through a series of screens with restricted openings to reduce the size of the

RAW MILK: DON'T BE NAÏVE

First off, I believe in raw food, I eat raw food, and I was raised on raw milk and confidently feed it to my family. While raw milk has some amazing innate properties that make it capable of protecting itself from a certain level of contamination, it is not magical. In my opinion, the biggest concern with consuming raw milk is not bacterial contamination from the environment but the passing of disease from animals to humans. In the past the concern was mainly tuberculosis (TB) and brucellosis (also known as Bang's disease). Now, most herds are tested for these two biggies, and cattle are vaccinated for brucellosis.

But even with that, there are still major concerns as new diseases make their way through commercial herds, into wild populations of deer and bison, and back into domestic herds. The latest threats include things like Q fever, which *Goat Medicine*, by Smith and Sherman, states was present in 24 percent of the dairy goat herds tested in California. Q fever is caused by the bacterium *Coxiella burnetti*. It is passed in birth fluids (becoming airborne during the birth process and potentially infecting humans and animals within a fairly large geographical radius). It can cause flu-like symptoms and in the worst cases endocarditis (inflammation of the heart muscle).

Johnes (pronounced yo-knees), a milk-transmitted disease of growing concern, is being studied for a possible link to Crohn's disease in humans. So if you are a consumer or purveyor of raw milk and raw products, do remember that animal health is of equal importance to sanitation and handling of the milk after it leaves the animal.

fat molecules and change some of the proteins that help milk fat rise to the top. The fat molecule is altered by this process and becomes more vulnerable to breakdown by enzymes. Between the damaged fat molecule and the changes in the protein network in the milk, homogenized milk becomes a poor choice for cheesemaking. If you must use store-bought milk for making cheese, it will work, but expect a weaker, softer curd.

> Between the damaged fat molecule and the changes in the protein network in the milk, homogenized milk is a poor choice for cheesemaking.

STANDARDIZATION AND THE BALANCE OF PROTEIN AND FAT

There are several milk components, such as fat, protein, calcium, and sugar, that are sometimes standardized (increased or decreased to an identical amount in the milk) in industrial cheesemaking. While the artisan will rarely be performing any of the standardization processes, it is important to have an awareness of the *balance of protein to fat* and how it helps or hinders the production of certain cheese types. Remember that cheese recipes evolved around the ability of that milk and its uniqueness—whether species, breed, or seasonally specific—to produce a cheese with certain properties.

In large-scale cheesemaking a goal protein-to-fat ratio is determined and the milk is *standardized* to that ratio. This is basically done either by removing or adding cream or by adding skimmed milk to reach the goal protein-to-fat ratio. The main point to remember is that each cheese variety has an ideal protein-to-fat ratio. So when you are setting goals for the type of cheese you want to produce, you must keep in mind that the ratio that exists in your milk will greatly influence the ability to attain the goal cheese type.

A FORMULA FOR DETERMINING PROTEIN-TO-FAT RATIO

To determine the protein-to-fat ratio, divide the protein percentage by the fat percentage. For example:

Milk fat = 4.25
Protein = 3.25
3.25 ÷ 4.25 = 0.76
Ratio = 0.76:1

There is another reason you might have to turn to some level of standardizing milk, even if you are an artisan commercial producer. The US federal government, as well as many other nation's governments, maintains a list of cheeses that are required to have certain qualities, including fat percentage (the federal government maintains a list called the Federal Standards of Identity, which is part of the Code of Federal Regulations). So as a commercial producer, if choosing a name for your type of cheese might bring certain restrictions on your product, this is something to consider before putting your cheese on the market.

While most cheese milk will have more fat than protein (since milk naturally has a higher percentage of butterfat than it does protein), when milk is skimmed to make cheeses such as Parmesan or part-skim mozzarella, the percentage of protein becomes higher. For example, milk for cheddar cheese might have a ratio of protein to fat of 0.8:1 (meaning, for every 1 part of fat there are 0.8 parts protein), while milk for part-skim mozzarella might have a ratio of 1.46:1 (meaning, for every 1 part of fat there are 1.46 parts protein).

Standardization is also a way to maximize cheese yield. As you learn about the cheesemaking process, you are probably starting to understand that the milk fat cannot be captured in the cheese without a strong protein-calcium phosphate network in which to trap the fat. If the fat isn't surrounded by the network, it

HOW TO ESTIMATE PRESSED CHEESE YIELD

Here is a handy formula for making some pretty accurate estimates on just how much cheese you can make from your milk. These calculations are based on components—butterfat and protein. So you will either need to have your milk tested or base the calculations on breed/species averages. Either way, this can be a useful tool if you are hoping to make a career of cheesemaking and want to try to predict how much cheese you can make given a certain volume of milk.

Semihard to hard cheese formula:
1.63 × (Fat % + Protein %) = Yield

Example: 1.63 × (4.5 + 2.8) = 1.63 × 7.3 = 11.899, or about an 11% yield

Soft cheese guideline:
Soft cheese yield will usually be 20 to 40 percent, depending on moisture content of final product

Keep in mind that there are many different formulas for doing these calculations and almost all of them are much more complicated (but potentially more accurate) than what I am giving you here. These simplified formulas do not take into account such things as protein lost in whey, a too high butterfat-to-protein ratio, or the different moisture levels some cheesemakers desire in their soft cheeses. But I have tested these here at our facility, and honestly, if we made them any more complicated, most of us wouldn't try them!

will be lost in the whey. In general, the closer the ratio, the higher the yield. If the fat content is too high in comparison to the protein content, there will not be enough protein to form a network capable of capturing the extra fat. (This is only true for any cheese where the curd is cut and stirred in the vat—curd that is not cut but is ladled directly into molds can have extra fat; a double crème Brie is an example, as is cream cheese.)

Milk can be standardized for fat (and ideal protein-to-fat ratios) in several ways: for example, by the removal of cream, to decrease the total fat content; second, by the addition of milk or cream with higher fat, to increase the total fat content; and last, by the addition of skim or low-fat milk, to reduce the fat and increase the protein.

CULTURING AND RIPENING

Now the real cheesemaking begins! Selecting the best culture, determining the right amount to use, knowing the correct temperature of the milk, and deciding how long to let the milk sit (or ripen) once cultured are decisions that are usually dictated by the recipe you are using. As you advance in your cheesemaking skills, you will learn that many other factors might affect the amount of culture, the ripening temperature, and the length of ripening. Remember that a recipe is based on what works well in *most* circumstances but may need to be adapted as needed. In this section we will cover some of the possible conditions under which an adept cheesemaker might alter the usual parameters and steps.

FINDING AND BUYING CULTURES

A little note of caution regarding culture shopping: Some companies are not very good at properly describing the dosages and what exactly is in the package. In addition, some are priced so high that using them would cause the cheese to be very expensive to make. I have not included any of these companies in my recommendations, even though some of them are great sources for other items. If you find that cultures are too costly, shop around and also learn how to buy multiple-dose packages and repack them as I describe a bit later in this chapter. Repackaging will bring your costs down significantly!

Fortunately for the small artisan and avid home cheesemaker, there are now more options for purchasing a wide array of cultures than ever before. The revolution began with New England Cheese Supply in 1978, which for close to 25 years was the main supply source for noncommercial cheesemakers in North America (they are still at the top of my recommended source list). Now many local stores, including food preservation, wine making, beer making, sporting goods, kitchen supply, and even livestock feed stores carry cheesemaking kits and cultures. While many of these cultures are a bit limited (no rotation of strains or variety of choices), they are reliable and accessible and are certainly better than using just store-purchased yogurt (for high-heat cheeses) and buttermilk (for medium-heat cheeses) as your two primary cultures.

Two mainstay culture suppliers in the United States and Canada have been Dairy Connection, out of the state of Wisconsin, and Glengarry Cheese Supply in Canada. Another cheesemaking supply company, whose updated and improved website is set to launch while this book is in production, is Fromagex also in Canada. They offer a full line of cultures from the two biggest producers in the world, Chr. Hansen and Danisco (Choozit). The product lines of these two culture houses are so extensive that you might need some help navigating the choices.

If you happen to be perusing the Fromagex catalog and are feeling a bit overwhelmed, I have a few tips to get you started. First, take a marker and cross out all the frozen cultures (these are for larger producers and can only be shipped within the Fromagex). From there, start lining out cultures that are only available in sizes that are too big for your use (for most of us that will be anything over twenty to fifty units). Then grab the culture charts in chapter 1 and the recipe chapters and start looking for culture blends and varieties that are what you are looking for. Soon things will start to make sense, trust me! (For contact information for all of these companies, see appendix B.)

SELECTING CULTURES

In chapter 1 you met some of the many types of starter cultures available to the cheesemakers of today. In part II I have included more specific charts of recommended blends and dosages for the categories of cheese being made. As you become a more accomplished cheesemaker, you will begin to see how to make alterations and substitutions.

I debated quite a bit about whether to include instructions on how to make a mother culture and how to use bulk starter culture. Mother cultures are those that are continually recultured and maintained at the cheesemaking facility to provide fresh starter culture liquid for each batch of cheese. "Bulk culture" can refer to either that produced by the mother culture and added to the vat or to purchased frozen starter produced by a culture manufacturer. Since bulk starter is frozen (as opposed to freeze-dried), it "comes to life" more quickly in the vat, so it offers some advantages to the cheesemaker. In reality, though, these methods are rarely used by the home-scale and small-market producer because of the likely inconsistency and issues presented by the process.

One of my favorite books (and the first written to explain the science of making cheese for the general population), Paul Kindstedt's *American Farmstead Cheese*, explains in detail how to produce mother and bulk cultures, but he then goes on to recommend that home and small artisans not even contemplate working with mother cultures, because of the labor-intensive aspects, the high risk of the mother's becoming contaminated, the variation and inconsistency of the starter produced, and the requirement for precision equipment. If you are interested in either of these methods there are several online websites (in addition to Kindstedt's book) that cover the process in detail, including New England Cheese Supply's website and, at this writing, a great photo essay on the Cheese Forum (links to both of these are in appendix B).

THREE MUST-OWN CULTURE BLENDS

There are three core culture blends that you can use to make pretty much every cheese in this book. So if you are on a "real person's budget" and also don't want to have a bunch of packages of culture that you won't use very often, stick with these three:

- Fresh, feta, and surface-ripened cheeses: Flora Danica (Chr. Hansen)
- Semihard to hard aged cheeses whose curd is cooked to only 102°F (39°C): MA 4000 series (Choozit)
- Italian style and any cheese whose curd is cooked over 102°F (39°C): Thermo B (Abiasa)

You will no doubt want to order other cultures later, but this will give you a good start. If you buy ten-unit packages, then repack them into smaller doses, they will be relatively inexpensive.

MEASURING AND STORING CULTURE

Many recipes, including the recipes in this book, give measurements for adding direct-set culture in teaspoons. This will work but is less than ideal. If you are making very small batches and using prepackaged doses purchased from a home cheesemaking supply company (in other words, little packets meant to culture one 2-gallon batch of milk), then stick with the instructions. But if you are making batches of over 4 gallons, you will probably start purchasing larger packets of culture that you will need to learn how to measure.

If you are making cheese in very small batches of no more than 4 gallons it is quite convenient and fairly economical to purchase premeasured packets from a home cheesemaking supply company. These culture packets may not offer quite the variety, but since they are premeasured they are easy to store, and they maintain the culture's quality. Large multidose packets are at risk of humidity damage—every time you open the packet you let in some air and humidity that will start rehydrating the culture, eventually leading to deactivation. If you decide to venture into the world of multiple-dose culture packages, you will need to learn how to properly

EQUIPMENT FOR WEIGHING AND REPACKAGING CULTURE

- Gram scale reading to the tenths (should read to 0.0) or, even better, to 0.00 or 0.000
- Small, sealable baggies (from lab supply, food service supply, or craft store)
- Sanitized, dry measuring spoon

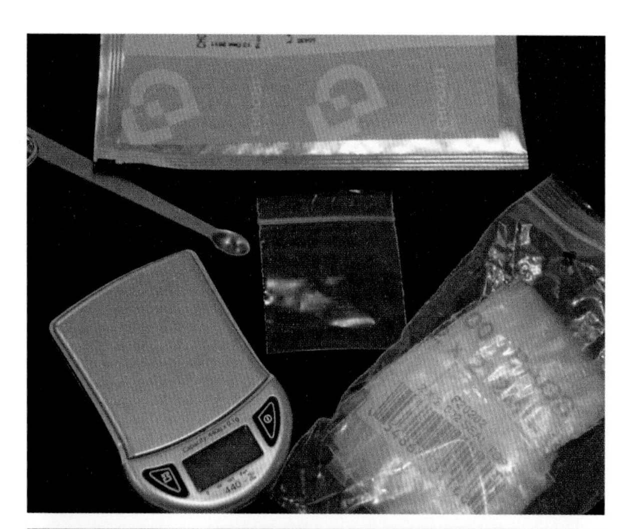

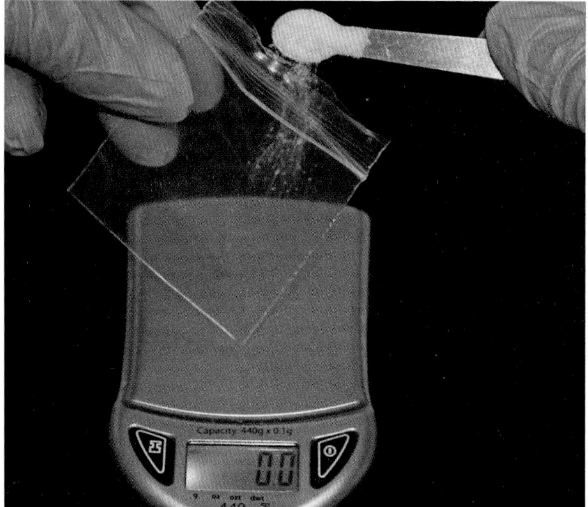

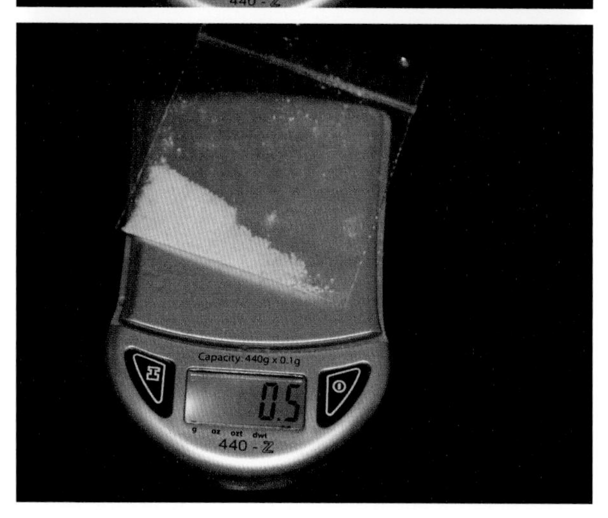

Equipment and steps in dividing a large culture packet into smaller doses

measure them into smaller, individual packets as well as how to measure proper amounts for each vat of cheese.

While there is no doubt that it is always better to weigh and measure culture doses carefully, the reality of this being practiced in most home and small artisan situations is a bit unlikely. After studying all the variables that go into choosing the correct amount of culture—such as whether the milk is late lactation, high in protein, and so on—you start wondering how you could ever get it precisely right anyway. The fact is, there is a bit of wiggle room for measuring. So come to terms with the goal of precision, then work with a method you can live with, whether that is weighing each dose on a tiny gram scale or purchasing a set of very small measuring spoons and using a meat skewer tip for super tiny doses (see pages 117 and 178 for more on measuring with a skewer).

MULTIPLE-DOSE CULTURE PACKETS

The first thing to learn about multiple-dose culture packets is that they are almost always measured by the number of units (a.k.a. doses) that the packet contains. Packets usually come in ten- and twenty-unit sizes, but some do come as small as five units (doses). Culture is sold by the number of units because when culture is manufactured it frequently varies in strength and in its ability to acidify the milk. So each batch is tested, then each packet is filled with the same amount based on strength instead of on weight.

Doesn't sound too confusing yet, right? Well, unfortunately, one dose does not culture the same amount of milk for every type of cheese. A rule of thumb is one unit (1U) per 50 pounds (22 L) of milk. But this doesn't apply across the board. In the charts listed in each cheese recipe chapter, I have included common dosing rates when that information is available.

Once you know how many doses are in the packet, you need to find out how much the contents weigh. Many packets now include weights. As an example of the varying strengths of the same type of culture, one blend that we regularly use can vary in weight by as much as 20 percent (for the same number of units). If the packaging doesn't say the weight of the contents, then you will need to empty the culture into another container whose empty weight is known. If you are quite clever, you might want to set the empty culture packet on the scale and make a note of its empty weight for future reference. Take the weight of the culture and divide it by the number of doses/units.

Then place an empty baggie on the scale, and note its weight. Add the weight of the baggie and the weight of the dose. This will be your total weight for each 1U bag. (You can divide them into 2U doses, or more, if that suits your scale of production better.) Then, using the sanitized, dry measuring spoon, fill each baggie to the total weight. Label a larger storage bag with the culture type and doses per bag, and place all of the baggies in the larger container. (I also cut off the portion of the original packaging that contains vital information and place it in the bag, as it includes the lot number and manufacturing date—all things that the commercial producer should be keeping track of!) Push or vacuum (using a special, resealable bag and handheld vacuum, see photo on page 120) the air out of the bag, and store it in a freezer. It may seem like a lot of trouble to do this, but it will keep the culture in great shape, will be convenient to use, and makes buying the larger packets more economical than the premeasured cultures from smaller suppliers.

EXAMPLE STEPS TO DIVIDING MULTIUNIT (TEN OR MORE) CULTURE PACKET

1. Culture weight = 4.5 grams
2. Dosage per package = 10 units
3. 4.5 divided by 10 = 0.45
4. 1U = 0.45 grams

OTHER FACTORS INFLUENCING DOSAGE

There are a few other factors that should be considered when determining how much culture to use. Raw milk, high protein and/or fat content, and late-lactation milk (with a higher somatic cell count and lower lactose) are a few of the instances when culture volume should be adjusted. If you are working with raw milk, remember that adventitious bacteria will help develop some acid during ripening, so you can generally use about half the amount of starter culture as when you are using pasteurized milk. By contrast, milk with higher than average protein content, milk with high somatic cell counts, and milk with lower than usual lactose content will have a negative affect on acid development, so you might need to use a bit more culture. This is why it becomes important to know your milk *and* to monitor acid development through the ripening process (see chapter 3 for an in-depth exploration of learning to measure acid development).

WHEN AND HOW TO ADD THE CULTURE

The usual temperature at which culture is added and milk is held during ripening is between 86 and 90°F (30 and 32°C). Some acid-coagulated cheeses are ripened at much lower temperatures, as is yogurt (not covered in this book), and some surface-ripened cheeses are ripened and coagulated at higher temperatures. Knowing the goals of the recipe and the temperatures at which the bacteria in the particular starter culture used will grow will help choose the correct temperature. Mostly, though, you will depend upon the directions given in the recipe you are following.

Here is one trick that I learned from Peter Dixon (my first, official in-person instructor): When working with raw milk, add direct-set starter culture at 80°F (27°C), then continue to raise the milk temperature to the goal ripening temperature. This will allow the freeze-dried culture time to rehydrate and "wake up" so that when you reach the ripening temperature it will be ready to go. Otherwise, the native bacteria might get a foothold from the time you add the freeze-dried direct-set culture to when it is ready to start populating the milk. This can be done with pasteurized milk as well. It will reduce your ripening time by a bit for either type of milk.

If you are working with milk that you have just pasteurized, you can add the culture at 100°F (38°C), then cool it to the goal ripening temperature—the reverse of what I suggested for raw milk, but it accomplishes the same goal: to give the direct-set culture a chance to rehydrate and get to work.

Direct vat set (DVS) cultures are the only kind we are working with in this book; they are added by sprinkling them on top of the warm milk, letting them set for a few minutes, then stirring them thoroughly into the milk. Letting the powder sit on top of the milk allows it time to absorb some moisture and begin to rehydrate. If it is stirred in immediately, sometimes the powder will form a clump in the milk that won't dissolve completely. Not only will this lead to an inadequate amount of bacteria in the milk, but the clump can cause problems later in the cheese. Some DVS cultures have more of a pellet shape than a powdery texture (such as Flora Danica); these will need to be stirred for longer to make sure they completely dissolve.

PREMATURATION

Prematuration, also known as preripening, is a nifty technique not practiced widely in the United States with the goal of introducing a limited population of lactic starter bacteria along with a concurrent drop in pH (a rise in acid level) that "protects" cold stored milk until it can be made into cheese. Remember psychrotrophes, the cold-loving bacteria that we learned about in chapter 1? Well, before the advent of on-farm high-efficiency cooling equipment, the natural bacteria in milk had a bit of time to grow before the milk was adequately cooled. This milk (which if tested in a lab would show fairly high total bacteria counts) was essentially inhospitable to the growth of the cold-loving bacteria. So psychrotrophes were less of an issue for milk quality than they are today.

On the other hand, the milk collected on many farms has the potential to be "cleaner" than in days past, thanks to modern techniques that rapidly chill the milk and strong chemical cleaners that help prevent bacterial populations from permanently residing in milking equipment, so there is no guarantee that if left to sit and cool more slowly (than is required), the milk would grow a population of "good guy" bacteria.

Prematuration attempts to mimic what used to occur naturally in these on-farm situations. After the milk is collected, a small amount of starter culture is added to the milk and the milk is held at 46 to 50°F (8–10°C) for 12 hours (basically overnight). This milk can then be combined with the morning milk for cheesemaking or chilled and stored. The goal drop in pH is 0.1 to 0.2 (see chapter 3 for more on pH). In some cases prematuration is followed by a heat treatment to kill the bacteria so that another culture can be added without further (competitive) growth of the prematuration culture. If you are going to try this technique, be sure to have the ability to check pH or you will be working with some unknown factors that might really wreak havoc on your cheesemaking process!

RIPENING TIMES

Ripening times can last a number of hours or be extremely short. (Keep in mind that curd will continue to ripen during the coagulation phase, so even though the "official" ripening period might be very short, in reality it will continue through the next phase.) Remember what is happening during ripening: The bacteria in the starter culture begins to populate the milk, consume milk sugar (lactose), and produce lactic acid. Depending on what else occurs during the make process, the ripening time must be of a length that helps, not hinders, the desired acid production during the rest of the make *and* the goal pH or titratable acidity (TA) (more on both pH and TA in chapter 3) at draining. It may sound complicated, but it is one of the secrets of becoming a great cheesemaker. You have to know that *everything* you do, once the culture is added, will influence how that cheese turns out.

Here is an example: Gruyère-type cheese has a very short ripening period at 90°F (32°C) of 5 to 10 minutes, not even really time for the direct-set culture to wake up. Why would that be? Well, Gruyère uses thermophilic (heat-loving) culture, so even if you held it for 30 to 40 minutes, the thermophilic culture wouldn't do much, but any native bacteria would. So you only want to hold it at this lower temperature for long enough to let the freeze-dried culture rehydrate. Then you start bringing the temperature up into the more ideal growth range for the thermophiles: 104 to 113°F (40–45°C).

So think of ripening times as a fluid parameter. Learn to think about what is coming later (what is your final goal) in the make process as much as what is happening at the moment. Above all, take notes, check pH or TA, record the results, and learn what works for your milk!

ADDITIONAL INGREDIENTS

Technically speaking, one could refer to these additional ingredients as "additives," but it makes me picture such things as monosodium glutamate and dicoraglucomioapolysaccharides (I just made that last one up, but you get the idea). Anyway, there are ingredients you might want to add to the cheese milk at this point in the process—after ripening and just before coagulating. For the artisan cheesemaker these extras will most likely be natural coloring, calcium chloride, lipase, or flavored liquid from herbs, seeds, smoke, or hot peppers. If you choose to color your cheese, then annatto, made from the seeds of the achiote tree, *Bixa orellana*, is the likely choice.

As far as chemicals go, there are many that are, or can be, used, but they are choices for the industrial cheesemaker, not the artisan. The only chemical that is used commonly by the artisan is calcium chloride ($CaCl_2$). Adding calcium chloride is a good choice in many circumstances (see chapter 1), so don't let the word "chemical" make you feel bad (remember, humans are made of chemicals). I have included calcium chloride dosages for each recipe. Flavored liquid from the boiling of seeds or other solid seasonings (that will be added to the drained curd) is also added while the milk is still in a liquid state (see chapter 5 for more on adding flavors).

Annatto, should you choose to use it, is added before the coagulant and is not diluted. Amounts added range from 15 to 60 milligrams per 2.2 pounds (1 kg) of milk. So for each gallon of milk, you would use 60 to 240 milligrams of annatto. You will need a metric scale that weighs milligrams to properly measure the annatto. Triple-strength solutions can be added at the rate of 0.5 to 2 milliliters per 26 gallons (100 L).

If you are using $CaCl_2$ that is already in an approximately 32 percent solution, it will come with usage instructions on the container. You should not add more than is recommended, as it will have a reverse effect and slow coagulation. The maximum amount that can be added (legally in the United States) is 0.02 percent by weight. See appendix C for a chart with dilution rates for liquid and calcium chloride crystals.

Calcium chloride should be diluted in a small amount of cool water; say, twenty to forty times the volume. (So if you are adding ½ tsp [2.5 ml] of $CaCl_2$, dilute it in ¼–⅜ cup [50–100 ml] of water. Add it to the milk before adding rennet, and stir in thoroughly.)

THE PROCESS OF COAGULATION

Once the ripening goal has been reached (whether determined by time or acid-level change), it is time to

FOUR NICE THINGS YOU CAN DO FOR YOUR COAGULANT

1. Keep it in the dark; light reduces its activity.
2. Keep it cool (store at fridge temps, never freezing).
3. Don't dilute with water until just before adding.
4. Keep it fresh; older rennet will be more likely to cause bitterness.

add the coagulant. Rennet, whether it is from an animal or a microbial source, comes as either a liquid or a tablet. While tablets have a longer storage life, liquid rennet is much easier to measure accurately. If you are making batches larger than 2 to 4 gallons or making cheese frequently, using liquid rennet will be the best choice.

DOSAGE

Liquid veal rennet comes in single strength, while microbial and fermented rennet usually come in double strength (if you are using rennet tablets, follow the instructions on the packet). *You must know which one you are using before calculating the correct dosage for your cheese recipe.* Typical single-strength dosage is 0.07 to 0.09 milliliters per pound of milk. Double strength is half that at 0.035 to 0.045 milliliters per pound. So for example, if you have 100 pounds of milk (about 12 gallons), you will use 7 to 9 milliliters of single-strength rennet or 3.5 to 4.5 milliliters of double-strength rennet. In the recipe sections all measurements are for single-strength liquid rennet. You can adjust them as needed.

> Remember that every step is fluid, and to become a master cheesemaker, you will have to learn how to adapt and adjust as the process changes.

As you read through the recipes, you will probably notice how the suggested amount of rennet will vary greatly, even between two similar cheeses. There are several reasons for this:

- Temperature of milk at renneting: Coagulant will work faster at slightly warmer temperatures, making flocculation time faster.
- The pH at renneting: The more acidic the milk, the faster the coagulation time.
- The goal total coagulation time: This is a combination of desired texture at cutting and acid development during coagulation.

When any of these factors varies, you will have to adjust the goal. For example, if the recipe says to add the rennet at 90°F (32°C) and you add it at 92°F (33°C), you should expect a shorter total coagulation time. You also may have to adjust the time in the vat or the length of time for cutting to make up for the shorter coagulation time. Remember that every step is fluid, and to become a master cheesemaker, you will have to learn how to adapt and adjust as the process changes.

DILUTION

Liquid rennet must be diluted in *cool, nonchlorinated water* before use. If the rennet is added to chlorinated water, the chlorine will attack the rennet and either neutralize it or greatly diminish its potency. Heat will also deactivate it; hence the need to use cool water. Don't mix the rennet with the water until just before use, as light exposure and the minerals in tap water will have a negative effect on the coagulant as well (some people use distilled water, but this can be costly when making large batches of cheese). *Dilution rates* are 20x for single strength (e.g., 20 milliliters of water for every 1 milliliter of rennet) and up to 40x for double strength. It is important to use the right amount of water to help the rennet disperse evenly through the milk. If not, once you stop stirring, the coagulation will not take place evenly.

To dechlorinate tap water from a municipal (city) source, add enough drops of milk to give the water a slight clouding. The chlorine will "attack" the organic compounds of milk and be neutralized. This is a normal practice in larger cheesemaking facilities where tap water is almost always from a water source that is chlorinated.

Water that is to be mixed with rennet should have an ideal pH of under 7.0. If you don't know the pH of your water, it is worth checking. If the water is at all alkaline (over 7.0), it will have an adverse effect on the coagulant. You can lower the pH by adding a drop or two of white vinegar or citric acid. Once diluted, rennet will not be stable for long, so you should mix it just before adding it to the milk.

ADDING THE COAGULANT

The style of stirring and the length of time you stir the coagulant into the milk are vital. As soon as the enzymes in the coagulant make contact with the milk, they will start working, so the more evenly and effectively that the diluted solution is incorporated into the milk, the better. Once mixed, it is critical that the milk stop moving in the vat and remain as quiet and still as possible until coagulation is complete. For these reasons the rennet solution should be dispersed over the surface of the milk by pouring it over a perforated ladle or using a "garden" sprinkle can (when adding to large vats of milk). The stirring motion should be "up and down" instead of circular. When any fluid is stirred in a circular motion, it tends to keep moving for a long period of time once stirring ceases. If this happens during coagulation, you will see swirled fractures in the coagulated curd. Stirring up and down does not create this whirlpool effect. Once you stop stirring, you can use a flat ladle to help still the milk. In rectangular vats a flat metal plate can be inserted into the milk to help quiet it.

Dispersing rennet over the surface of milk with a cheese ladle

Ideally, the milk is kept moving until just before the protein clusters start forming a network. This is more important with cow's milk than with goat and sheep milk, which do not "cream" as rapidly. Remember, the protein network is meant to trap as much fat as possible. Stirring for just the right amount of time both keeps the fat incorporated in solution and evenly disperses the coagulant. So for cow's milk you may need to stir for 4 to 5 minutes. For goat and sheep milk, 1 to 2 minutes is adequate. When you stop stirring, note the time you added the coagulant in your logbook. This is essential for the next step.

> The success of great cheesemaking depends on the total sum of all steps. There is no one step that will be accomplished perfectly every time, but you must try, because if you don't, all the little omissions and shortcuts add up to poor-quality cheese.

CUTTING THE CURD

Okay, now the artisan part of the process escalates (and will continue to do so throughout the rest of cheesemaking, so brace yourself!). Most recipes tell you to add the coagulant, then wait for a set period of time or "until clean break" to cut the curd. This works fine when the milk and all other factors are standardized, but it's not so simple with artisan hands-on milk from naturally fed seasonal animals. In other words, *artisan often equals variation*. So the artisan cheesemaker must learn to be in tune with these changes and follow them, *not* the instructions! But enough of my ranting. . . . Let's talk about why the firmness at cutting should be based on the goal size of the curd and the goal texture of the cheese. Then we'll talk about using flocculation factors as well as "clean break" checks to help determine when to cut the curd.

FIRMNESS AT CUTTING

The firmness of the coagulated milk when you cut it "fixes" the type of curd you will end up with later on in the make. So if you know the goal, you will know at what firmness to cut and what size to cut the curd. Here is an example: If you are making a thermophilic (high cook temperature) cheese, such as Gruyère or Parmesan style, the curds must be cut smaller than if you are making a Camembert or a feta (cheddar- and tomme-type curds fall somewhere in the middle). For a curd to be cut smaller, the coagulum should usually be a bit softer than if the curds are to be cut larger. As the coagulum sits for longer, the protein structure will become progressively more firm and granular, so when you attempt to cut an extra-firm mass into small pieces, it will create curds with rough, granular sides with more surface area. Imagine if you tried to run a closely spaced wire whisk through a bowl of extra-thick Jell-O; it would tear into ragged pieces. But if you ran the same whisk through a softer set, the Jell-O would cut before it tore.

So why is this a problem with tearing? After all, doesn't cutting curds small essentially increase the surface area? Yes, it does, but when the curds have a rough surface, they will lose more butterfat into the whey through the jagged parts that stick out from the main part of each curd. So the goal is a *smoothly cut curd of the desired size*.

SIZE OF THE CURD

The goal of cutting the coagulated milk into evenly sized pieces, otherwise known as *curds*, is to expose an equal amount of surface area so the curds can shrink, by losing whey, evenly from curd to curd as well as for each individual curd. So the more even in size you can cut the curds, the better. That being said, it is almost impossible to get them all the same. And just so you know, they are rarely perfect cubes. I always tell my students that the success of great cheesemaking depends on the total sum of all steps. There is no single step that will be accomplished perfectly every time, but you must try. If you don't, all the little omissions and shortcuts add up to poor-quality cheese.

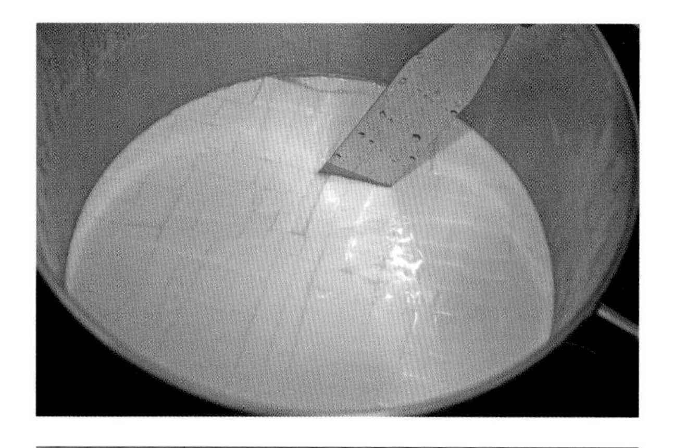

Cutting curd in a small vat

An Italian-style curd-cutting tool called a *spino* being used at Bellwether Farms, California

When trying to determine the goal curd size (usually indicated in the recipe), think about what you will be doing to the curds after they are cut. The temperature at the end of the cook and the rate at which the temperature is reached are the main factors in determining curd cut size. Remember, the goal is to lose whey evenly from inside the curd. Think about a large curd being immersed in hot whey: As you might imagine, the surface of the large curd will lose whey more quickly than the interior and will form a less permeable skin, which will limit the loss of whey from inside the curd. When this occurs it is sometimes referred to as *case hardening*. The retained whey in these larger curds (which includes lactose, the milk sugar) will continue to ferment after pressing, leading to pockets of crumbly cheese and off-flavors—not what you want in your cheese! Cheeses whose final temperature is lower—say, about 100°F (38°C)—and is attained over a longer period of time typically have a curd size a bit larger than a high-temperature-cooked cheese. On the other end of the curd-size spectrum, cheeses whose curd temperature never goes much beyond that of ripening, like bloomy rinds (Camembert, crottin, and so on) and brined cheeses (such as feta), have curds cut larger or not at all.

The size at which the curds are cut also influences fat retention. No matter what the original fat content of the milk, if the processing does not help keep that fat in the curds, you will lose a percentage of it in the whey. The smaller the curds, the more fat is lost. It cannot be helped. Remember that protein network that is trapping fat? When you cut the curds, areas of the network are exposed, and the fat trapped within leaks out into the whey. This is one reason that you only find a "double or triple crème" in a bloomy rind style where the curds are cut in large, ladle-size pieces.

DETERMINING WHEN TO CUT

We've covered the goal firmness and curd size. While most cheese recipes have an ideal total coagulation time (that you can make adjustments for by using more or less rennet in future batches), here are some tricks to helping you know when to cut the curd. The first method involves observing the milk for the point at which the enzymatic phase ends (when most of the κ-casein has been cut) and the coagulation phase begins—in other words, observing the flocculation point, then using that time to determine the total coagulation time. The second approach is to watch for the ideal texture for cutting, or the "clean break."

I originally learned the flocculation technique from Peter Dixon and subsequently from two instructors from France. So just what is flocculation? A floc is simply a clumped particle. (Some examples of flocking in other situations would be fabric that is flocked with fuzzy dots or a holiday Christmas tree sprayed with a textured, white coating.) If you remember back to chapter 1 and what is happening when the coagulant is working in the milk, you might remember that the first phase involves some enzymes knocking off the outside, water-loving kappa-casein projections (that keep the clusters from further clumping together). Flocculation occurs when this "enzymatic" phase is 75 to 80 percent complete and clumping begins to be noticeable.

There are several techniques for checking for flocculation. You can dip a dry knife or curd blade into the milk (below the cream line for cow's milk), draw it back out, and observe the surface for little flecks (those are flocs); float a lightweight lid (such as you would find on a jar of jelly) on the surface of the milk and note when it will no longer spin freely; or remove a small

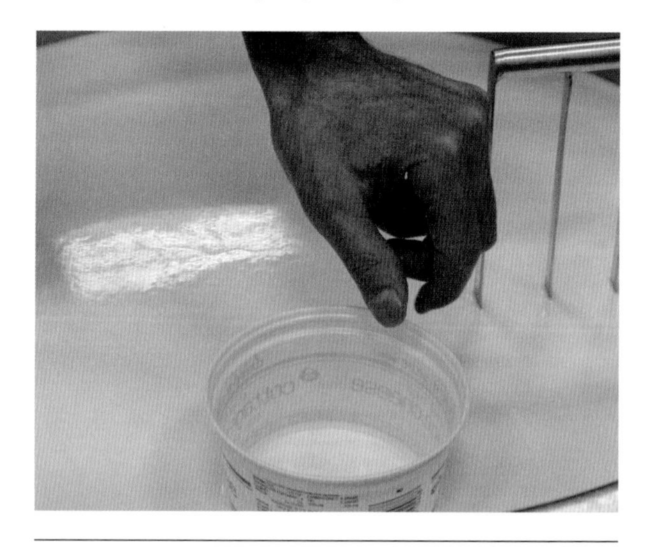

Peter Dixon demonstrating a popular technique for determining flocculation by noting when a cup or lid will no longer spin freely on the surface of the milk

HOW TO TEST MILK FOR COAGULATION POTENTIAL BEFORE MAKING CHEESE

A few years ago we lost a lot of vats of milk (translated: lots of future income) to slow, or no, coagulation. I won't tell you all the things we ruled out, as that's a long story, but in the end Dave Potter at Dairy Connection in Wisconsin (whom we buy our cultures from) told me how to do a simple test on the milk to determine its clotting ability. This has been very helpful to us, so I would like to share it with you.

1. Mix 24 milliliters cool water with 1.25 milliliters double-strength rennet.
2. Warm ¼ cup milk to 92°F (33°C) in a clear measuring cup.
3. Add 10 milliliters of rennet mixture to warm milk while swirling milk.
4. Time the seconds it takes for small flocs to form on sides of measuring cup.

If it takes more than 20 seconds to flocculate, there is an issue. Repeat the test, but this time add ⅛ teaspoon calcium chloride to the milk. If flocculation time improves, then add calcium chloride to the cheese milk at the recommended rate.

sample of milk in a clear container, swirl it, and see if flocs form on the sides of the container. By watching for flocculation you can judge the total cut time better than by just watching for a "clean break" (we'll talk more about that in a moment).

Seasonal and other factors will change the milk's ability to coagulate. Also the pH and temperature of the milk will influence coagulation times. By watching for flocculation time you can note how receptive each batch of milk is to the coagulant.

TABLE 2-2. COMPARISON OF COAGULATION AND CUTTING FACTORS*

Cheese Type	Single-Strength Rennet Rate for 2.5 Gal (10 L)	Flocculation Factor	Goal Coagulation Time	Clean Break Type	Curd Size
Semisoft, washed curd	2 ml	1.1	25–30 min	Soft-medium	¼ inch (6 mm)
Stabilized bloomy	2.5 ml	2.5	15–20 min	Medium	¾ inch (2 cm)
Extra hard, hard	1.8–2.5 ml	2–2.5	20–30 min	Medium	¼ inch (6 mm) or smaller
Semihard to hard	2.5–3 ml	3–3.5	30–50 min	Medium-firm	⅜ inch (1 cm)
Feta types	2–3 ml	4	50–60 min	Medium	¾–1¼ inch (2–3 cm)
Traditional bloomy	1–2 ml	5–6	50–60 min	Medium	Ladled

*Examples and guidelines only; many recipes will differ.

Once you see flocculation, the time is noted, then multiplied by the specific factor (see Table 2–2, Comparison of Coagulation and Cutting Factors) being used for the type of cheese you are making. The resulting number is the time from adding the coagulant to cutting. For example, when I make our types of hard cheeses, the flocculation time is almost always (if things are going well) 12 minutes. I multiply that by 3.5 and come up with a cut time of 42 minutes total. If you are following the flocculation method, you must watch the milk closely after adding the rennet, for the first 15 minutes or so, but after that you can go do other things, as you will know precisely when to return for the cut.

If you forget to watch for flocculation, or if you prefer to go by feel rather than formula, you will need to learn how to determine when a "clean break" is achieved. In fact, even if you use the flocculation formula, checking for a proper clean break is a good secondary method to ensure proper cutting time. So just what is a clean break, and how do you check for it? First, you must know that a clean break is not a singular moment in the vat; instead, it is a curve during which different qualities of cuts and curd size can be attained. Remember what is happening during coagulation: The proteins are knitting together to form a network. As you wait this network will continue to grow in strength. If you attempt to cut too soon, the structure will mush rather than cut, and you will lose fat into the whey. On the other hand, if you wait too long, it will become so firm that it will resist the knife and tear rather than cut cleanly. So think of the optimal cutting time as a range, not a moment.

There are several techniques for checking the curd mass for a clean break. The most common one is to crook your index or middle finger, insert it into the curd up to the second knuckle and lift slowly up and forward. I prefer to use a knife, cut a slit, insert the knife—flat side up—at an angle just behind the slit and under the uncut portion in front of the cut. Then I slowly lift the knife and see if the cut seam expands forward. As the curd mass gives way to the pressure below, it will part. As it parts, I observe the sides of the tear and the liquid that escapes. Are the sides of the tear smooth or jagged? Is the whey white or yellow/green? If the sides are smooth and the liquid is more yellow/green than white, the curd is ready to cut. Remember, though, that some clean breaks should be softer than others, depending on the size of the curds to be cut.

Checking for a clean break: A slit is cut in the curd mass, the blade is inserted through the slit, and the curd is lifted while the slit is observed as it expands forward.

For each cheese there is an ideal coagulation time that should be monitored with each make. If you find that you're having to cut earlier (based on flocculation time and clean break) or later than the goal time, then adjust the rennet volume in the future. Also remember how temperature and pH dependent the action of rennet is. Be sure your process is adhering to the goals in this regard before you adjust the rennet volume.

COOKING AND STIRRING

Some cheeses, such as soft fresh cheeses and many bloomy rinds, are made *without* the curd going through

Cheese curds being stirred in a 30-gallon steam kettle vat

a phase of heating and stirring. For most of these types, the curd is held at the ripening temperature until the entire mass has achieved the goal acid development and texture, at which point it is ladled (sometimes with some cutting just before it is ladled), scooped, or drained right into the forms (but more on that in a bit).

Almost all other hard cheese types, though, will go through an often time-consuming phase during which the curd is heated and stirred. In some books this is called "scalding" and in others "cooking" (I think this extra work is the real reason they are called "hard" cheeses). The goal of this treatment is to attain the right curd texture, moisture level, and acid development within the curd.

A helpful way to think of curds is as little fermentation factories. Each has an internal supply of lactose, starter bacteria, some trapped whey, calcium, and (of course) a network of protein. The curd factory is enclosed by a semipermeable membrane and floats in a bath of whey. The whey also contains starter bacteria, lactose, and calcium. To achieve the right texture and acid development, you need to understand how these little fermentation factories work and what you can do to influence and assist them.

HEAT

Increasing the temperature of the whey will cause the curds to shrink and lose more whey. If this happens too quickly, it will cause the curd surface to dry prematurely, thereby trapping whey inside the curd. Consequently, temperature increases are usually done in slow increments, with the initial temperature increases being the slowest, so the inside of the curd can have a chance to lose whey at the same rate as the surface.

As the temperature in the vat changes, the bacteria will respond by producing more or less acid, depending on the type of starter bacteria used and the ideal growth temperature of that particular bacteria. You can adjust the rate of acid development by adjusting the temperature in the vat. For example, if your curds are not developing the correct texture quickly enough, but the acid level is increasing a bit too fast, you can increase the vat temperature toward the maximum cook temperature more quickly. At this higher temperature the bacterial activity will decrease, since you are now at the limits of the temperature range in which the bacteria will be active, and the curd will also shrink more quickly.

As the curd shrinks and loses whey, it also loses lactose. As lactose decreases in the curd, acid production (fermentation) decreases as well. If the curd is held at a temperature that limits shrinkage, lactose can move into the curd and replenish the supply for the starter bacteria. This method is used in cheeses such as feta, where it is stirred at a lower temperature to encourage acid development and whey retention in the curd. By your having a wetter curd during drainage, more fermentation can occur over a longer period of time. All this helps create the desired tanginess and texture of feta. But we'll talk more about this when we get to the recipe section.

STIRRING

Recipes (including the recipes later in this book) typically tell you to stir the curd gently, but just what exactly does "gently" mean? The best rule to follow is this: Stir the curd as gently as is needed to prevent the curds from "shattering," or breaking into tiny pieces. Some curd will be sturdy enough for you to move the cheese ladle in a gentle swirling pattern through the curd without excessive breakage, while some other curd may need to be initially agitated by shaking the pot gently. You will only know the precise "gentleness" needed after some trial and error. Pay close attention when you first begin stirring, and you should be able to adjust your technique to suit the curd in your vat at that moment.

The rate at which the curd is stirred will affect the rate at which whey is lost. And you now know that the rate at which the whey is lost will change how acid develops. In most cases curd is only stirred as vigorously as is needed to prevent the curds from matting back together. If curd is stirred more rapidly, whey will be lost more quickly, and the resulting curd will be drier. If curd is stirred more gently, whey loss will be slower, and the resulting curd will be moister. But remember, too much retained whey can cause problems, too!

THE CALCIUM CONNECTION

Much of a cheese's final texture is due to the amount of minerals, especially calcium (mostly as calcium phosphate), retained in the curd and final cheese. Low calcium content equals crumbly, and high calcium equals pliable (think about the difference between feta and cheddar). If you remember, calcium is the "glue" that holds the protein matrix together, and as calcium leaves the curd, the protein networks collapse, creating a fractured texture.

So what causes curd to lose or retain calcium? I like to have people think about mineral deposits on their coffee- or teapot and what they use to dissolve this type of buildup. Acid (usually in the form of vinegar), right? Calcium inside the protein clusters and network responds the same way to an increase in acid. So as the pH of the curd decreases, the calcium will become soluble and leave the curd along with the whey. If the acid level drops too low, too much calcium is lost. This loss can occur in the vat or during drainage. It can even occur after pressing. Remember when we talked about whey retained in curds that were too large or that did not shrink enough during cooking? This retained whey will provide extra lactose for excess acid production, resulting in calcium being removed from the protein structure and lost in the whey. You can probably imagine that this is not ideal for most cheese but is desired for some.

WASHING THE CURDS

Did you ever wonder why some aged cheeses are so pliable, moist, and tender? Like that creamy Monterey Jack oozing from your chili relleno or those luscious slices of Havarti served with crisp green apple slices? Well, these cheeses fall into the category called *washed curd* cheeses.

Not to be confused with *washed rind* cheeses, washed curd cheese has a portion of the whey removed during the cook phase. The same, or a lesser, volume of water is then added back into the vat. When whey is removed, a good portion of milk sugar is also removed, giving the starter bacteria a little less nourishment, thus slowing their growth and acid production. This lowered acid production helps the curd retain calcium, which will lead to a more pliable texture in the finished cheese. When water is added back into the vat, it does several other things: First, it keeps the volume of liquid the same so the curds can still move freely during stirring; second, it creates a situation in which more water is absorbed into the curds (especially when cool water is added to warm curds), leading to a higher-moisture cheese; and last, it can help moderate the cooking of the curd—depending on the temperature of the water.

If you are following a recipe for a washed curd cheese, it will tell you how much whey to remove, when to remove it, what temperature water to replace it with, and how long to cook after that. Since many traditional cheeses are washed curd cheeses, guidelines are fairly accessible for this information (see Table 2-3 for a few examples of the way washed curd techniques can vary with cheese type). But remember, the artisan cheesemaker can employ this technique both to develop his own unique cheese and as a stopgap measure to control acid development in the vat.

In general only 25 to 30 percent of the whey should ever be removed. If more is taken away, you risk removing too much lactose for the curd to continue to develop the proper amount of acid. But that also depends on how much acid has already been developed, the final moisture content of the curd, and the amount of pressure during pressing. The point at which the whey is removed is up to the cheesemaker, but it must be done with the rest of the process in mind—especially the goal pH at draining, temperature of the room during draining, the amount of weight that will be applied

TABLE 2-3. WASHED CURD CHEESE EXAMPLES

Cheese	Amount of Whey Removed and Temperature	Temperature of Water Added
Edam	⅓ at 86–87°F (30–31°C)	122–140°F (50–60°C) to goal vat temp 98°F (37°C)
Gouda	⅓ at 86–87°F (30–31°C)	140–176°F (60–80°C) to goal vat temp 98°F (37°C)
St. Paulin	⅓–½ at 95°F (35°C)	95°F (35°C) weak brine (0.02%)
Monterey Jack	⅓–½ at 102°F (39°C)	Cold water to goal vat temp of 86°F (30°C)

USING WASHING AS AN EMERGENCY MEASURE

While there are many cheeses that have curd that is washed as a regular step in their production, you can also use washing as an "emergency measure" during almost any cheesemaking process. As you learn to better monitor acid development and understand the finished goals of each cheese you make, you will see that you can use the removal of whey, with or without the addition of water, to slow pH development and/or increase moisture content and mineral retention of the finished curd. While this may not make a lot of sense right now, keep it in the back of your mind for later. It's one of those great tips and secrets that a master cheesemaker can choose to use.

to the draining curd, and the goal pH at salting. This may seem impossible at the moment to figure out on your own, but as you follow existing recipes for washed curd cheeses, think about the point at which the curd is washed and about the temperature of the water (and sometimes a weak brine solution) that is added—how does this change the texture in the vat? How does it change the finished cheese?

CHECKING CURD FOR READINESS

There are several approaches to determining when curd is finished shrinking and is ready for the next step—either "pitching" (more on pitching in a bit) or draining. Some people are able to judge readiness by how the curd feels as it moves in the vat, others by tearing curds open to check for even texture. I even saw one cheesemaker that likes to throw a few on the floor to see how they respond (like people who throw cooking pasta at the wall to see how well it will stick). Keep in mind that different cheese varieties will require different textures!

My cheesemaker friend Pablo Battro (who supplies a blue cheese recipe later in the book) taught me a great method that I find makes it very easy to check texture for many semihard to hard cheese types. To use this method, scoop a small handful of curds from the vat, flatten your hands, and gently press the curds between your fingers. Without uncovering them, turn your hands so the curd mass can drain from the top and bottom. Then separate your hands, and turn the side with the curd patty on it upside down. If the curd mass sticks to your hand, it is ready.

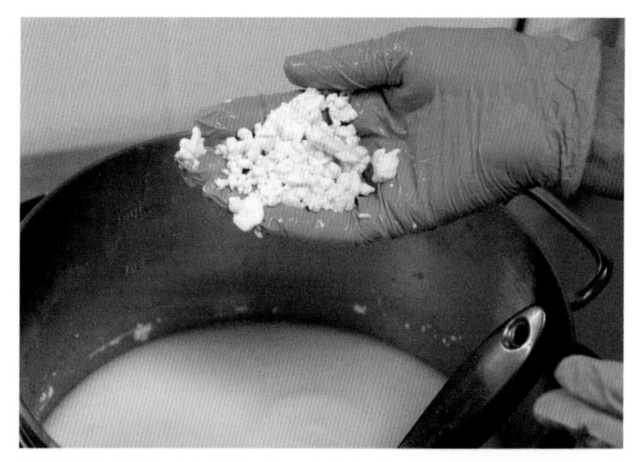

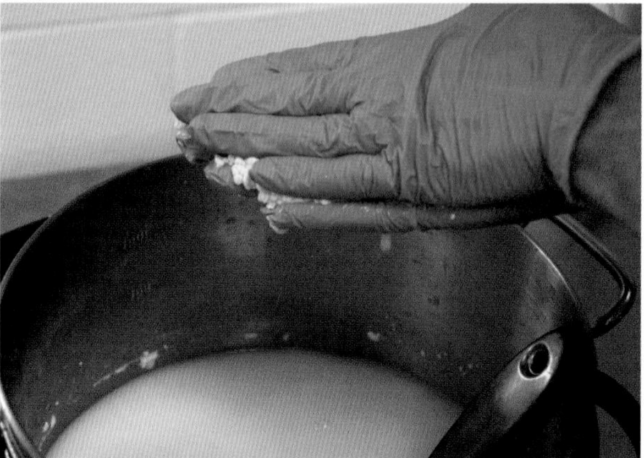

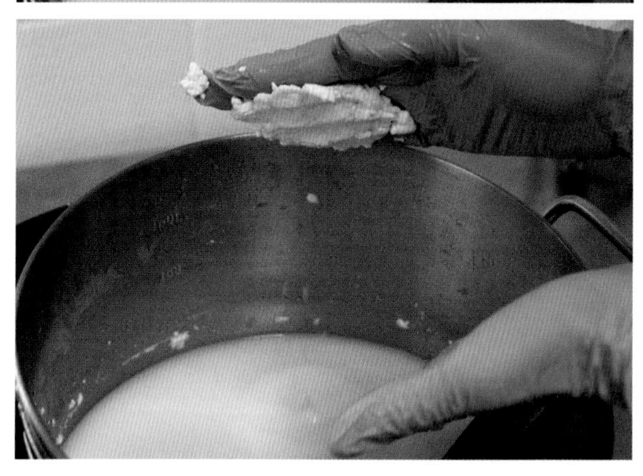

Checking for curd texture readiness for semihard to hard cheeses: First, remove a handful of curds from the vat. Second, gently squeeze the curds together by pressing your flattened fingers over the curds, squeezing until no more whey runs between your fingers. Last, lift the hand on top and turn the curd patty upside down. If it remains stuck to your hand, the curd is ready.

PITCHING THE CURD

Here is another term for you, and no, it doesn't mean throwing a ball of curd at someone holding a baseball bat! (Remember, these terms are going to come in handy when bantering with your other fromage-focused friends.) "Pitching" refers to letting the curds settle in the vat, sometimes just before draining but often as a technique for allowing a bit more acid development in the vat without continuing to shrink the curds through stirring. Sometimes pitched curds are stirred intermittently and sometimes not. If a recipe calls for letting the curds settle, great, but this is yet another technique that can be employed once a cheesemaker better understands what is happening in the vat and what their goals are for the cheese.

Many recipes call for the curd to be pitched, then pressed under the whey. This helps consolidate the curd into a mass that will assume its final shape and texture in the mold with more ease. In larger facilities the curd is pressed under the whey using perforated, stainless steel plates upon which weights are set. At home or on the small scale, you can use a cheese mat and your hands, or just your hands, to slowly and gently press the curd.

> The main goal of pressing is to achieve the right shape, a closed rind, and proper whey drainage from the interior of the cheese.

DRAINING AND PRESSING

Finally, you are ready to remove the whey from the vat—or, conversely, remove the curds from the whey. You've gotten them this far safely, right? Just a few more steps and you'll have something that is really starting to look like cheese. Here is actually the best piece of advice I can give about draining and pressing (or draining in forms without pressing): The main goal of pressing is to achieve the right shape, a closed rind, and proper whey drainage from the interior of the cheese. There are no hard and fast rules for just what amount of pressure is needed. Each cheese; different form, size, and shape; temperatures during draining; and so on will influence just how much pressure and time is needed to achieve the before-mentioned goals. (For more on choosing presses, see chapter 6.)

DRAINING IN BAGS

This technique is almost strictly used for soft, spreadable cheeses such as chevre, fromage blanc, and cream cheese. But some surface-ripened cheeses are briefly drained in bags before being ladled into forms. Bags and cloths meant for draining the mushy curds of a

RINSING FORMS, BAGS, AND CLOTHS BEFORE USE

No matter the method of draining, the receptacle should be sanitized (see chapter 6 for ways to sanitize) just before use. To help rinse away any chemical residue, as well as prepare the form to receive the curds, the form and liner should have some warm whey from the cheesemaking vat (or saved during draining of the vat) poured over it. This not only removes or neutralizes any chemical residue, but it also warms the form and brings the surfaces to a similar pH and acid content. This creates a "friendly" environment that is less likely to shock the curds when they are first placed in the form. By not shocking them you will help the initial drainage potential of the curd mass.

Bags of chevre draining at Black Mesa Ranch, Arizona

fresh, soft cheese must be of a very fine weave or you will lose valuable product in the whey. In fact, take a good look at the whey collecting below the bags; it should be relatively transparent and a yellow-green color, not white. If it's whitish, you know you are losing some butterfat and product through the cloth.

Bags can be formed by using large pieces of cloth whose ends you gather together, knot, and tie with a string, but bags actually designed for draining soft cheeses will be less frustrating (no chance of their popping open while being filled). You can even use a sanitized pillowcase (remember, any bag you use should be sanitized).

DIPPING AND LADLING CURDS

If you have made a soft, high-moisture curd, such as for Camembert or any of the lovely French goat cheese varieties, you will be removing the curd from the whey. The curd mass is often cut just before ladling, not in the cube shapes into which a cooked curd cheese is cut, but in columns. This helps the curd drain quicker once in the form and helps fill the form evenly and create the right shape.

Once you start ladling the curd, you must try to get it all out of the vat and into the forms in a reasonable amount of time. Just what does that mean? Well, think about what is going on while you are blissfully ladling your future Pouligny-Saint-Pierre style cheese into its pyramid-shaped forms: The curd waiting in the vat is continuing to acidify and lose calcium to the whey. So you must try to size your batches to match the speed at which you (or you and your assistant) can get the curds into the forms; you don't want the batch to vary greatly. A general guideline is 20 to 30 minutes. If you aren't sure, just check the pH at the beginning and end of filling the forms; if it varied too much, flag one of the cheeses from the beginning of the ladling and one from the end. Compare them when they are ripe, and see what you think.

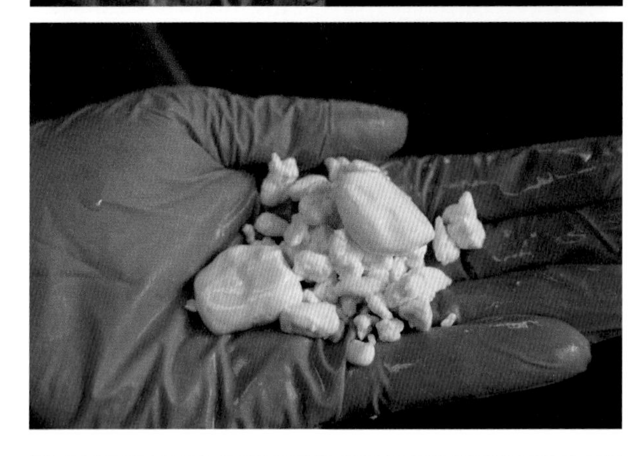

Curds of wheat grain size—about ¼ inch (0.6 cm), curds of hazelnut size—⅜ to ½ inch (1–1.3 cm), and undesirable irregularly sized curds

WHEY OFF

If you have made curd that is meant to be pressed, you will be removing a large portion of the whey from the vat before removing the curd. This is referred to as *whey off*. Usually, whey off occurs after the curd is pitched. In larger vats a stainless steel screen or a food-grade plastic mesh is placed over the drain at the bottom of the vat, and the whey is released. I make our cheese in a 30-gallon steam kettle. Most steam kettles (also called soup kettles) have a pouring mechanism that allows you to tip the entire pot and pour whey off the top. While my little vat has its drawbacks, this is a feature I really love—it's fun and feels quite official. . . . For home-size batches you can use a pitcher to remove the whey or pour it off the top of the pot.

TEXTURING

If you do some in-depth cheesemaking reading, you will come across the terms *textured cheese* or *textured curd*. While your mind's eye might immediately see a cheese with a pattern on the rind, this term really refers to an internal change of structure that alters the final texture of the cheese. A couple of good examples of textured cheeses are cheddar (both traditional and stirred-curd varieties) and mozzarella. Cheddar curd is textured during the cheddaring process, while stirred-curd cheddar and several other cheeses (such as another British favorite, Cheshire) are textured by stirring the curds for a period of time in the drained vat. Cheddar is also milled (recut into fairly uniform sizes), as are some stirred-curd varieties. Mozzarella and all so-called *pasta filata,* or kneaded-curd cheeses, are textured when the curd is heated after reaching the goal pH, then stretched and shaped. I'll give you the specifics about how each cheese type is textured when we get to the pertinent recipe chapters.

CHOOSING FORMS

It is amazing how the details can sometimes derail you. Choosing a form (also called a hoop or mold—not to be confused with cheese mold, which also goes on the outside of cheeses but is fuzzy and smelly) can be confusing. If you happen to come across one of the cheese supplier websites that offers many different forms for sale, it is hard to know where to begin—or to stop! So let's go over some basics. (For more on forms, including how to make your own, see chapter 6.)

Many cheeses have an ideal diameter-to-height ratio or proportion. So if you are trying to make a certain variety of cheese, you must choose a form that will duplicate these proportions—or you might be disappointed with the resulting cheese. For example, about the only cheeses that can be taller than they are wide are bandaged cheddars and blues. A bloomy rind must have dimensions that allow ripening to occur evenly, so they are typically very short for their diameter.

Marks created by cheesecloth during pressing can usually be avoided.

Here I go again—you must know the goal of your cheesemaking! A small cheese will age more quickly but will also lose more moisture during the same period of aging. A tall cheese could fracture (develop interior fissures and cracks) if not formed of curd with the proper moisture content. A thin cheese, meant to become a hard cheese, might invite some wild red surface-ripening bacteria to populate its exterior and, since it is so skinny, soften and grow into a lovely little stinky cheese instead. (Just like children, you can't always control what they grow up to be.)

Once you decide on the proportions of your cheese, you might have a few more choices to make when choosing a form. If you are using a mechanical press, you may be limited to certain sizes and shapes, such as offered by Kadova forms. This brand of form is made with a built-in mesh liner and is meant to stack for mechanical pressing. But if you are able to make cheeses that do not require mechanical pressing (more on that in a moment), you will be able to choose from a larger selection of forms.

DRAINING AND PRESSING IN FORMS

Now it's time to help the cheese attain the desired shape, moisture content, and acid level prior to salting (or if you are making cheddar or one of its cousins, pressing will occur after salting).

MOISTURE LEVEL OF THE CURDS

Soft, ladled, high-moisture curds will not need more than their own weight and the right temperature room (very important!) to drain and shape up, but the drier the curds, the more pressure will be needed. Cheddar and other stirred-curd cheeses that have had their curds salted in the vat are very dry, thanks to the dehydrating effect of the salt, and therefore need a whole bunch of pressure to properly knit into a wheel.

TEMPERATURE OF THE CURDS

When curds are over 75°F (24°C), the butterfat is still mostly in a liquid state and therefore is less stable in the curds. So warm curd must be treated very gently, or you'll squeeze the fat right out of the cheese! Watch for white, greasy whey, and you'll know you are using too much pressure. By the same token, if the curds cool too much, they will repel each other and not knit well—unless you apply the right amount of pressure.

The room temperature can help or hinder drainage. Generally, a 72°F (22°C) room is ideal for draining most cheese types—just below the butterfat melting temperature. The room temperature can be dropped later in pressing, once the goal shape and moisture level have been reached, to slow acid development if needed. Why would you need to slow its development? Well, the usual scenario involves the right pH being attained and the cheese needing to go into the brine at 3:00 a.m., unless you can slow the development down so it is ready at a more civilized hour.

THE ROLE OF CHEESECLOTH

You might think that cheesecloth is just to hold the loose curds in the form. While it is true that this is its first job, it performs an equally important task by wicking draining whey to drainage holes in the form. The fewer holes in the forms, the more important this job becomes. If the cloth has a weave that is too open for the texture of the curd, curd bits will plug the holes and make cleaning the cloths difficult to impossible.

Besides choosing the right weave and openness of cloth (more on selecting cheesecloth in chapter 6), it is important to prepare the cloth before placing the curd in the form. This is best done by rinsing the sanitized cloths with some warm whey from the vat. This not only rinses any chemicals away (or neutralizes them), but it puts the cloth at roughly the same pH as the curd. With the warmth, moisture, and ambient pH, the curd will be less likely to close up and not drain properly.

Toward the very end of pressing, the cloths may be removed from some types of cheese so that any pattern from the form (such as a basket-weave form) can be embossed on the surface. But don't take the cloth off too soon or you are liable to end up with pockets of whey inside the cheese and little pockmarks all over the surface of the cheese where the whey escaped from the inside of the cheese but was then trapped by the solid sides of the form. Removal of the cloth and repressing, even when embossing a pattern isn't the goal, will help prevent marks being left by the cheesecloth wrap.

HOW MUCH PRESSURE IS ENOUGH?

As much as it takes! Seriously. That being said, many recipes will give you some guidelines, such as how many pounds of pressure to apply (although most don't clearly state whether this is actual weight or weight per square inch) or how many cheeses to stack on top of each other. Remember what I stated above about the goals of pressing—shape, moisture, and acidity? You need only apply as much pressure as it takes to attain these goals. Once you understand the other things that influence pressing—the moisture and temperature of the curds, the desired texture—you will be able to choose the appropriate amount of pressure to be applied at the right time. Until that time, by observing the cheese at each flip, you can observe the progress and adjust the weight accordingly.

THE TIME AND TEMPERATURE CONNECTION

The amount of time it takes to press a cheese properly is closely tied to the temperature of the cheese during pressing. There are several factors that need to be considered when choosing the right time and temperature for draining: wheel size, pH at the beginning of draining, the goal pH by the end of draining, and desired texture of the cheese.

Large wheels, especially those made from higher-temperature cooked cheeses, such as Parmesan types,

After the first flip curd shapes are still evident, after the second flip the rind is closing nicely, and after the third flip the rind is mostly closed.

will stay warm for longer than small wheels made from cooler curd. The pH at the beginning of draining, compared to the goal pH at the end of draining, will influence how warm you will want the cheese to stay. The cooler the room, the slower both the drainage and the pH development. You can sometimes use this to your advantage; if you need to delay the time when the cheese needs to be moved to the brine (or otherwise salted), you can cool the room or move the cheese to a cooler location. Where texture is concerned, if the cheese cools too rapidly, the curd will not knit as well, and mechanical openings will be more likely.

I have included goal room temperatures during draining and pressing in the recipes. As a rule of thumb, 72°F (22°C) is an ideal room temperature during this phase. If you have only one or two wheels and cannot maintain an even room temperature, you always have the option of setting them in an ice chest or other heavily insulated space to help stabilize the temperature.

THE ESSENTIALS OF SALTING

Salt does amazing things for cheese. (In this section when I say "salt," I mean specifically sodium chloride—NaCl, a.k.a. table salt.) In chapter 1 we covered the multiple important functions of salt: the slowing and halting of acidification, its assistance with drainage, the control of bacterial and mold growth, its influence over texture, and its contributions to flavor. Now let's go over types of salt and options for properly incorporating salt into cheese.

WHAT KIND OF SALT?

There is salt, and then there is *salt*. Sea salt and rock salt (the former produced by evaporating sea water and the latter by mining salt deposits from the earth) are not pure sodium chloride. (Throughout the recipe section I will be referring to salt used as an ingredient as "pure salt.") This does not mean you can't use them (unless they contain over a certain percentage of other minerals that are not legally allowed or desirable in the manufacture of cheese), but you will need to keep in mind that the goal amounts of salt listed in Table 2–4 are for pure sodium chloride. Also, sea and mined salts often contain visible elements that are not desirable in your cheese. Just things to remember before you order a bag of Dead Sea salt, which might be "100% natural, visible purified," as some ads say, but also might be composed of more potassium, magnesium, and calcium chlorides (also salts) than sodium chloride (there is a reason it is used in bath-care products instead of cooking!).

The texture of salt matters, too. Rock and large-crystal salts will not dissolve evenly on the surface of cheeses. Extra-fine ground salt, such as that used in butter making, is too fine for most cheeses and will dissolve and run off the cheese along with the whey before it has a chance to be absorbed into the rind. The best texture for most cheesemaking applications is a flaked salt, such as you find in kosher salt. In fact, kosher salt got its name not because it is considered kosher as defined by Jewish food laws (almost all salts are) but because it was originally manufactured for use on meats being made for kosher consumption by salting to remove blood. The flake form of kosher salt allows it to coat meats and cheeses more easily than other forms of salt.

Unfortunately, all kosher salts are not created equal. For some time I blithely used the most common brand available, Morton kosher salt, but then found out that it contains an anticaking agent that is safe to eat but is not a good choice for commercial cheesemaking. Another brand, which is a bit harder to find but is completely free of additives, is Diamond Crystal kosher salt. It can be found in kitchen and gourmet stores, and I have heard that some larger or specialty grocery stores carry it as well. We like Sonoma kosher (I purchase it online through SaltWorks; for more information see appendix B) in a 30-pound bag, which brings the cost down to less than $2 a pound (and sometimes you can even get free shipping). If the recipe calls for fine salt, as does one of our blue cheese recipes later in part II, you can choose canning salt (free of additives) or keep a salt grinder on hand and grind coarse kosher salt to the desired texture.

METHODS OF SALTING

There are two main methods of adding salt to cheese—*dry salting* and *brine salting*. Some cheeses can be salted using either method. Often the method used is the personal preference of the cheesemaker, but at other times one method will be preferred over another.

TABLE 2-4. AVERAGE IDEAL SALT LEVELS IN CHEESE

Cheese	Ideal Salt Content
Emmental (Swiss type with eyes)	0.5–1.2%
Crottin (bloomy rind, soft ripened)	1–3%
Cheddar	1.5–1.95%
Gouda	about 2%
White mold	about 2%
Limburger (stinky, washed rind)	2.5–3.5%
Provolone	about 3%
Parmesan	3.5%
Feta (rindless, brine aged)	3.5–5%
Blue	4–7%
Romano	5–7%

How to Dry Salt Curd and Cheese

Dry salting is done by sprinkling or rubbing salt onto the curd or formed cheeses. Soft, fresh cheeses such as fromage blanc and chevre have salt added to the drained curd and mixed in by hand or by machine. Bloomy rinds, such as the little crottin mentioned in the table above, have salt sprinkled (either by hand or with a small sieve) onto the still-draining but unmolded (out of the forms) cheeses. While bloomy rinds are almost always dry salted, they can be brined. Blue cheeses may be partially salted in the vat and partly salted as finished wheels. Textured cheeses, such as cheddar, are salted after milling or stirring of the curds, when the final goal pH has been reached. When salt is added to curd, it must be given a period of time to melt and begin to be absorbed. This is called *mellowing*; otherwise, dark lines surrounding the curds might be present in the mature cheese, and texture flaws also might occur.

When dry salting, ideally, you need to know how many pounds of curd you are working with or the

WHAT ABOUT LOW-SODIUM CHEESE?

My temptation here is to say don't even waste your time thinking about low-sodium cheese, but the truth is that sodium is an issue for many people. To make great cheese sodium chloride *must* be used. As you have learned, there is a small range where that amount can be varied and you will still end up with a tasty, safe, and recognizable product. Cheeses can be made with other "salt substitutes" (which are really just other types of salt, such as calcium chloride), but these commonly lead to flavor defects, which the industry has addressed by adding flavor enhancers.

But a 1-ounce piece of natural (meaning made from dairy milk to which culture, enzymes, and salt have been added) cheddar cheese is *naturally* 50 percent lower in sodium than 1 ounce of processed *reduced* sodium cheese! A 1-ounce piece of natural cheddar has about 170 milligrams of sodium, while 1 ounce of regular processed cheese has about 450 milligrams.

The point is that natural cheese is the better choice in a reduced sodium diet than so-called "reduced sodium" processed cheese products. Of course, it must be eaten in proportion to the rest of the sodium intake and the level of sodium allowed in the person's diet.

weight of each cheese. If you don't want to weigh the cheeses, you can weigh the drained whey instead. If you know how many pounds of milk you started with, you will know that the curd weight is the difference between the starting weight and the whey weight. You also need to keep in mind how much moisture is still retained in the curd—the wetter the curd, the more salt will be washed out along with the exiting whey. So a bit more salt is needed to obtain the same level in the final cheese. In the recipe chapters I'll be a bit more specific about particular cheese types and their salt requirements.

How to Brine Salt Cheese

Brine salting is my favorite way of salting most cheeses. First, the cheeses float supported and suspended in the solution, instead of being salted on a mat (this means no mesh marks). Second, the cheese holds its shape instead of wanting to bulge or spread out while still on the draining table (although you can use this bulging, spreading effect to your advantage in making some cool new shapes for your cheeses). I also like the fact that I can go by time in the brine versus measurement of dry salt.

Ideally, cheeses are put into the brine at the same temperature as the brine (usually, 46 to 59°F [8–15°C]). This helps prevent shocking the exterior of the cheese and reducing the permeability of the rind to the brine (kind of like closing up your pores by splashing ice water on your face). Temperature also affects the uptake of salt into the cheese. In general, the warmer the brine, the faster salt diffuses into the cheese. Keep this in mind when following a recipe that gives you a range of brining times. On the other hand, if the brine and cheese are too warm, fat may leak out of the cheese into the brine, causing a slower uptake of salt. In addition, to maintain microbiological safety, brine should be stored below 55°F (13°C).

Just before placing wheels in the brine, it is a good idea first to agitate the solution to help ensure an equal distribution of minerals and contents. This will also give you the chance to check for salt saturation. Over time other components such as proteins and fats will also accumulate at the bottom of the tank, so don't assume that the white solids are salt; feel them to see if they are actual salt crystals before assuming that the brine is still fully saturated (if that is your goal concentration).

Cheeses should not be so crowded that the brine cannot flow around them. If your tank is occasionally

An overcrowded and poorly maintained brine tank

overcrowded, make sure to move and turn the cheeses a bit during their time in the brine so each side gets equal exposure time. By that same token you should always sprinkle dry salt on top of each cheese after it is floating in the brine, unless you have the capability of totally submerging the cheeses. This can be accomplished using a rack that presses them down into the brine (but expect mesh marks if you use this method). During brining it is always best to agitate the solution, but to be honest I never do, and it doesn't seem to affect the process negatively—or at least not to a degree that is noticeable.

The higher the fat content of the cheese, the slower the uptake of salt will be. This is especially true if you lose some fat from the curd during pressing. The oozing fat basically forms a water-repellent (and also brine repellent) coating on the surface of the cheese and curds within the cheese. So allow more time for brining, and maybe brine at the higher end of the acceptable temperature range, where the fat will be more soluble than when it is colder.

TABLE 2-5. AVERAGE BRINING TIMES BASED ON CHEESE TYPE

Cheese Type	Average Brine Time (23% salt brine) (wt is lb/kg per wheel)
Bloomy rind	30 min
Semisoft, washed curd	2–3 hrs/lb/0.5 kg
Semihard	3–4 hrs/lb/0.5 kg
Piquant hard (Romano type)	8–12 hrs/lb/0.5 kg
Feta types	8–9 hrs/lb/0.5 kg

Making and Maintaining Brine

A well-made and maintained brine can last almost indefinitely. The home and small artisan cheesemaker, however, will probably not have access to some of the purification and filtration methods that larger producers will, but we also won't have such a large volume of brine that we cannot dispose of it without concern (yes—because of chloride's reactiveness, brine is considered a pollutant by many regulatory bodies).

Fresh brine is made using water, salt (sodium chloride), calcium chloride, and vinegar. Calcium chloride is added to mimic the mineral balance in the cheese. If the brine is low in calcium ions, ions in the cheese will migrate out and into the brine in an attempt to balance the solution (diffusion to balance the osmotic pressure). This migration of calcium softens the rind and can even dissolve part of the surface of the cheese into the solution. Vinegar is added for a similar reason, to balance the pH and the hydrogen ion concentration (more on that in the next chapter). As with calcium, if the acid isn't balanced, there will be a movement of ions and a softening of the rind. The amount of vinegar (or other food acid) added will depend on the cheeses being brined. The goal is to maintain the brine pH so it matches that of the cheeses (see the reproducible chart in appendix C for common brine concentrations and the proportion of salt that should be added). In general, for each gallon (4 L) of water, add 1 tablespoon (15 ml) of calcium chloride (of a 32 percent premixed solution) and 1 teaspoon (5 ml) white vinegar.

Some cheesemakers use whey instead of water as a brine. Since whey has calcium in solution and is similar in pH to the cheeses, this works well. I prefer using water, however, to create brine, for several reasons. First, if you work with raw milk, the whey will have a population of starter and nonstarter bacteria; second, whey contains a lot of protein that can break down in the brine and lead to quality issues; and last, it is possible to be more consistent with each batch of brine when it is made from scratch. But many traditional recipes harvest the whey both for initial brining and for a lighter storage brine.

If you want to experiment with using whey to create a brine, first heat it to 185°F (85°C) or until the whey proteins precipitate out of the solution (you are basically making whey ricotta). Allow it to cool a bit, then skim the proteins and filter the whey. This heat treatment will not only help you remove the excess proteins but will also rid the whey of any bacterial activity.

If the water you have available has a high calcium level, you might not need—or want—to add calcium chloride to the brine. Too much calcium in the brine can mean a migration of calcium into the surface of the cheese, leading to a hard, hornlike surface. If you aren't sure, make your first brine without any or no more calcium than the recipe calls for, and see if you get any sliming on the surface of the cheeses. Remember that if cheeses start to slime, you can usually salvage them by scraping off the slimy part and adding more calcium chloride to the brine before placing the cheeses back into the solution. Don't forget to check the brine pH at the same time.

> If you need to remove brine from your tank to keep it from overflowing, be sure to stir the brine well first, so you are leaving a balanced solution.

A well-maintained brine should keep its mineral and pH balance fairly well. Each cheese that is brined will lose whey, and calcium, into the solution. If you are consistently brining similar types of cheese, your brine is more likely to remain stable. If you brine cheeses over a range of finished pH levels, however (such as feta, with a low pH, and Gouda with a higher pH), your brine will fluctuate in response. It is a good idea to routinely check not only salt concentration (more on that next) but also pH. The brine volume will increase by the loss of whey from the cheeses. If you need to remove brine from your tank to keep it from overflowing, be sure to stir the brine well first, so you are leaving a balanced solution. (As the brine sits, it will have some areas that are of different concentrations from others.) If your brine becomes too acidic, you can neutralize it a bit by adding some baking soda.

Brine saturation levels will vary depending on the use of the brine. For initial salting of cheeses, the usual level of salt to brine is about 23 percent (fully saturated brine is closer to 26 percent salt). When talking about the percentage of salt in the brine and the level of saturation, remember that the amount of salt the brine can hold will vary based on the temperature of the solution. Brine concentration is almost always measured at 60°F (16°C). Since most brine is stored at a lower temperature, this must be taken into account. A brine that looks fully saturated at 55°F (13°C)—you might see some salt crystals at the bottom of the tank—might be less than saturated at the standardized temperature of 60°F. For this reason many small cheesemakers keep their brine fully saturated at the cooler temperature and are content with this level of salt.

Brine must be maintained to prevent bacterial growth, keep proper salt and calcium balance as well as proper pH, and to prevent off-flavors (you are salting the cheese, not marinating it!). In commercial-scale cheese manufacturing, brine is often continuously passed through a filter and ultraviolet (UV) light purification system to ensure that the brine remains free from bacteria, yeasts, and molds. (UV purifiers are a chemical-free solution for purifying water sources, as well as brine. The water passes through the strong UV light, which kills microorganisms.) At home you can periodically filter the brine through a finely woven cheesecloth to remove particles and "sludge" that will accumulate at the bottom of the tank. The sludge is made up of proteins and fats that have leaked out of the

BRINE QUALITY MAINTENANCE QUESTIONS

1. Do I see salt at the bottom? If not, add more salt (at 55°F [13°C]).
2. Does it smell nice or like strong cheese? If it's too strong, it's time to make some fresh.
3. Are there little chunks of cheese floating or a white sludge at the bottom of the tank? If so, it's time to filter it through a cheesecloth.
4. Have you added calcium chloride lately? If you use the brine frequently, you will need to refresh the calcium level periodically.
5. How is the pH? If it is too low, you can make a partial new batch or add a bit of baking soda. If you haven't added calcium chloride lately, try that first, as it will have a slight buffering effect.
6. How long has this brine been in use? If it has been over 6 months and it still smells nice, consider putting it in a pot on the stove and pasteurizing it. If it doesn't smell so good, well, you should have caught that under question 2!

Measuring brine strength using a salometer and homemade PVC vial

cheeses. Brine maintenance is easy to forget, but as long as you pay attention to your senses when putting a cheese into the brine, you should be fine.

A more accurate way to monitor salt concentration is by using a salometer (also spelled "salinometer"), also known as a brine hydrometer, to determine the salt content of a sample of whey. A salometer is a type of hydrometer that measures a fluid for salt concentration. You can purchase digital salometers, but the basic floating-bulb type works very reliably and is inexpensive. The first thing that confused me about using a salometer was understanding the difference between a brine's being 23 percent salt and at the same time 90 percent salt by saturation on the salometer (at 60°F [16°C]). I am going to give you a minute to reread that and mull it over. . . . Ready to go on? The first measurement of 23 percent salt (also called 23° Baume, after the Frenchman that developed the method) means that in 100 pounds of brine there are 23 pounds of salt. The 90 percent saturation means that of that 100 pounds of brine, 90 percent of the solution is saturated by salt.

A salometer's scale is read as degrees, so 90 percent is read as 90° SAL. If you are really clever, you may want to buy the kind of salometer with a scale calibrated based on saturation by weight, instead of the percentage of salt, so that when you add 23 percent salt, the salometer will read 23 percent. When purchasing a salometer, you will also have the choice of plastic or glass. Though glass is less expensive, it is worth spending a bit more for the plastic version and not having to worry about shattering glass in your cheesemaking space. Most salometers are meant to be floated in a clear cylinder to read the concentration (if your brine tank is deep enough, you can float the salometer inside the tank), but we built an inexpensive vial out of PVC. Since it isn't clear, we have to fill it to the top with brine so the salometer will float at a readable level, but it still only takes less than a cup of brine to get a good reading. Salometer readings will vary a bit based on the balance of other components in the brine. As the brine ages, look for a higher degree measurement than with a fresh brine.

An Unusual Brine Problem

Slimy, ropey brine is not too common, and if it occurs it is not an issue for safe food, but it is quite visually unappealing. There are some bacteria (both mesophiles and thermophiles) that are natural producers of compounds called *exopolysaccharides* (EPS), which have a viscous, raw-egg-white-like consistency. Some of these types of bacteria are purposely used in some yogurt cultures to help create a product that doesn't separate when stirred. Many of these bacteria are also environmental and can only be avoided by proper brine maintenance—including maintenance of acid levels and even periodic changing, pasteurization, or other methods of bacteriological sanitation. Avoiding the use of yogurt cultures as a substitute for thermophilic cultures is a good idea to help prevent these EPS producers from inhabiting your brine.

3: THE FUNDAMENTALS OF ACID DEVELOPMENT AND MONITORING DURING CHEESEMAKING

It's tough to know at what point the new cheesemaker should go beyond the basics regarding the topic of acid development during cheesemaking. In the beginning of a cheesemaker's education, I believe it's better to discuss acidity in fairly general, unintimidating terms. But at some point mastery of the subject will become more important, and any cheesemaker worth their cheese salt will need to better understand how to read and interpret acid levels. There is really no other tool or skill that a cheesemaker can use that will make the difference between erratic, unpredictable results and quality, safe, consistent products.

In this chapter we'll cover some basics about pH and how it is a measure of acidity in milk and cheese; we'll explore the somewhat complicated subject of buffering; we'll follow the path of acid development during cheesemaking; and then we'll go into more specifics about measuring acid using the two main standards, pH and titratable acidity. I've included quite a bit about how to work with pH meters and their moody and needy dispositions.

THE ESSENTIALS OF ACIDITY AND PH

In the first chapter of this book, you learned that the milk sugar consumed by bacteria produces lactic acid. Without acid production, cheese does not exist. So let's go over some basics about understanding pH and acid.

You may remember when beauty and hair product advertisements began using the term "pH neutral." If you do, you may recall that this implied that the shampoo (or whatever they were trying to sell) was gentle, not harsh. So think of pH neutral (which is 7.0, by the way) as gentle. The farther you go from 7.0, up or down, the harsher the solution. Things that have a pH below 7.0 are sour, or acidic. Things above 7.0 are bitter, or basic (also called alkaline).

TABLE 3-1. FAMILIAR SUBSTANCES AND THEIR PLACE ON THE PH SCALE

Examples	pH	Hydrogen Ion Concentration Compared to Distilled Water
Lye, caustic soda	14	1/10,000,000
Bleach	13	1/1,000,000
Detergent	12	1/100,000
Ammonia	11	1/10,000
Milk of magnesia	10	1/1,000
Toothpaste	9	1/100
Baking soda	8	1/10
Distilled water	7	1
Milk	6	10
Cheddar cheese	5	100
Feta	4	1,000
Orange juice	3	10,000
Lemon juice	2	100,000
Stomach acid	1	1,000,000
Battery acid	0	10,000,000

Do you also remember learning about the number line in math? There is zero (think neutral) and the positive and negative sides. If you combine the same number from each side—say, −3 added to +3—you end up with 0; in essence, they neutralize each other. The same happens with substances from opposite sides of the pH scale. If you combine equal strengths and measurements of vinegar (whose pH is 4 points below neutral) with ammonia (whose pH is 4 points above neutral) guess what you end up with? A solution with a pH of 7.0—neutral!

While the pH scale covers things on both the acidic and basic side of neutral, the cheesemaker is only concerned primarily with the acid side. For this reason it sometimes sounds as though the terms "acidity" and "pH" are synonymous, but I hope you will now understand that they are not quite the same thing.

In chapter 1 you were introduced to *ions*, atoms or molecules with a positive or negative charge. You might remember from basic high school chemistry that positives and negatives are opposites that attract. Solutions below neutral (acids) contain positively charged hydrogen ions (+H) while solutions above neutral (alkaline) contain negatively charged hydroxyl ions (−OH). It is the combination of these two opposites that brings about neutralization.

Hence, any solution with a pH below neutral will have hydrogen ions in varying amounts, and above-neutral solutions will have fewer hydrogen ions and more hydroxyl ions (−OH). Since these ions carry an electrical charge, they can be measured with a meter capable of reading the electrical potential of a solution. *Titratable* acidity, on the other hand, measures the amount of an alkaline solution it takes to neutralize the acid and uses that measurement to define the amount of acid present (more on measuring titratable acidity and reading pH with a meter later).

> Once the pH nears the goal, even tiny decreases beyond the target will make huge differences in the amount of acid present in the curd.

There is one more thing you should understand about the pH scale: It does not increase or decrease in an equal fashion; it's what is called a *logarithm*. If you are familiar at all with the Richter scale, by which earthquake magnitude is measured, you might already understand that logarithms increase (and decrease) exponentially. Just like the Richter scale, the pH scale is exponential (meaning, multiplied by an exponent). The pH scale ranges from 1 to 14, with pure water being neutral at a pH of 7. Every point lower than 7 on the pH scale is ten times more acidic than the number before it. So a pH of 6 is ten times more acidic than 7, a pH of 5 is 100 times more acidic than 7, and a pH of 4 is 1,000 times more acidic and so on. I know, that might be a little bit too much math, but stop and think about the implications of this type of scale. For the cheesemaker it means that once the pH nears the goal even tiny decreases will make huge differences in the amount of acid present in the curd.

CHEESEMAKER MOTIVATIONS:

THE HERDKEEPER

Because making a pound of cheese takes over a gallon of milk, it is an excellent way to use a surplus milk supply. Folks who milk a couple of goats or a family cow often turn to cheesemaking as a way to use up that fluid, white bounty; after all, you can only drink so much! Turning this surplus into cheese gives small herdkeepers a way to capture the nutrition of the milk for use during the season when their animals are dried off (not milking) in preparation for birthing. It also often leads to sharing and trading (we won't worry about legalities of barter right now) with friends and neighbors. Many herdkeepers turned cheesemakers find themselves contemplating "going commercial"; indeed, this is probably the primary incubation grounds for small licensed farmstead cheesemakers.

BUFFERING CAPACITY EXPLAINED

Buffers are substances that have the ability to absorb acid (in the form of hydrogen ions: +H) and resist pH change. Solutions with good buffering capacity contain a lot of these substances. A solution with good buffering capacity will *attempt* to prevent pH changes when acid is added or created.

The buffers in milk include casein, calcium phosphate, and other minerals. Each of these has a different capacity to absorb +H. To help appreciate milk's buffering ability, consider how drinking a glass of milk—or swallowing a Tums or Rolaids tablet made of calcium carbonate—will help neutralize acid in our stomachs (in fact, many acidic medications recommend you "drink with a glass of milk" at the same time as the medication).

Buffering capacity is the ability of a solution to resist pH changes, but that doesn't mean that the pH doesn't change, it means that the solution *attempts* to keep it from changing. A solution with a strong buffering capacity will still become acidified; it just takes more acid to make that happen than with a solution without any buffering ability. Buffering includes resisting acidic and alkaline changes, but for our purposes I'll only refer to changes in acid level.

Buffers in a solution can be any number of molecules, proteins, or mineral salts that can bind +H by combining and creating a new compound or molecule. In milk the primary buffers are phosphate (usually bonded to calcium) and the protein complex, micelle, of caseins. With caseins (brace yourself for a science-packed statement), it is acid and base protein segments, called amino acid subunits, that act as buffers, while phosphate combines with hydrogen to form an acid salt.

The buffering capacity of milk will change seasonally and by species and breed of the milking animal. In general, the higher the protein content of the milk, the more buffering capacity. We covered this in chapter 2 when discussing measuring culture—that high-component milk, whether from late-lactation animals or a breed with naturally high components, may need more starter culture than called for in the recipe. The extra starter culture will help create more acid to overcome the additional "acid sponges" in the high-component milk.

ALKALINE MILK

What if the starting pH of milk is higher than 7.0? Can milk be alkaline? Yes, when an animal has mastitis—an udder infection—her body does its best to fight it off by forcing white blood cells into the udder. Blood is slightly alkaline. These blood cells, also called *leukocytes*, can be counted in milk as somatic cells (when milk is tested for somatic cells, it is called a somatic cell count or SCC). Their presence will cause the pH of milk to rise. Not only do somatic cells change the pH, but they are bad news for cheesemakers.

What do white blood cells do, after all? They attack foreign bacteria. When you add starter culture to milk, you are adding foreign bacteria. Somatic cells therefore attack the added culture and hinder the cheesemaking process. If that weren't enough, mastitic milk is also higher in enzymes that break down protein; often, this breakdown occurs before you have made the milk into cheese. Whether you are buying milk or have your own dairy animals, you can check for high somatic cells by using one of several inexpensive tests, such as the California Mastitis Test or PortaCheck.

It is important to understand the properties of these buffers, as the entire process of cheesemaking is one of manipulating the buffering capacity of milk through the production and absorption of acid by the curd.

MORE ON ACID DEVELOPMENT DURING CHEESEMAKING

As acid develops throughout the cheesemaking process—from the warming of the milk, the ripening time, and the cutting of the curd to cooking the curd and pressing the formed cheeses—each step will behave differently because of changes in buffering capacity. Building on what you just learned about milk's buffering ability, let's take a look at how acid develops throughout the cheesemaking process.

CALCIUM PHOSPHATE

As you now know, one of milk's main buffers is calcium phosphate. In milk calcium phosphate resides inside the protein clusters and in the milk. It is very sensitive to temperature changes and can move in and out of the protein structures depending on temperature. When it is cold it moves out of the clusters; this is called becoming soluble. When it moves out of the clusters, it is more available to buffer the pH of the fluid portion of the milk. When the milk warms up, it moves back into the clusters. When this happens, the pH of the milk will read a bit lower. If you take the pH of the milk when it is cold, it might read about 6.66, and when it reaches the ripening temperature, it might read 6.56. I used to blame this on native milk bacteria already producing lactic acid, but it is more a function of the calcium phosphate. You can take some of the warm milk and cool it back down, and the pH will go back up. This is your first lesson in learning why you have to take pH readings at the same time for each vat of cheese! **It is not necessarily the actual pH that is important; it is the rate and proportion of change.**

Temperature Compensation

Most decent pH meters have a built-in feature called *temperature compensation*. Until recently, I thought (and I know many cheesemakers who think this, too) that the temperature compensation is supposed to account for milk's natural response to warming. But that's not completely correct. The feature is designed to make adjustments based on how the electrolyte solution in the meter's electrode responds to temperature changes. Now that I know this, it seems obvious: How could they build a feature that would be able to predict the response to the great variety of solutions and substances on which people are using the meter? You can still maximize this feature, however, by calibrating when the buffer solutions are at the same temperature as the solution you want to measure. I say this, but don't ask me if any of us are actually doing this. . . .

Acidity in Milk versus Curds versus Whey

Okay, now you are going to have to do some thinking outside the box. Some of the concepts dealing with buffering during cheesemaking might at first seem to contradict themselves. The first step in dealing with this is to remember that milk is a very different creature from curds and whey, and that curds and whey together in the vat are very different from curd being pressed. Let's start with milk.

When the milk is still fluid in the vat, the components that buffer the milk are dispersed throughout the fluid. Remember that the main constituent of milk is water; when you take the pH of milk, you are reading the acid level in the "water phase." Initially, there are fairly quick changes in the pH, such as the difference created by temperature alone, since the buffers are diluted by so much water.

Once coagulant is added and the curd is cut, you must start thinking of the whey and curd as two separate substances. The whey will have some buffering components, but the curd will have more—casein. You might think, given what I said about milk's showing more of a pH change initially because of its lower buffering ability, that the whey pH will drop quicker than the curd. But remember that most of the "sponges" are in the curd, and these will be able to better soak up the acid (the hydrogen ions [+H]). Consequently, the curd soon will contain more acid than the whey. You can see this phenomenon readily by comparing whey pH right after cutting the curd to that of the curd itself. The whey pH will be about 0.10 higher than the curd, depending upon the length of time the milk "ripened" (for a refresher on ripening, coagulating, and cutting, refer back to chapter 2). This difference will continue to increase as the cheesemaking process continues.

> You should attempt to measure and record pH under the same conditions; otherwise, the numbers will not give you meaningful information.

When you are checking pH during pressing, the amount of whey leaking from the wheel will influence the reading. The temperature of the curd or wheel is another factor causing readings to fluctuate. When you are developing goal pH readings, it is a good idea to record the curd temperature at the same time. Consistency is the key! You should attempt to measure and record pH under the same conditions; otherwise, the numbers will not give you meaningful information.

I have included as many benchmark/goal pHs as possible in the recipes, but even when you have this type of information, you can probably understand by now that it is important to develop your own standards. The more I learn about cheesemaking, the more I realize that so much is really in the hands of the cheesemaker—both the interpretation of results and the implementation of control measures.

MASTERING MEASURING ACID DEVELOPMENT

In the early part of this chapter, I talked a little about both pH and titratable acidity (TA). Before the availability of accurate, affordable pH meters, using TA and other tests to determine acid development was the norm. When I first started making cheese, TA was still very common, especially with European cheesemakers. But now pH is definitely becoming the standard method for measuring acid development. I am going to cover TA at the end of this section, though, since it is still in use, and you will no doubt find recipes and directions that only include this type of acid measurement.

One important fact is that pH and TA are not directly related; pH measures the number of "free" hydrogen ions (+H) in the solution, while TA measures all the ions present whether they are free or bound by buffers (those acid "sponges" we talked about earlier). Since the concentration of unbound hydrogen ions is what affects other things going on during the cheesemaking process, pH is the most direct measure of what is concerns the cheesemaker. For each cheese type you can develop parallels, but they will work only for that type of cheese and will change slightly if the buffering capacity of the milk changes; for example, if the protein level is high, such as at the end of the animal's lactation. In other words, as I said, TA and pH do not measure the same thing.

MEASURING ACID USING PH

This is my favorite way to monitor acid development in cheese. It is not perfect, but it is quick, it allows you to check curd and finished cheeses, and it is relatively easy to find pH parameters for different cheese types. Meters come with their own set of irritating issues, such as needing frequent calibration and cleaning, but in general, if you treat them well, they are quite reliable.

Choosing a pH Meter

pH meters come in all sizes, from a simple, handheld instrument, not much larger than a big felt-tip marker, to larger so-called "benchtop" models meant to remain stationary on the work table. Handheld models cost just around $100 and are usually quite reliable. Be sure to check new units immediately upon arrival. If there is a problem, it's important to troubleshoot it immediately for warranty coverage. I advise purchasing a starter kit with your first meter. This will ensure you have all the proper calibration solutions (also called

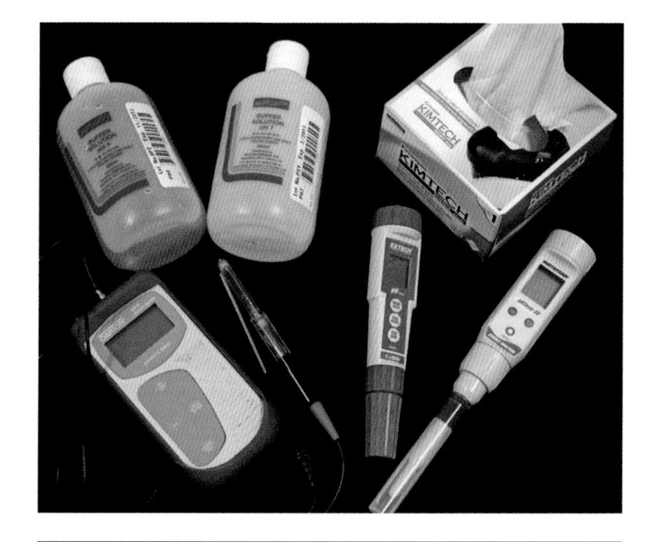

Various pH meters, buffering solutions, and Kimwipes: Meters pictured from left to right are the Oakton Acorn portable with spear-tip electrode (about $350 as a kit), the Extech Ex Stik handheld with a flat electrode (about $110), and the Oakton pH Spear (about $300).

buffer solutions because they will hold their specific pH, making calibration an accurate process), cleaning solution, and storage solution. Meters should come with instructions for rehydrating the electrode before first use, calibrating, and cleaning. Be sure to follow these instructions carefully before and during use. Satisfaction with any pH meter's performance usually directly correlates with its proper care. Must-have features in any pH meter are as follows:

- Readings to two decimal places; in other words, 0.00 instead of 0.0
- Range of 0.00 to 14.00
- Two- or three-point calibration
- Waterproof

My favorite meter so far is the Oakton Acorn portable. It looks a bit like its expensive cousin, the benchtop meter, but it is about half the price at under

A DAY IN THE LIFE OF A PH METER:
WHEN AND HOW TO CHECK PH

So you are learning all of these in-depth details about pH, acidity, and pH meters but you are still wondering about some basics, such as when the heck do you actually use the meter? Let's go over the use of a pH meter during the making of a simple semihard cheese.

After calibrating the meter I like to check the milk pH as soon as it goes into the vat or pot. I scoop a sample of milk into a plastic vial (you can skip this for home cheesemaking) and insert the electrode and temperature probe (handheld varieties have both of these parts situated side by side in the tip) and move the electrode gently and slowly in the solution. I watch the digital readout until it stabilizes—after about 1 minute—then write that reading on the cheese make sheet.

When the milk reaches the right temperature, I add the culture. After the recommended ripening time, I take another milk sample and repeat the process above to get a pH reading. That number is recorded on the log or make sheet, and I keep going with the process.

I usually don't check the pH again until curd has reached the right texture and I believe that it is ready to drain. Sometimes I don't even do it then, but that is only because I am usually making a cheese that I make so frequently that its behavior is pretty predictable. For any new cheese I take more readings than ever, just to learn and develop some benchmarks for future makes of the same cheese. When checking at this stage, you are usually checking the whey, which will have a different pH from that of the curd. You can form a little ball of curd and check it instead. It doesn't really matter which one you check, as long as you record the number and develop a consistent pattern.

When I didn't have a meter with a sharp "spear" tip that I could insert into whole cheeses without leaving much of a mark, I would save a small amount of curd and press it separately. This small cheese was a sacrificial clump that I could check pH on without ruining a whole wheel. Granted, this small amount developed pH at a different rate from the large wheel, but by checking both a few times, I was able to establish parallel numbers. In my opinion, the pH level at the *end* of draining, right before pressing, is the most important number, other than maybe the starting pH.

Whatever you do regarding taking readings, be consistent! This is the only way you will begin to collect information that will help you become comfortable with the process and make decisions that improve your cheesemaking.

$400. It has been the most reliable, satisfying meter that we have worked with so far. The glass spear tip easily checks pressed curd and cheeses without leaving much of a mark in the cheese.

Using and Maintaining a pH Meter

When checking milk or whey pH, it is a good idea to pull a small sample out of your pot or vat and check it separately from the entire batch. There are a couple of reasons for doing this: First, most meters have a glass bulb/tip, and keeping glass away from your batch of cheese is always a good idea; second, checking pH in small samples can help prolong the life of the electrode; and last, any buffer, storage, or cleaning solution residue will only contaminate the small, disposable sample.

How pH Meters Work

A good pH meter for the cheesemaker consists of four main components, even if they are combined in one handheld device: (1) the meter body that gives the readout; (2) a glass electrode; (3) a reference electrode; and (4) an automatic temperature compensation (ATC) probe. Without going into too much detail (as varying meters and electrodes are made differently), let's look at what each part of the meter does.

The glass electrode contains a thin metal wire surrounded by an electrolyte solution. Electrolyte solutions got their name because of their ability to conduct electricity. For example, salt water will conduct an electrical current (from, say, an electric eel) better than tap water. When the glass electrode is placed in a solution that contains hydrogen ions (+H), the ions interact with the glass

WHAT ABOUT PH STRIPS?

Using pH strips is an inexpensive and easy way to check the approximate pH of liquids. Strips are impregnated with a sensitive material that changes color in the presence of acid or base liquids. They range in sensitivity increments from single point (such as 6.0, 5.0, 4.0) to quarter point (0.25, 0.50, 0.75, etc.). They can come in a roll (with a single color to read) up to slightly rigid strips with two or three color indicators. As you might imagine, the finer the point change and the more color tabs, the better chance of an accurate measurement.

It isn't a bad idea for every cheesemaker to have a few pH strips on hand, but they are limited in range, *and* since you can't accurately check curd and finished cheeses, they are also very limited in application. That being said, there are a couple of times where they are useful:

1. If you can't yet afford a meter
2. If your meter is broken or malfunctions
3. If you are only making soft fresh cheeses where the whey can be checked
4. If you are making quick mozzarella (also often called 30-Minute Mozzarella) and want to see if you have added enough citric acid

You can order pH strips online or purchase them at a science or educational store and at most places that supply products for wine, beer, and other fermented food making.

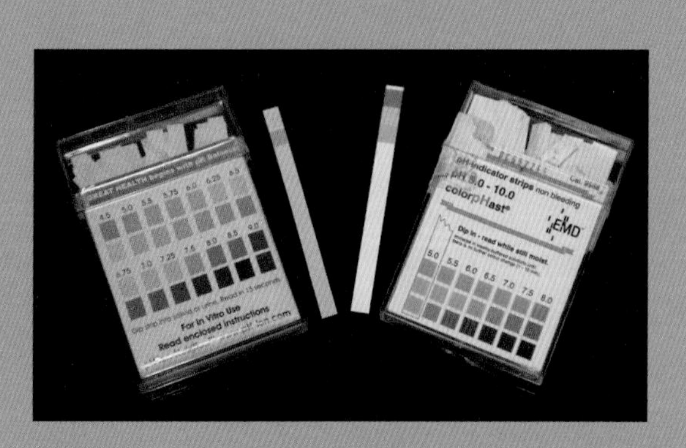

Examples of pH strips that can be used with milk and whey

surface and create a tiny voltage (a current of electricity) that the metal strand picks up and sends to the meter.

Meanwhile, the reference electrode (usually in the tip of the handheld unit) serves to complete the electrical circuit (much like you complete the electrical circuit of a hot-wire fence when your body touches both the ground and the fence). The meter body interprets the electrical signal and converts it to a pH reading. Knowing how pH meters work will help you understand why they do not last forever and need to be properly treated and maintained. The electrodes are sensitive instruments with a life span.

TWELVE-STEP PROGRAM FOR A GOOD RELATIONSHIP WITH YOUR PH METER

1. Never allow electrode to dry out.
2. Always store electrode in proper storage solution (never leave in milk or whey).
3. Calibrate (using 7.00 and 4.01 buffers) before daily use and as needed.
4. Rinse tip in distilled or deionized water and blot dry between calibrations and before use.
5. Use only Kimwipes or other lint and static-free cloth to dry electrode.
6. Never use fingertips to rub electrode.
7. Use dish soap such as Dawn to clean electrode.
8. Monitor and refill electrode (if refillable).
9. Use small amounts of buffer and storage solution, and change regularly.
10. Always take the milk, whey, or curd sample out of the vat, test, then discard.
11. If meter has a separate temperature probe, be sure it is in the sample.
12. Depending on use, expect to replace electrodes every year or two, sometimes more often.

In case you still aren't feeling comfortable working with a pH meter, let me walk you through our cheesemaking day and how we use our meter. When we first come into the make room, we take the pH probe out of the storage solution vial, rinse it with clean distilled or deionized water, wrap a clean microfiber cloth or Kimwipe around the tip to dry it (never rub the electrode with a cloth because you can disrupt the charge inside by creating static electricity), and set the tip in a vial filled with 7.0 buffer solution. Then we go about getting everything else ready—sanitizing the vat, pouring the milk, preparing a make sheet—you get the idea.

After about 15 minutes we turn the meter on and get a reading in the 7.00 solution. If the meter reads in the 7.00 to 7.02 range, we are happy and proceed to rinse and dry the tip, then check it in the 4.01 solution. If either reading does not stabilize within a very close range, we follow the instructions for calibrating (these vary with every meter, so I won't cover them here). If the meter still will not take an accurate reading, we clean the tip using Dawn dish soap (or another gentle but slightly alkaline dish soap) and warm water, rinse, dry, and recalibrate. If the meter is still not behaving the way we like, we soak it for about 30 minutes in the 7.00 solution and start all over.

Are you starting to see that success with a meter requires patience and persistence? During the cheesemaking process we rinse the tip after each reading, dry it, and leave it setting in the 7.00 solution. At least once or twice, we clean the tip with dish soap, since our Nigerian goat milk is so high in fat that it tends to clog up the surface of the electrode.

MEASURING ACID USING TITRATABLE ACIDITY

Before pH meters became as reliable as they are today, *titratable acidity* (TA) was the gold standard for measuring acid development during cheesemaking. A titration is a way of measuring the unknown concentration of a substance that is dissolved in a liquid. First, a color reactor is added to the solution, then tiny, measured amounts of a substance with a known chemical composition are added. The known substance is chosen for the predictable way it will react with the substance whose measurement is being determined. In the case of measuring acid in milk and whey, you use sodium hydroxide (a base or alkaline substance) to slowly neutralize the acids in the sample. (Remember the example earlier of combining ammonia with vinegar to create a neutral sample.) The added color indicator used for TA will turn pink when the milk crosses the line from acidic to basic. The amount of sodium hydroxide that is added correlates to a number that represents the amount of acid in the sample.

The important thing to remember is that TA measures *all* the acid in milk (I like to remember this by thinking of the abbreviation TA as meaning "total acid"). So while lactic acid is what you are hoping to measure, with TA you are also measuring the other acids in the sample, such as fatty acids that might have been released and citric and phosphoric acid. Still, TA is an excellent way to measure acid in milk and whey and is still the preferred method for many cheesemakers.

Titratable acidity can be measured with a fairly simple kit available from home cheesemaking, beer, wine, or other fermentation supply companies. It can also be done using a more expensive device ($90 to $300, depending on quality) called an *acidimeter*. Both instruments use the same chemicals and basic techniques, but if you are not going to be checking TA often, the inexpensive kit (under $20) is fine, and it's safer and easier to use.

For either method the milk or whey sample (9 ml) is mixed with 5 drops of phenolphthalein (a color indicator that will turn pink as the solution turns from acidic to basic). Then sodium hydroxide (NaOH), a strong basic, is added drop by drop while the solution is swirled or stirred. As each drop combines with the sample, it will briefly turn pink. When the basic has essentially

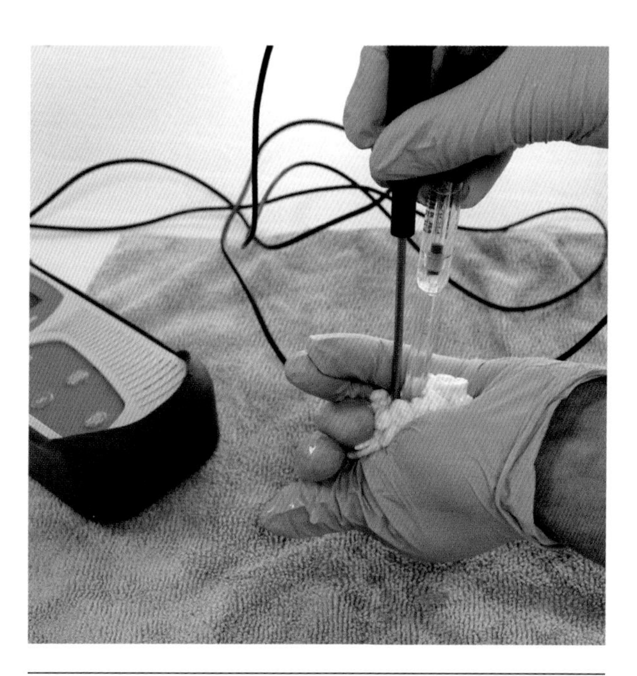

Testing the pH of the curd, instead of the whey, will result in different readings and is an important technique for some cheese types, such as mozzarella.

THE HOT IRON TEST

You might be one of those who say, "Well, they have been making cheese successfully for centuries without pH meters and chemical titrations, right?" Well, you'd be right, but cheesemakers often used a very basic technique that told them if the cheese was ready for salting, which they called the "hot iron test." If you have ever made mozzarella, you might know that the curd stretches thanks to two critical points having been attained: temperature and acid level. When the curd drops to a pH of 5.1 to 5.4 and reaches about 160°F (71°C), it will stretch. You can check a piece of curd this way by immersing it in very hot water (about 170°F [77°C]) and seeing how well it stretches.

But whole wheels cannot be checked by immersing them in hot water, so an iron rod was heated to a dull black heat (not glowing red), then touched briefly to the side of the wheel. When it was drawn away, if a small string of cheese pulled and stretched away, too, it indicated that the wheel had developed enough acid and was ready to salt. So much for no science being used in "the old days."

neutralized all the hydrogen ions, the solution will remain pink. At that point the amount of NaOH required to complete the change of color is noted, then converted to either a percent scale or a degree scale.

One of the most common ways to read TA is by interpreting the number as percentage of acid; for example, converting 1.5 milliliters to 0.15 percent. To make it simple you move the decimal point to the left one place. While TA percent (also referred to as "lactic acid" percent) is the most common way of interpreting the results of titratable acidity, there are also three different ways to read the results in "degrees." The takeaway lesson from this knowledge is to always understand the type of measurement you are taking or attempting to interpret from a recipe or instructions.

Whenever people ask me for help troubleshooting a problem with their cheese recipe, my first question is almost always, "Did you track the pH?" When the answer is no, the likelihood of my being able to help them figure out what went wrong is significantly reduced. But when pH is monitored and the recipe includes benchmark acid goals, then usually the cheesemaker is able to answer their own questions about "what went wrong." In fact, I would now like to award those readers who have mastered taking acid and pH levels with a pHD!

4: AGING CHEESE GRACEFULLY—THE ART OF AFFINAGE

Affinage is French for the art of aging cheese. And let's face it, without as much effort, talent, and art applied to the aging of cheese as there is to the making of cheese, most of it wouldn't be worth eating.

In this chapter I will walk you through some of the main considerations surrounding the craft of affinage. We'll start with setting up a cheese-aging space, then we'll go over the many ways you can protect and finish cheese rinds, and finally, we'll go through a large section on troubleshooting. I will attempt to adequately answer the most frequently asked questions about affinage and even give you some tips and answers that will turn you into the consummate *affineur*.

THE HOME CHEESE CELLAR

Figuring out how to age small amounts of cheese is often the biggest roadblock to making aged cheeses at home or on a small commercial scale. In my first book, *The Farmstead Creamery Advisor*, I devoted an entire chapter to the proper design of an aging space meant to accommodate several hundred or more pounds of cheese, so if you need one with that capacity, you might want to refer to my first book. For this book we'll be focusing on space for the smaller-scale artisan.

When I first got started making cheese, I used a small wine cooler. Then I moved up to a refrigerator outfitted with an external thermostat. This served me quite well until we became licensed and got a larger, walk-in space. Along the way we researched using upright and chest freezers and various refrigerators. Let's go over those options and their pros and cons. Then we'll talk about creating a small room cellar.

WINE COOLER

These are appliances designed to keep wine at proper storage temperatures of between 50 and 57°F (10 and 14°C). In other words, also the perfect cheese-aging temperature! Unlike most food refrigerators, they don't have fans, so they also tend to keep more moisture in the air (also good!). One of adequate size—say, that will hold twenty to fifty bottles of wine—can usually be purchased for under $200 new. The shelves are not designed for cheese, but they can be replaced with baker's racks (the kind you cool cakes on). I aged our first cheeses in one almost a decade ago, and we are still using it, but for wine storage. My first cheeses that won top awards were aged in this type of cooler.

While maintaining the correct temperature is no problem in these units, you will still have to monitor and maintain humidity (more on monitoring humidity later in this chapter). A simple, effective way to maintain humidity in a small space is to hang a moistened terry cloth towel over the lowest shelf in the unit or on shelves in the door, monitor the humidity, and remoisten as necessary. Remember to open the door daily to provide some fresh air for your living, breathing cheeses (if you are checking the humidity and turning the cheeses as you should, the air exchange is accomplished at the same time). These coolers are usually a bit too small to include a fan for air movement (except some very small personal fans) if that is required for your cheese type, so if you are making cheeses that need more airflow during aging, consider a larger unit or a modified refrigerator.

> If any ammonia odors or mustiness develops, you will know you are not getting enough fresh air to your cheese.

REFRIGERATOR

There are two basic types of home refrigerators: one that has a freezer compartment and one that doesn't. If at all possible, choose one that doesn't have a freezer compartment. They are a bit more expensive and often harder to find, but they will be worth the time and cost. First, you won't give up any of those precious cubic square feet to freezer space; second, they use less energy (instead of holding the unused freezer compartment at freezing and blowing cold air from that space to the refrigerator compartment); and finally, the lack of a fan can help reduce overdrying of the cheeses.

Whatever type you choose, be sure to get one with coated-wire, not glass, shelves! Glass shelves will not allow adequate air circulation around your cheeses and will also accumulate condensation on their undersides that could drip on the cheeses below, possibly ruining those wheels.

So here's the biggest question, how do you get a refrigerator to "not refrigerate"? When we first got ours (an all-refrigerator unit purchased used for a hundred bucks), we called different cooling repair guys, expecting to be able to have some sort of temperature control override performed, but the answer was simpler—and cheaper! For about $60 you can buy a special thermostat (designed for beer brewers wanting to convert a fridge or freezer to a fermentation unit) that you can set to keep the appliance at the temperature you want for your cheeses (I still use mine, but now it keeps a chest freezer operating at refrigeration temperatures for our milk can storage). To operate it you plug the appliance into the thermostat, plug the thermostat into the outlet, set the dial to the desired temperature, then insert the temperature probe into the appliance (usually through the door, which will still close without any problem). The thermostat then turns the appliance on and off as needed to maintain the temperature. By the way, operating this way causes no harm to the appliance.

As with a wine cooler, you can simply increase humidity by hanging a moistened towel or sheeting on a lower shelf, so it doesn't drip on cheese. You can even let this towel wick additional water up from a pan. Simply setting a pan of water at the bottom of the unit rarely provides enough of a humidity increase. If you

A refrigerator outfitted as an aging space utilizing an external thermostat

are monitoring the humidity carefully, you will know if it is being maintained at the correct level. If the passive methods mentioned don't work well enough, you can use a small electric ultrasonic humidifier. Be sure to watch for moisture buildup on the walls and any dripping.

While refrigerators have a greater volume of air than a wine cooler, you will still want to make sure that you open the door daily and allow for air exchange. If any ammonia odors or mustiness develops, you will know you are not getting enough fresh air to your cheese. If air movement is required (in addition to air exchange), a small fan can be placed on one of the shelves. Prevent the fan from blowing directly on any cheeses by providing a baffle of thin cloth or creating an air sock.

FREEZERS

While it might seem as though a chest freezer, modified with an external thermostat, would make a good aging space (as they are relatively inexpensive and don't

WHAT IS RELATIVE HUMIDITY?

Relative humidity is fairly tricky to explain. While you don't need the full meteorologist's level of understanding, it's good to be able to grasp a bit more of the concept so you can deal with certain issues that will come up when trying to maintain the proper level of humidity in your aging space.

To explain it simply, first you must know that the total amount of water vapor in the air (if you could measure its mass) is called *absolute humidity*. The amount of absolute humidity relative to the maximum amount that same air could hold is called *relative humidity* (RH). (To be clear, the vapor does not become a part of the air but instead occupies empty space between air molecules.)

The amount of water vapor carried by the air at any given time is very dependent upon the atmospheric pressure and the temperature. For example, if you trap air in a jar and change the temperature, the amount of water vapor would remain the same—and because of the constraints of the jar, the air pressure would change. Consequently, the ability of the air to hold the vapor would stay about the same. But if the air is trapped in a flexible container, such as a balloon, and the temperature changes, the amount of water vapor stays the same, but the ability of the air mass to hold the water vapor—the relative humidity—would change.

It is also important to understand the effect that temperature has on evaporation. For example, a pan of water placed in a cool aging room will have almost no ability to increase the relative humidity. Why? Warm air and warm water contain more thermal energy than cool air and cool water. When more energy is present, the amount of evaporation—the changing of water to water vapor—increases. The cooler the air and the cooler a body of water, the slower the molecules move and the less evaporation occurs. This is one reason that tropical climates are also so humid: More heat and more moisture in the same place equals more evaporation, so more water vapor in the air.

The amount of water vapor needed to saturate the air increases as the temperature increases—when all other factors, such as atmospheric pressure, are equal (since the potential for evaporation increases and the potential for the air to carry more vapor increases). By the same token the amount of water vapor needed to saturate the air decreases as the temperature decreases (since there is less potential for evaporation to occur). Remember that container full of trapped air with the consistent amount of water vapor? If we cool it, the relative humidity would increase, and if we warm it, the relative humidity would decrease.

Another useful term to understand is *dew point*. When the absolute humidity is the same as the relative humidity (meaning the air is fully saturated with water vapor), condensation will occur—this is the dew point. You can see the dew point in action and observe the effects of temperature on relative humidity by placing that same closed jar in the freezer. Remember, you are not changing the absolute humidity (the amount of water vapor in the air). As the temperature decreases, the relative humidity rises. At the point that the RH reaches 100 percent, water vapor will condense out of the air and freeze as ice crystals. The dew point has been reached. If you take the jar out and warm it, the dew point will rise and the ice crystals melt and, if it is warm enough, become vapor again.

use much electricity), the actual reality of creating cheese shelving space that is easily accessible is unlikely. Upright freezers, usually frost-free, have some issues as well. First, frost-free units maintain their frost-free conditions by utilizing a heater that runs on a timer (I was unable to verify if this timer works regardless of the temperature at which the unit is set). Manual-defrost uprights have the condenser coils (the parts that create the cold air) running through the shelves, so not only are the shelves not adjustable, but they will have condensation dripping, which is never good for cheese!

SMALL-CLOSET CONVERSION

This is just about the holy grail for the serious hobby cheesemaker: their own "real" cheese cave/cellar/closet. And it isn't that hard to do. The first thing, when creating a small room for aging, is to think about insulation and waterproofing.

> If you are waxing or otherwise coating your cheeses (or vacuum sealing), humidity is not important for their proper aging—you don't need to create a waterproof space.

If you can choose a space that naturally maintains a steady temperature, or at least does not fluctuate too much, you will be ahead of the game. If you are lucky enough to have a basement or a "real" cellar, that is the place to start. One of the cheesemakers profiled later in the recipe section, Jos Vulto, created a wonderful cheese in a nook accessible from a work space that

Jos Vulto's home cheese-aging cave beneath a sidewalk in Brooklyn, New York. Photo courtesy of Jos Vulto

was literally underneath a sidewalk in Brooklyn, New York. Whatever the space you choose, provide as much insulation as possible. Insulation is available based on thickness and what is called R-value. Think of R-value as the material's ability to *resist changes in temperature.* The higher the number, the more that material will keep the temperature from fluctuating. There are thinner materials with higher R-values and thicker materials with lower values, so don't base your selection simply on how thick the material is. Of course, as you might imagine, you will spend more on the thinner material with the higher R-value. If you have space, you might be able to build thicker walls and spend less.

When it comes to waterproofing, I like to start people out by suggesting they visualize creating a space that could hold up to being a bathroom shower. If you start there, you know you will be able to build something that can withstand the humidity and cleaning needs of a cheese-aging space. Showers are far more humid than a cheese cave, after all. If you are waxing or otherwise coating your cheeses (or vacuum sealing), humidity is not important for their proper aging—you don't need to create a waterproof space, but it will still need to hold up to regular wipe-downs of the walls (and you will have a versatile space for future "naturally" rinded cheeses).

Cooling of your cheese closet can be accomplished very economically by using either a "through-the-wall" wine-cellar cooler (what we use here at our small commercial creamery) or the very popular choice of a properly sized air conditioner combined with an aftermarket controller called a "CoolBot." This controller overrides the air conditioner's thermostat, allowing it to drop the temperature to ideal cheese-aging numbers, instead of the ideal people-cooling temperatures. Wine-cellar coolers are a bit more expensive than air conditioners but are easily set to maintain ideal cheese-aging temperatures. The CoolBot has an advantage in being adaptable to any easily purchased room air conditioner. When we were building our room, CoolBots were not available; now they are fairly widely used. I have not heard of any complaints about their performance, provided that the air conditioner itself is sized properly for the room size. Your best bet for wine-cellar coolers is online. There are many choices; just be sure to choose the right size unit for your space, keeping in mind that larger is always better than too small.

A wine/beverage cooler makes a wonderful aging unit.

Whether you use a wine-cellar cooling unit or a room air conditioner, you will need to think about the space into which the cooling unit vents. In other words, what's on the other side of the wall that the cooling unit protrudes through? It should be a well-ventilated space that will not be too warm. Remember, the cooling unit will be sending hot air into that room, so you should have a way to remove the hot air, even if it is an exhaust fan or window (try to keep in mind that all cooling units are basically removing heat from the air, and they have to send it somewhere else). Some units, such as the wine-cellar coolers, can only work well if the room they are venting into does not go over a certain temperature (ours works best below 70°F [21°C]), so you don't want to vent into a space that is too warm (large commercial cooling units have a remote compressor that usually vents to the outdoors).

As with wine coolers and refrigerators, ideal humidity can be accomplished in several ways: passively through wet sheeting or actively through ultrasonic, or other, humidifiers. Because most home aging spaces will not have a floor drain, I don't recommend some of the more aggressive ways that commercial folks might humidify their spaces. Again, for the more commercial-scale information, see my previous book, *The Farmstead Creamery Advisor*.

> A musty, moldy-smelling cheese cellar is not a good place for young cheeses to grow up.

Don't forget about air exchange in your cheese closet. As with smaller units, simply opening the door daily is usually adequate. Remember that cheese "breathes"; it needs oxygen for optimal aging processes to occur. A part of breathing is exhaling unwanted gases, and with cheeses these are often ammonia. A musty, moldy-smelling cheese cellar is not a good place for young cheeses to grow up.

THE INS AND OUTS OF MONITORING HUMIDITY

In chapters 2 and 3 I mentioned that I think the single most important step in making superior cheese (given equal-quality milk and ingredients) and preventing problems in the vat is monitoring pH. Well, once the cheese is out of the vat and into the aging space, the issue that many cheesemakers fail to monitor properly is humidity. If you read the sidebar on page 77, "What Is Relative Humidity?" you, I hope, understand the difference between moisture in the air and relative humidity. When cheesemakers talk about and monitor humidity, it is really relative humidity (RH) that is being measured, but for simplicity's sake we call it (and I do the same in this book) simply "humidity." Gauges for reading relative humidity are called *hygrometers*. You can purchase simple dial versions, digital varieties, or a wet-dry bulb type also called a *psychrometer*.

Dial hygrometers are usually the least expensive, but I have not found them to do well in the continually very humid environment of the aging room. They operate by utilizing the moisture-absorbing properties of a fiber, traditionally a human hair; the fiber is attached to a lever that moves the hand on the dial. Quite basic.

Digital hygrometers, also called thermohygrometers, come in inexpensive models ($20 to $30) on up to more expensive types. I recommend having two, even in a small aging space, as a way to correlate the readings for accuracy. If one shows a very different reading from the other, you can catch a problem. Two will also give

CHEESEMAKER MOTIVATIONS: THE FOODIE

Whether you are an accomplished home chef or simply an avid diner, you might be a part of the growing segment of the population often referred to as "foodies." While I usually avoid the adding of the "y" sound to the end of words, I do like this word better than its more presumptuous counterpart, gourmet or gourmand. Foodies make an avocation of all things related to the appreciation of real food. Fortunately, cheese is growing in status in the pantheon of the foodie's worship.

Foodies that also make cheese represent a huge segment of the cheesemakers in the United States, and I am constantly amazed at the talent of some of these extreme hobbyists and the depth of their cheesemaking abilities. For these folks the pure joy and thrill of making their own divine cheese creations drives them to build their own cellars, write blogs, and support other budding cheesemakers via online forums. I find this group of cheesemakers to be one of the most inspiring for writing this book. In fact, it is from many of them that I have learned and been pushed in my own craft!

you a way to monitor humidity differences in different parts of your aging space.

Finally, my favorite tool for measuring humidity; the old-fashioned, reliable psychrometer. This instrument is composed of two thermometers, one with a dry bulb and one with a wet bulb. The wet bulb is wrapped in an absorbent wick that runs to a small reservoir of water; the water travels up the wick and keeps the bulb moist. The water will evaporate from the bulb at different rates depending on two main factors: the temperature and the dryness of the air. If you read the sidebar on relative humidity on page 77, you know that the warmer the air, the more evaporation will occur, but as the air becomes more saturated with water vapor, the evaporation will slow. As the water molecules evaporate, they cool the bulb and lower the temperature of the wet-bulb thermometer. When this temperature is compared to the dry bulb (which is only affected by water vapor in the air), it correlates with a relative humidity percentage.

Since relative humidity is temperature, air pressure, and water vapor pressure dependent, you can't simply take the two numbers and consistently compare the difference. In other words, 2° difference at 55°F (13°C) will not equal the same RH as at 40°F (4°C). For that reason charts exist to tell you how to compare these two numbers to determine the RH. Since the range of temperatures at which the cheesemaker is concerned with RH is fairly limited, I have prepared two tables (one for Fahrenheit and one for Celsius) with some common goal temperatures and some information showing how bulb temperature differences correlate with RH. You can find these charts in appendix C. Please feel free to copy them and keep them handy in your cheesemaking area.

From these tables you can see that you can develop a consistent rule of thumb based on your aging-space requirements; for example, my goal in our aging room is 50 to 55°F (10–13°C), so I watch for a 2.5°F difference between the bulbs. This difference keeps it in a nice, acceptable range. Other than needing to clean the wick once in a while and refill the reservoir with clean water, the psychrometer does not need recalibrating and is reliable. Besides, I think they look cool.

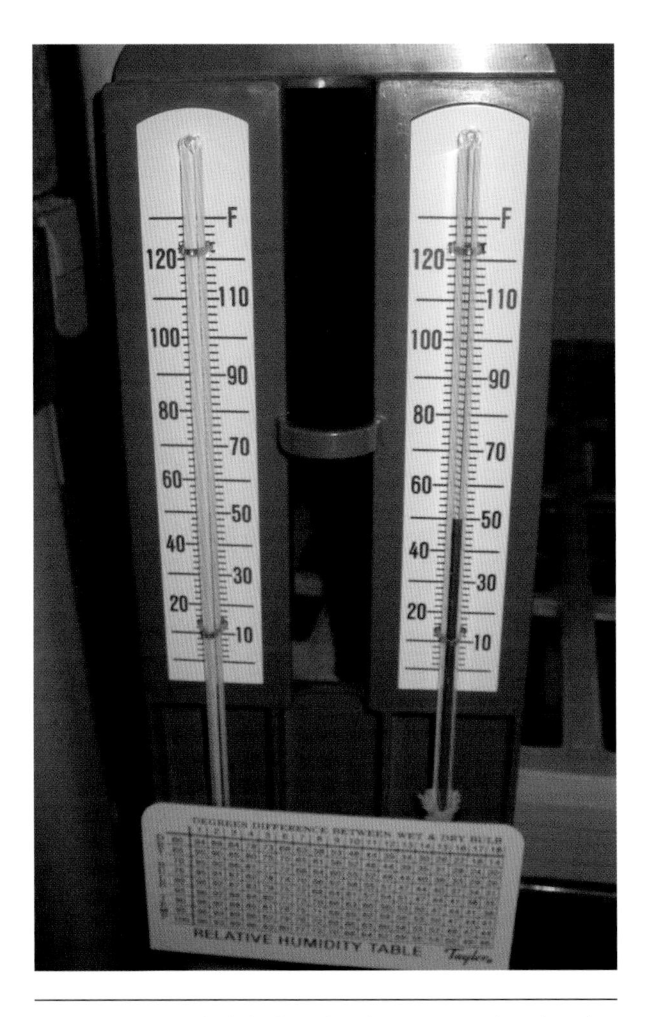

A psychrometer provides fail-safe readings for monitoring relative humidity.

A digital hygrometer offers an easy and quick way to monitor both relative humidity and temperature.

CHEESE WITH SPECIAL NEEDS

It is possible to do some clever choreography to be able to age cheeses with different needs in the same space. My goal is to help create flexibility so you are not thwarted in your passion as a cheesemaker. The following tips will, I hope, help you age special-need cheeses alongside the less needy ones. By the way, I'll cover all of these special needs in detail in the recipe chapters about these types of cheese.

HIGH HUMIDITY NEEDS

Some cheeses need a bit more moisture during their early stages, or later, such as washed rind, stinky cheeses, and bloomy rinds. If you have a small, shared space, you can segregate these cheeses into a higher, moister environment, such as a plastic tub or behind a shower-curtained partitioned area that is kept moister. Be sure to give each space its own hygrometer to monitor the humidity.

HIGH TEMPERATURE NEEDS

If you are making a Swiss mountain–type cheese, such as Emmental, where large eye formation (from the activity of propionic acid bacteria) is critical, you will need to be able to move these cheeses to a warmer space (60 to 68°F [16–20°C]) for a 2- to 4-week period. On a small scale this can be accomplished using an ice chest, wine cooler, or small refrigerator outfitted with an external thermostat. With any of these choices, be sure that you are monitoring the temperature correctly. Humidity during this time is also critical. These kinds of cheeses can be quite high maintenance in their affinage stage but are worth the effort in the end.

> You can't expect stellar results unless you respect the needs of the cheese!

AGGRESSIVE MOLD ISSUES

If you have ever tried to age blue cheese side by side with, say, a naturally rinded tomme type, you know how aggressive blue molds can be. I actually didn't learn well enough my first time and tried it again for the sake of research for this book. We lost many wheels

Using an ice chest placed on its side to age an Emmental-type cheese during its warming phase

of one of our most popular cheeses because the blue found entry points in the rind and made its way deep into the cheese. So if you want to give making blue cheese a try—and I highly recommend it—do your best to keep it totally separated from any other cheeses. Remember, mold spores can inoculate even the milk in the pot and bloom later, so ideally, blues are made and housed entirely separate from other cheeses.

LOWER TEMPERATURE NEEDS

Some cheeses are aged at lower temperatures than most. I like to remind myself that traditional cheeses are very much a product of evolution and developed these special needs because they were made where these conditions were natural. But as with so many things, now we must try to mimic nature. If you are making blues, a good part of their aging is done at lower temperatures. So you will probably need to consider a separate refrigerator for this purpose. The same is true with Emmental (Swiss-type cheese). You can't expect stellar results unless you respect the needs of the cheese!

SELECTING SHELVING

On a small scale, shelving is more open to options than for the commercial cheesemaker, in that no inspector will be shaking their head at your lovely, hand-planed wood shelves. When choosing what to set your beautiful handmade wheels of cheese on, just keep in mind the goals: air circulation, the right amount of moisture, and a neutral surface that will not impart any off-flavors, toxins, or other undesired elements. Wood is the historical embodiment of the ideal cheese-aging shelf for several reasons: First, it was all that was available for centuries; second, it helps maintain humidity without condensation; and last, it can impart its unique stamp of locality—*terroir*, as it is most often referred to today (another lovely French word eagerly coined by North Americans).

If you are choosing wood, look for thick planks (a full two inches thick is perfect) of a tight-grained, pitch-free type of wood. This usually means a hardwood without a lot of knots (markings from where branches grew from the tree), sometimes called "clear." You want the tight grain produced by a hardwood to give you a smoother surface that is easier to clean. That being said, I know plenty of cheesemakers who age cheeses on pine boards (pine is a soft wood and often high in pitch production).

Wood planks can be cleaned by washing with warm water and dried in a clean place in the sun (the ultraviolet rays will help sanitize the boards). Commercially, they are often kiln dried or sanitized in the heat of an oven. Don't use chlorine on wood, as it will eat away at the softer portions of the wood and make it more difficult to clean in the future. Dish soap and detergents can leave a residue that may impart off flavors to cheeses. When we were using wood shelving, we washed them in warm water, scrubbing with a medium stiffness brush, allowed them to dry, then brushed them with vinegar. We also tried brushing on a solution of water and dissolved starter culture (to inoculate the boards with good bacteria). Both methods proved free of bad bacteria when swabbed and tested at a lab after we'd aged cheeses on them for months.

You can set cheeses directly on the plastic-coated wire racks of a refrigerator or on baking racks (if you have modified these to serve as shelves in a wine cooler). If you're aging on coated wire racks, it is a good idea to also put down a mat of plastic mesh to help even out the pressure and reduce the chance of creating impressions from the shelving on your cheeses.

If you have a cheese closet or other walk-in aging room, I recommend freestanding shelving units made of epoxy-coated wire. You can use either mesh mats or wood laid on top of the coated shelves. My personal favorite is MetroMax shelving, which consists of a high-quality epoxy-coated framework with snap-on plastic shelving for ease of cleaning.

The author's personal favorite choice for cheese-aging shelves: MetroMax

AGING-SPACE HOUSEKEEPING

Aging rooms present two cleaning challenges: One, they are warm, humid environments, with a food supply, that are appealing to pests such as rodents and insects; and two, unwaxed or packaged cheeses will leave particles and share their molds and flora with the environment.

To prevent mildew and molds from growing on the walls of your aging space, a periodic wipe-down using a chlorine solution or other sanitizer is adequate (see Table 6–1 in chapter 6 for a guide to sanitizers and proper dilutions). If there is evidence of molds or debris, first use a damp cloth or a mild detergent to remove any visible buildup (remember the axiom: "You can't sanitize dirt"). Make sure to not use excess sanitizer, as this will not only be wasteful, but it could impart unfavorable odors and flavors to your aging space, and therefore your cheese. You can either schedule your cleanups or do them at the first signs of need. Remember, by the time you see the mildew and mold, it has been there for a while. If you are making particularly delicate cheeses, such as white mold ripened varieties, you will want to clean *before* you see any contaminants, or they will grow on your cheeses as well as on the walls.

Many French farmhouse varieties of surface mold ripened cheeses, such as Pouligny-Saint-Pierre, have a multitude of molds growing on the surface. If this cheese variety is your goal, you don't have to be quite so vigilant. (In fact, you may have to work hard at inoculating the aging space with the right molds, then be careful to *not* remove them completely from the environment.) But if you are trying to keep the surface of the cheeses lily white, you will need to be more attentive.

If the space is tightly sealed, you won't have problems with large vermin, such as rats and mice, but you might still get the occasional fly. You should also watch for fly egg cases (they will look like black, brown, or whitish grains of long-grain rice) being laid in nooks and crevices. If the space is large, you can use fly strips or black light fly traps to help with pest control. Small spaces generally don't have as much trouble with flying insect pests. Cheese mites, on the other hand, will likely find their way to any space where cheeses are being aged for more than three months and have a natural rind. Mites are such a big topic that I am giving them their own section later in this chapter under "Affinage Troubleshooting."

USING A CHEESE TRIER/IRON

Before you start cutting into wheels and finding out that they are not ready, invest in a handy little tool called a cheese trier or cheese iron. Triers consist of two parts: one for removing a core sample from a whole cheese and the other for tapping the outside of a cheese to listen for air pockets.

By learning to sample cheeses, you will be able to check the progress of aging cheese without cutting into it. You can make notes on flavor and texture changes that will greatly help you determine when a cheese is ready. Remember that as cheeses go through phases in their ripening, they might be bitter. Making notes on when a bitter phase occurs and when it ceases will be very helpful in deciding ideal aging lengths for each type of cheese. A core sample can also help determine when a blue cheese is ready to be wrapped or moved to colder storage.

Using the iron side to tap a cheese can help determine when eye development is adequate or when a cheese is suffering from early or late blowing (see "Affinage Troubleshooting"). And it's fun to turn a cheese into a percussion instrument.

OPTIONS FOR PROTECTING AND FINISHING A CHEESE

How to take care of rinds is one of the most mysterious parts of affinage, and the one for which I receive the most questions. In this section we will cover everything from waxing to washing. As with most things regarding cheesemaking, the right choice for rind care is a matter of your individual goals.

VACUUM SEALING, CLEAR COATING, AND WAXING

While some cheese snobs may frown upon a waxed or otherwise sealed cheese, many fine cheeses are made in this fashion, including some of the world's best blue cheeses. Cheeses aged with a coating that prevents air from contacting the surface of the cheese are considered "rindless." Think about a lovely slice of blue cheese that has a soft, white surface, as well as the 2-pound block of cheddar from the grocery store.

You should consider sealing your cheese with wax or by vacuum sealing if you cannot guarantee that the humidity in the aging environment will be ideal throughout the cheese's affinage. Why would you want to spend hours making a cheese, just to guarantee its doom in a dry room? There is no shame in ensuring that your cheese stands a chance at a future! Cheese wax is made of a paraffin base (a by-product of petroleum production) with other additives to prevent cracking. Some cheesemakers use beeswax as a substitute. Beeswax works but is more expensive and softer than cheese wax, and it imparts a lovely honey flavor to the cheese (which might be just what you are going for).

Vacuum sealing, while less traditional and less visually appealing, can be done as an alternative to waxing, which can be messy and painstaking. Vacuum sealers come in two versions: small household units that will handle only smaller bags and pull the air from the bag, and larger "chambered" units that create a vacuum

HOW TO CORE SAMPLE A CHEESE

1. If the cheese is semisoft, prechill it to refrigeration temperature.
2. Sanitize the trier and dry well. You may want to use a warm to hot sanitizing solution to help the trier more easily penetrate the cheese.
3. Insert the trier about two-thirds of its length (if the cheese is large enough) or halfway through the diameter of the cheese.
4. Rotate the trier in several complete rotations.
5. Tilt the handle back slightly, and slide the sample out of the cheese.
6. Observe the sample for eyes, flaws, texture, and aroma.
7. With a sanitized knife, cut a small portion from the tip of the sample.
8. Slide the trier with the cheese sample on it back into the hole and push from the rind side to seat the sample back into the cheese.
9. Use a tiny bit of the cut sample to smear over the seams.
10. Return the cheese to aging, and note on the make sheet the date of the sampling.
11. Analyze the sample for texture and flavor.

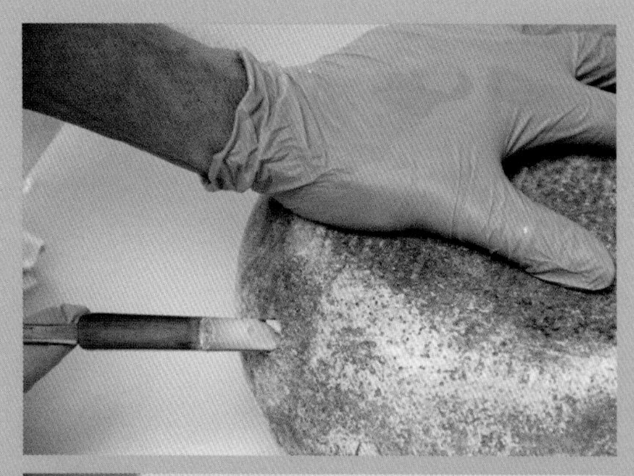

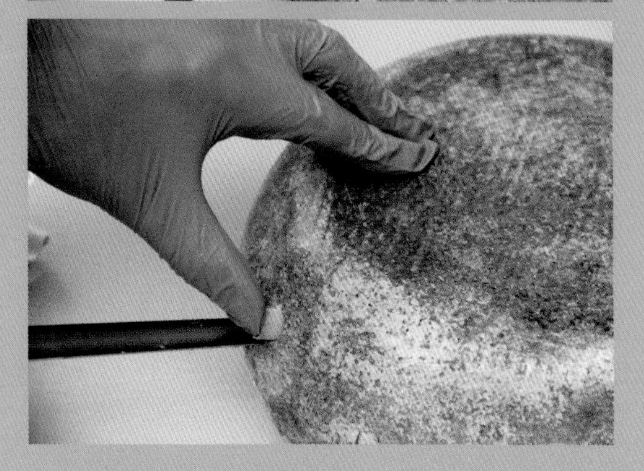

Steps in core sampling a cheese: First, the trier is inserted and rotated several times; next, the trier is pulled gently from the cheese using a slight downward pressure on the handle, a small sample is taken from the tip, and the plug is returned to the wheel.

within the chamber that seals the product between two layers of plastic. These larger units (Cryovac is probably the most well-known brand name) are much more effective and versatile, but of course more expensive. Some cheesemakers find them indispensable for other uses as well, such as packaging cheeses for sale, shipping, and display.

Clear coating, also called "cream wax" or "plastic coating" (brand name Paracoat) is another option and creates a more natural-rind look. Clear coating is, indeed, a plastic polymer product (smells a bit like a popular brand of white glue). It is semipermeable, meaning the cheese will be able to breathe a little bit. It is often used as a preliminary coat before wax sealing. In that case it serves two purposes: First, if there is mold inhibitor in the clear coat, it will help prevent mold from forming under the wax; and second, it will serve as a "primer" (like that used when painting a car, wall, or piece of furniture) that will help create a surface to which the wax will better adhere. Many commercial cheesemakers use the clear coat as the primary treatment of their rinds. Since the coating is somewhat breathable, you can still achieve a more artisan taste, but without quite the maintenance work of a completely natural rind. It is worth remembering that not all clear coats contain mold inhibitor and some do come with coloring.

NATAMYCIN: THE MOLD INHIBITOR

While the name of this approved food treatment sounds as though it is straight from a doctor's office, *natamycin* (brand name Natamax and also known as pimaricin) is an antifungal (or antimycotic) produced by fermentation of a natural, unmodified bacteria called *Streptomyces natalensis*. While some data will define it as an antibiotic and/or antimicrobial, it has no effect on bacteria, just on molds and yeast. For this reason humans have not shown any reactions or resistance from its use. It is widely used in the food industries of a large number of nations. It is approved for use in certified organic foods. It is the inhibitor used in the clear coating applied to many cheeses and is also used in other forms on cheese. For the production of some cheese on the industrial scale, it has allowed a more naturally aged cheese, while at the same time protecting it from molds known to cause human health problems. While most home artisans aren't going to be using it other than in the clear coat, I think it is important to know of it and its use.

When to Wax, Coat, or Seal

The most important issue when you are using one of these techniques to protect your cheeses is *timing*: The cheese must be ready to be sealed. If the cheese was not first properly pressed or air-dried, it will weep whey after being sealed, creating loose pockets and possibly areas of rot.

The cheese is ready to seal when the surface of the cheese is sufficiently dry, usually after a week to ten days of aging at 50 to 55°F (10–13°C) at 85 percent RH. Larger wheels will take a bit longer to dry properly than will small wheels. Many instructions call for air-drying at room temperatures, which I don't recommend, as gas-producing bacteria that might be present in the cheese will have an opportunity to grow at these temperatures.

During the drying period be vigilant about mold growth on the surface. At the first signs of mold, take a cloth moistened in a light brine (a solution of 1 tablespoon of salt and 1 tablespoon of vinegar [optional] to 1 cup of water) and gently rub the surface of the cheese. Be sure to not leave the surface of the cheese wet.

How to Wax and Coat

Depending upon the temperature of the wax, the temperature of the cheese should be adjusted to anticipate certain issues. In general, the hotter the wax, the colder the cheese should be. If the wax is at the low end of the melting temperatures, a cold cheese will thicken the wax very quickly when it is applied to the cheese. But for very hot wax a chilled cheese will help prevent the heat of the wax from bringing fat out of the cheese. If you are waxing blue cheeses, a wax that is too hot (and therefore thinner) might penetrate the piercings in the cheese.

The cheese must also be prepared for waxing. The surface should look firm and smooth and be free from mold. If mold is present, even if it's a faint white bloom, follow the procedure above to wipe away the mold with a brine-moistened cloth, then place on a rack or cheese mat and allow to air-dry at room temperature.

For cooler wax temperatures allow the cheese to come to room temperature (the same as the clear coat) before waxing. If it is a hot day and the room temperature is above 70°F (21°C) or so, wait until the room cools (you can use a fan to help speed up this process).

For hotter wax applications clean the cheese as above, then chill thoroughly to refrigeration temperatures.

If you are clear coating before waxing, follow the instructions above. After clear coating, the cheese can go back to aging for another week or two before applying the hard wax or can simply be aged with the clear coat alone. If the clear coat contains mold inhibitor, this will help prevent mold growth under the wax.

Waxing can be done over a large temperature range, from the low, safe temperature of about 120°F (49°C) to the more dangerous high temperature of up to 220°F (104°C). The high-temperature method can scald skin

NATURAL RINDS:
CAN YOU EAT THEM?

One of the questions we often answer for customers is, "Can we eat the rind?" Natural, unwaxed cheese rinds are completely edible, but not always eatable. If a rind is quite thick and dry, with a certain amount of cheese mite residue, it may not be palatable, although I have often been at events cutting and serving cheese, only to have a customer ask for a large chunk of what I like to call the "heel" of the cheese wheel. So for some the rind is quite a delicacy. Cheeses with natural rinds that are less than palatable will probably not be labeled as such when packaged for sale. So take a tiny nibble if rinds are on your cheese menu, and see if they are flavorful. Some folks save dry, hard rinds, such as those from aged Parmesan, and use them to flavor soups and stock. At our house the dogs, and one cheese-crazed cat, consider the rinds better than any biscuit or treat on the market.

PRECAUTIONS TO TAKE WHEN WAXING

1. Be sure to work in a well-ventilated area, as the fumes from heated wax are not healthy and can ignite.
2. Wear protective clothing that will not stick to you, should hot wax splatter or spill.
3. Wear goggles when working with temperatures over 120°F (49°C).
4. Wear gloves.

Dipping a hard cheese in hot (200°F [93°C]) wax heated in a crock pot

easily and is a risk for fire, but it will kill mold spores on the surface of the cheese. It is not recommended for anyone who is not willing to be extra vigilant and prepared! Definitely don't try it with little children around or pets that might jump onto the counter or be splattered. You can use a double boiler to melt wax or a recyclable aluminum foil tray and set on very low direct heat. But better than either of these is a slow cooker or Crock-Pot. This eliminates open flames, hot burners, or the possibility of water (from the double boiler) splattering into the hot wax—especially if you find the higher temperature wax is easier to work with. Once you are done, you can simply turn the pot off and let the wax harden inside. Inexpensive Crock-Pots can be purchased from a thrift store.

Steps for Waxing a Cheese

When waxing your cheese, I recommend following this checklist:

1. Prepare cheese as mentioned above.
2. Dip half of the cheese into the hot wax.
3. Hold wheel above wax, and jiggle gently to remove excess wax.
4. Place wheel, unwaxed side down, on a mat or board.
5. When wax is dry, dip second side as before. Make sure that the two coats touch evenly around the cheese.
6. Allow to dry.
7. Repeat steps 2 through 6. (If dipping half of the wheel, instead of top or bottom halves, rotate the wheel for the second coat, so that the first wax seam is at a right angle to the last one.)
8. If you have any pinholes, use the tip of your wax thermometer to drop tiny wax drops over the holes to seal them.

The cheese will be dipped in a two-step process, with each coat needing to dry in between. Depending on whether you have a curved cheese rack or a flat mat, you should dip the cheese so that the waxed side will dry without touching anything. It is usually easier to hold a wheel by gripping the top and the bottom, so a curved drying rack can be very helpful for success in waxing. If you only have a flat surface to dry the cheese on, first dip the bottom half, let it dry, then turn and dip the other side. A second thin coat should be applied. Some people prefer to brush the wax on, as it is a simpler technique, but readily leaves rough brush marks, as the wax thickens very quickly as it cools. If you decide to try the brushing method, use a natural bristle brush, not a synthetic. I have seen photos of very clever setups that some innovative home cheesemakers have built that allow the entire cheese to be submersed for a single, smooth coat. If you are interested in waxing a lot of cheeses, this might be worth investigating.

BANDAGED RINDS

The old-time practice of bandaging is seeing a resurgence in popularity for the aging of artisan cheddar cheese. You will hear the resulting cheeses called either "clothbound" or "bandaged." You can bandage any semihard to hard cheese, but typically cheeses aging less than six months won't need the extra protection of a bandage wrap. Extra-hard (grana) cheeses are never bandaged, but you have a different goal: the driest of textures. Cheddar, which can age for as many years as a grana type, is typically not as hard and dry as its similarly aged cousin. It is, in part, thanks to the bandaging that cheddars have a different texture by the end of aging. Cheddars and other semihard cheeses aged under a year will usually achieve the desired texture without being clothbound.

The bandage and lard (or butter) wrap is embedded in the cheese's surface during an extra step of pressing. During aging the wrapping is gently brushed to limit mold growth but is never scrubbed or washed with brine. During very long aging bandage-wrapped cheeses are very susceptible to cheese-mite damage. In the past this was dealt with by using pesticide fumigants, but that, thank goodness, has been banned. Ozone-generating units now often do the job. Usually, the bandage wrapping is removed prior to cutting, but not always, depending on the aesthetics that the seller desires (see the "Affinage Troubleshooting" section in this chapter for more information).

When to Bandage

Bandaging is done after the final pressing of the cheese, when it would normally go to the aging phase. The

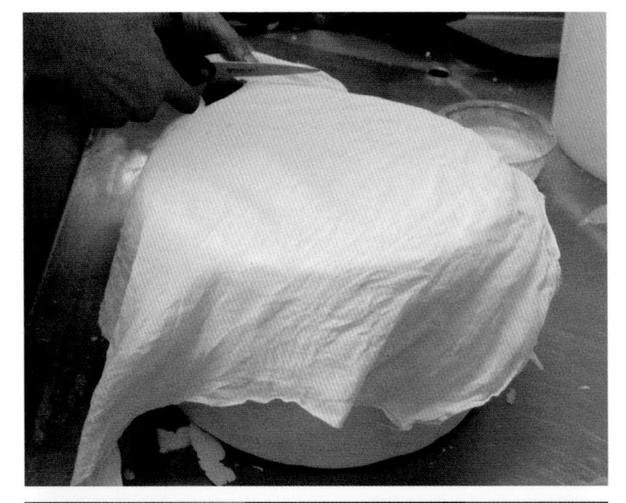

Bandage wrapping a cheese: First, the cheesecloth is cut to fit the wheel, then the top and bottom circles are soaked in butter and applied, and finally the side piece is soaked and smoothed around the wheel.

pressing room is usually cooled to a bit cooler than normal room temperature, so just under 70°F (21°C). If you are bandaging cheddar, the cheese will have already been salted, so acid development has ceased, but if you decide to experiment with bandaging a cheese other than cheddar (or other cheese in which the curds have been salted prior to pressing), you will need to take this into consideration. Options are to bandage after salting (but you may have difficulty returning the cheese to the form and press) or dry salting in the form, then bandaging. If you try to apply the cloth, *then* salt the cheese (obviously soaking in brine is not an option), the salt will not be able to penetrate well through the fat-coated bandages—and if it does, the whey it draws out of the cheese will soak the cloths. All in all, you are best sticking with cheeses whose acid development is complete before pressing and whose curds have been salted before pressing.

How to Bandage

When the cheese is finished with its initial pressing, it is time to apply the bandages. Remove the cheese from the press. Cut two circular pieces about ½ to 1 inch (2–3 cm) larger in diameter than the wheel of cheese. Cut a third piece the same height as the wheel and long enough to completely encircle (like a belt) the cheese with about ¼ to ½ inch (1–1.5 cm) overlap. Melt the butter or lard, and have it ready. Soak the circular pieces of cheesecloth in the butter or lard, and apply one to the top of the wheel. Smooth and press the cloth to adhere tightly to the surface, with the excess pressed over the edges and evenly down the sides. Turn the wheel over, and repeat the process. Then soak the side bandage in the oil, and with the wheel lying on its side, carefully apply the bandage, smoothing and tightening as you go. It should cover the top and bottom bandages evenly where they overlap the sides.

When the bandages have been applied and are smooth and snug, slide the wheel back into the press, and apply the same pressure as for the last part of the initial pressing. This repressing should continue for 12 to 24 hours. When complete, the bandage will be integrated into the surface of the cheese. The cheese can then be placed in the aging space.

Caring for the Aging Bandaged Cheese

After a few weeks you will start to see some mold growth on the surface of the cheese. If desired, you can carefully wipe or brush with a soft brush to limit the growth. Bandaged cheeses are not wiped with brine but are brushed and even vacuumed to limit molds and mites. The bandage will grow grey and look fairly unpleasant (to those who are uninitiated to the beauty that is being created beneath the wrapping!).

BRUSHED RINDS

One of the simplest ways to create a beautiful natural rind is to brush the rind periodically during aging. As with a bandaged rind, brushing does not eliminate the molds but instead limits their growth and "damage" to the rind. In the beginning use a very soft brush to prevent damaging the still soft cheese. It is a good idea to have a minimum of two textures of brushes, one soft and one medium, to maintain brushed rinds.

There are no hard-and-fast guidelines for when to brush a cheese. It will depend on your own goals and, frankly, your finickiness and diligence. I try to brush them before it is obvious that they need it. If the first brush swipe releases a large cloud of mold spores into the air, I probably waited longer than I should have. When you do decide it is time to brush the cheeses,

A bandaged cheddar with a brushed rind after 4 months' aging

take your clean, sanitized, and bone-dry brush and gently whisk it across the surface of the cheese. Doing this over a sink or dampened surface can help trap much of the mold spores and debris that might otherwise become airborne. Continue to gently brush the entire wheel until the rind looks smooth. Never use a scrubbing or circular motion, but instead think about brushing the molds off the cheese. Don't expect it to look like a brand-new cheese! Remember, your goal is to prevent the molds from growing so rampantly that they form a thick, felting dry rind—and attract cheese mites (more on those pests later).

When you are done, the brushes should be thoroughly washed, sanitized, and placed in a clean area to dry. I recommend having at least two of each texture of brushes so if one is still moist from being washed you will have a dry one ready when you need it. The brushes should be used only on cheese; they must not do double duty on dishes or other surfaces.

OILED RINDS

Oiling is another long-standing rind treatment. Grana-type cheeses, such as Parmigiano Reggiano, have an oiled rind that darkens somewhat and creates a lovely patina by the end of aging. The oil discourages most molds, as does the act of rubbing the oil onto the cheese. Cheese mites also are impeded by the oil simply because it is hard for their little legs to move on the

The rind of a Manchego-style cheese before and after brushing

A cocoa powder– and olive oil–rubbed rind

sticky surface. You can use any food-quality oil on an aging cheese, but olive oil and olive oil blends are favorites. When the cheese rind has dried off, usually 1 to 2 weeks into aging, you can apply the first coat of oil. If molds are already present, brush or wipe them off first with a dry, soft brush or dry cloth. New coats can be applied as needed when the cheese appears to be drying, when molds are present, or even if you simply feel like it! There are no hard-and-fast rules. After a few coats you won't need to apply more oil as often and the rind should start developing some depth of color.

Oiling can be combined with the application of other herbs, spices, or ingredients such as ash or cocoa powder. For the rind of one of our most popular cheeses, Hillis Peak, I attempted to copy a Spanish goat milk cheese called Majorero. I mix smoked paprika (usually Spanish, but only because it is the easiest to find in high-quality bulk) with a 75:25 blend of canola and olive oil and rub a wet paste mixture on the cheeses several times during their aging. Vella Dry Jack, from historic and well-respected Vella Cheese in Sonoma, California, has a mixture of cocoa powder, black pepper, and oil rubbed onto its rind during aging. When you combine things with the oil, the paste can help create a protective coating, and some of the spices provide a bit of abrasion when being applied that helps knock down any microflora growing on the surface of the cheese.

WASHED RINDS

The term "washed rind" really can be divided into two categories of cheese: those that develop a sticky, reddish, smelly rind from the activity of surface bacteria and those that are wiped or washed with other liquids to create a unique but firm and ultimately dry rind. The red-stinky variety is sometimes simply called "washed rind" cheese, so this category of rind treatment can create some confusion. I'll be covering the care of the soft, stinky cheeses in chapter 11, since the process that ultimately makes them into this type of cheese is very dependent on the rind care. In this chapter let's talk about the other methods for using liquids on the outside of cheeses during aging.

Some traditional cheeses that were once made with a bacteria/salt–washed rind are now washed with a light brine containing annatto (natural cheese coloring) to duplicate the orange hue reminiscent of the original, bacteria-reddened surface. Others are washed with a red bacteria brine to develop color and initial flavor changes, but because of the large format of these cheeses, the bacteria do not soften the cheese. Gruyère and the inimitable Pleasant Ridge Reserve from Uplands Cheese Company, Wisconsin (the only cheese, to date, to win the American Cheese Society competition Best in Show three times), are examples of large, semihard to hard cheeses that are washed, early in their aging, with coryneform bacteria.

A brine solution can be used to help keep molds down. Mix clean water with salt to 5 to 8 percent strength. See appendix C for brine recipes. Some people use a cloth moistened with vinegar; the vinegar can be any kind, but distilled white or cider vinegar is the most common choice (or a mixture of the two) to help keep molds down. If you try vinegar, be sure not to soak the cheese. Keep the cloth as wrung out as possible so the vinegar does not penetrate the rind.

Different types of drinking alcohol are popular choices for washing rinds. We use a locally made ale mixed with half water and a bit of salt to help maintain and develop the rind on several of our cheeses. Ale, wine, and spirits are all choices that can be tried on cheese rinds. They will each add a different look and bring their own characteristics to the cheese. Begin applying the washes when the rind has dried a bit and at the first sign of mold growth.

AFFINAGE TROUBLESHOOTING

In this section I will cover the major problems you might encounter when aging cheeses. As you become a more advanced cheesemaker and understand more about milk quality and the process, some of the answers to problems will become evident. But there are still many unknowns, even to the most advanced cheese microbiologist and chemist, so don't be too surprised if sooner or later something happens to your cheese that could be inexplicable.

PROBLEMS WITH WAXED AND SEALED CHEESE

When a cheese is waxed, or vacuum sealed, you are effectively trapping anything that is inside the cheese. If any mold or yeast spores are trapped, they might still grow; if any whey is still inside the cheese, it might weep out; and if any gas-forming bacteria are in the cheese, they will still grow and produce gas.

Mold spores trapped under the rind can still grow. If the growth is minimal, you don't need to do anything, but if it is extensive, it is a good idea to remove the wax, clean and dry the cheese, and reapply the wax. A precoat of liquid clear coat with mold inhibitor might be helpful if the problem is frequent. Be sure to always clean the surface of the cheese before waxing with a vinegar solution. Some cheesemakers use a hot wax (above 200°F [93°C]) application to help kill spores. For more on waxing options, see the section on waxing earlier in this chapter.

While cheeses may look dry and ready to wax, small amounts of whey can still be trapped within. If left with a natural rind on the cheese, this whey will evaporate through the rind, but when waxed, it will be trapped, leading to blisters or a moist surface to the cheese when the wax is removed. If moisture under the wax is a problem, you will want to address curd texture at draining and filling forms; pressure during pressing; time, temperature, and weight of pressing; brining and salting time (this helps remove moisture); and air-drying time after salting.

If you have used any gas-producing bacteria in your cheese make, or if your milk contains a gas-producing contaminant (this can be coliforms, clostridium, or propionic acid bacteria), gas production after waxing is a possibility. When this happens, the gas might expand the entire cheese and the wax, but it can also make its way to the surface of the cheese and create pockets or blisters. If gas production after waxing is a problem, you will have to look at its source to determine the course of action. Options include delaying waxing to allow desirable gas-producing bacteria to do their job; making sure temperatures don't reach the levels that would allow native milk propionic acid bacteria to thrive; and making sure that milk quality is such that contaminants such as coliforms and clostridium are not an issue.

CHEESE MITES

For some cheesemakers mites are a normal, expected, or even desired part of aging cheese, while for others they are persona non grata (or would that be insect non grata?). When we first encountered mites in our aging facility, it was almost impossible to find any information about how to deal with them. Now there is quite a bit of material online and in books. In fact, cheese mites are even getting a little respect for the role they play in aging such cheeses as the French Mimolette, which is purposely introduced to the little buggers during aging to help create a distinct rind and flavor profile.

So just what are cheese mites, and how do they find their way to your aging room? It seems that there are mites for every occasion and morsel—cheese mites, flour mites, mold mites, dust mites, and so on, and many of them are so picky that they won't dine on a variety of things. It is through this versatility that mites find their way to the aging cellar. Since most of us do a little baking or know someone who does, we become the unknowing host to some little hitchhikers. Mites are known to attach themselves to hair, clothing, and even flies in their quest for the "good life" of cheese and cured meats (charcuterie).

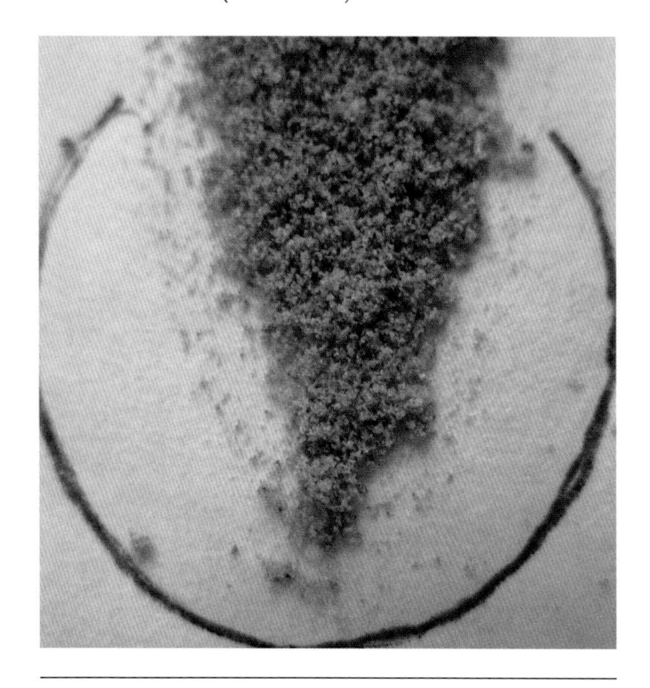

Cheese "dust" composed of living and dead mites, cheese rind, and mite excrement

A French Mimolette with a natrual rind showing the desired pitting created by cheese mites

You know you have cheese mites in your aging room if you start seeing a dry, fine powder on the floor or surface of the cheeses. This "cheese dust" is made up of dead mites, bits of cheese, and the mites' excrement. If you start seeing plugs of loose rind and craters on the surface of your cheese, you know that you have a large, very successful population of cheese mites.

So what do cheese mites do to the cheese? They dig in and tunnel through the rind of unwaxed and otherwise unprotected aging cheeses, including bandage-wrapped cheddars. It is now believed that they may not be after the cheese specifically but instead are eating the fungi (molds) on the cheese, digging under the surface for the parts of the fungus that grow there. It is even thought that the fungi and the mites have a symbiotic relationship—by eroding the surface, the mites help the molds find more area to grow.

Generally, they don't start to work on cheese that is under a couple of months of age. Cheeses that are being brine washed, rubbed with oil, or regularly brushed usually don't see evidence of mites until much later in their aging, or if rind treatment continues throughout aging, their activity is limited through the controlling of mold growth, as well as the inhospitable surface that such treatments create for the mites.

If you don't want mites in your aging room, the key factor in limiting their damage is early intervention. Once the mites work their way under the surface of the cheese, they gain a measure of protection from any attempts to disrupt them. If your cheese isn't intended to age for more than a few months, this may not be an issue. Mechanical means of limiting mites include brushing and vacuuming. If you are brushing, it is best to remove the cheeses from the aging room and brush them onto a damp floor or into a sink, so the dust is better contained. Vacuuming can be done with a small vacuum that you keep just for this purpose. Remember, the running motor will generate heat, which could be a factor for your cooling system. Both vacuuming and brushing will need to be done very regularly to have much effect, and again, remember that if the mites have already gone "underground" on your rind, these methods will slow the cheese rind deterioration but will not bring it to a halt.

Mites also seem to prefer a flat surface and darkness, so you will see more activity on the underneath side of a wheel that is aged on its flat surfaces (versus the curved sides). Wheels that are aged on their curved sides seem to suffer a bit less damage, as do those not aged on wood. (It isn't the wood itself that is the problem; it's just that the wood helps keep the mites dark and protected—they can't fall off the wheel as easily!)

Some data suggests that mites will not live at refrigeration temperature, so I tried chilling cheeses for a period of days before shipping. About all that did was slow the mites down. Some research says that mites can be controlled with temperatures less than 37.4°F (3°C), but that wasn't my experience with cheese already infested. And of course, most artisan cheeses would not be aged at refrigeration temperatures.

Waxing, clear coating, and vacuum sealing will of course prevent infestation, but bandaging will not. In fact, it is bandage-wrapped cheddars that seem to have the biggest problem. In the past, aging rooms were fumigated to kill mites, but this, thank goodness for the environment, is no longer acceptable or allowed.

Many cheesemakers find the best remedy to be the use of food-grade diatomaceous earth (DE). When dusted (using a fine-meshed sieve) on the surface of the cheese, the DE dehydrates the mites. Again, this technique must be employed early in the aging process, before the mites have gone below the rind's surface. Be sure to wear a mask when using DE; while not toxic, the fine, abrasive powder is not a good thing for your lungs.

Hydrogen peroxide can be utilized when mixed to 10 percent as a wash, either following vacuuming or before infestation occurs. I have not tried this method, but somehow it sounds effective yet unappetizing if your rinds are typically consumed by customers.

Ozone machines are being used to limit mite damage in larger aging facilities and some small ones. What is an ozone machine? Well, in a nutshell these machines take oxygen (O_2) from the room, utilize an ultraviolet light and electricity, and change it to triatomic oxygen (O_3). Ozone attacks organic compounds, such as mold, yeast, phage, and bacteria. It has been used for some time in hospitals, and other institutions use it to sanitize exposed surfaces. So how can this help with

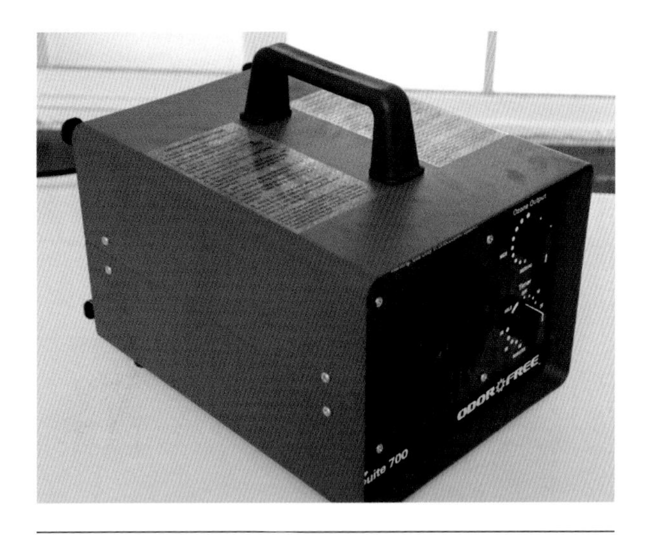

Small ozone generator

a cheese-mite problem? The research is unclear on the exact mechanics, but it seems that the mites don't appreciate the O_3 environment (it eventually eliminates them, through either sterilization or death), and as mold growth is also limited by ozone, the cheese surface becomes less appealing. At Pholia Farm we have a small ozone machine that we utilize for mildew prevention, air freshening, and general sanitation in the cheese make room. It has a daily setting and an "away" setting, during which living creatures should not be present. Remember, high levels of ozone can kill and sanitize for a reason; be sure to follow the instructions and recommendations for operating. (Note: Currently, most ozone generators cannot be sold or purchased in the state of California.)

I tried ozone for several weeks one fall in the aging room, and it did indeed rid it of mites. I can't use it in there most of the aging season, as we count on surface flora to help age some of our cheeses, but it was good to see that it did work.

So you get the idea that there are a lot of things still unknown about cheese mites and how to deal with them. There is some current research being done that should help shed more light on these little guys' eating habits and lifestyle. For now just know that you will likely encounter them if making any cheese that needs to age for a longer period of time and have a natural rind. For the commercial cheesemaker the FDA and inspectors still categorize them as filth and pests, but I believe as the awareness of the normalcy of their presence in aging rooms increases, there will be less concern by regulators.

BLOWING

The term "blowing" refers to a cheese's developing gas in a manner that causes the cheese to swell, or blow. There are two main types of blowing, early and late. Fortunately, these two terms are easy to correlate with the symptoms.

Early blowing is most often caused by the presence of coliform bacteria or yeasts in the cheese milk. In a soft, lactic acid–set cheese, the blowing manifests as a frothy, gassy curd. In a pressed or drained cheese, early blowing is seen as uncountable numbers of tiny, round holes in the cheese or yeast eyes that are two to three times larger than the tiny eyes that appear within a day or so after pressing. You may not discover these holes until the cheese is cut at a later date. Coliform eyes are tiny and elliptical and will be shiny (unlike mechanical eyes that are due to the amount of pressure used during draining). The most common cause of this type of contamination is poor milk collection and poor production cleanliness. Slow chilling of milk and poor acid development in the vat are also likely causes. Cheese with this defect should, of course, be thrown out. It can be fed to chickens and pigs if you happen to have any of these wonderful food recyclers on hand.

Late blowing is most often caused by one or more bacteria from the Clostridiaceae family. (Tetanus and botulism are also in this family, and if you remember much about these guys, you might recall that they are extra troublesome because they do well in anaerobic conditions and form spores that allow them to survive pasteurization.) Clostridia blowing creates a very easy-to-identify look in the mature cheese. Besides swelling and doming, the cut cheese will reveal a spongelike look often accompanied by cracks and fissures. Like the early blown cheese above, this is fit only for chickens and pigs to consume. There are three main subspecies of clostridia that cause blowing, each at different stages: one early in ripening (but later than coliform blowing), one at about a year, and one after a year. You don't need to memorize their names (*C. butyricum*, *C. tyrobutyricum*, and *C. sporogenes*), as the prevention for each is the same.

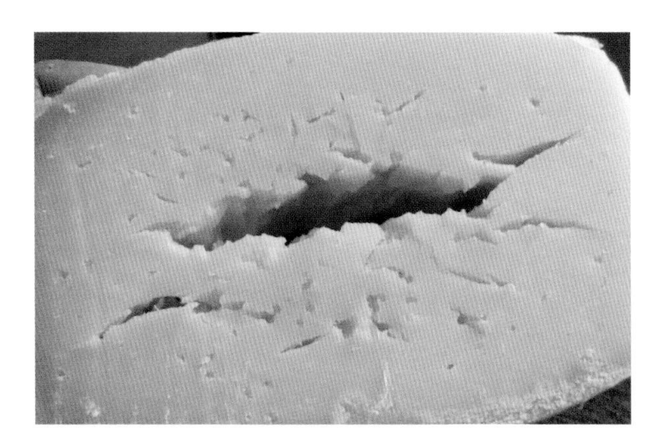

Late blowing caused by clostridium bacteria. Thank you to an anonymous cheesemaker friend for letting me photograph this cheese.

Superficial rind cracks caused by low humidity in the aging space

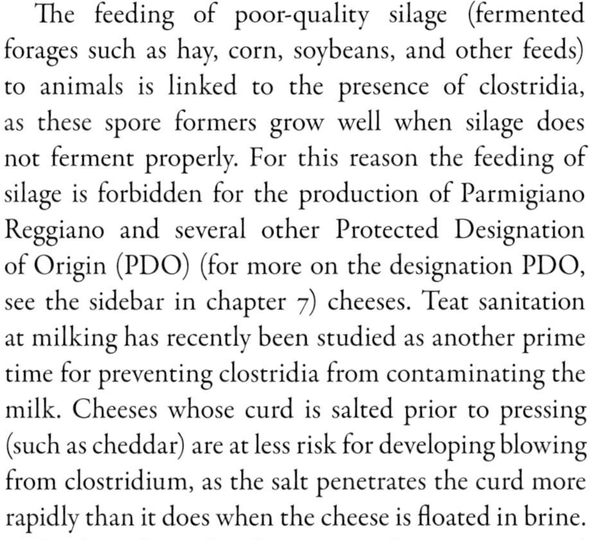

The feeding of poor-quality silage (fermented forages such as hay, corn, soybeans, and other feeds) to animals is linked to the presence of clostridia, as these spore formers grow well when silage does not ferment properly. For this reason the feeding of silage is forbidden for the production of Parmigiano Reggiano and several other Protected Designation of Origin (PDO) (for more on the designation PDO, see the sidebar in chapter 7) cheeses. Teat sanitation at milking has recently been studied as another prime time for preventing clostridia from contaminating the milk. Cheeses whose curd is salted prior to pressing (such as cheddar) are at less risk for developing blowing from clostridium, as the salt penetrates the curd more rapidly than it does when the cheese is floated in brine.

At the industrial scale *nitrate* or *lysozyme*, a natural antibacterial, can be added to inhibit growth of spore-forming bacteria. Also, the mechanical methods of *microfiltration* and *bactofugation* can be used: milk is put through a centrifuge and the fat and most of the spores are spun out; then that portion is ultrapasteurized and returned to the rest of the milk.

Another late-blowing culprit can be *propionic acid bacteria* (PAB). Of course, when making Emmental (a.k.a. Swiss cheese) PAB are purposely added and the cheese is nurtured to encourage these gas-producing bacteria to grow. But in other cheese types native propionic acid bacteria in the milk or from cross-contamination from the production of a cheese type such as Emmental cheese can cause some problems should conditions in the cheese shift to their range of growth (chapter 14 will tell you more about making cheese with eyes). Some European cheeses actually have nitrate added to prevent this problem! For the home and small production artisan, prevention is usually a matter of pasteurizing (or for raw production the prevention of cross-contamination of propionic acid bacterial cultures with other cultures, equipment, and milk). In addition, a primary preventive measure is maintaining temperatures at which PAB will not grow: 55°F (13°C) and under.

CRACKS AND CREVICES

Cracks in cheese can be on the surface of the rind, run throughout the cheese, or located only within the paste. There are various causes of cracks, some easier to prevent than others.

Superficial cracks in a pressed cheese are most often caused by low humidity in the aging room. If these develop early, it doesn't mean that the entire cheese will be lost but can mean a significant portion will have intrusions of mold. How much of the wheel is lost will depend on how many mechanical eyes (openings) are in the cheese. If there are many, the mold and other contaminants will likely work their way through the entire wheel. Don't confuse a crease from cheesecloth with a true crack. Creases may become filled with molds but will remain superficial in nature.

Cracks in nonpressed cheeses, such as mold-ripened varieties, can occur in places where curd does not knit properly during draining or from a late-draining

Internal crack (invaded by blue molds) caused by improper curd knitting during pressing

pocket of whey. To prevent these make sure curd size, or layers, is even in the form and that temperature during draining is optimal. You can briefly pierce the center of draining cheeses with a knife blade to help assist drainage of any whey trapped in the center of the wheel or shape.

Crevices that run throughout a wheel usually occur when large hunks of curd are put together in a form and then do not properly knit. This can be from too-low temperature at draining, too much initial pressure—causing the rind to seal and trap whey within the curd and seep into these joining pieces of curd—or pressure that is too low during the last stages of pressing. These types of crevices might include a streak of mold intrusion and discoloration from fermentations and proteolysis occurring within the crevice.

Crevices and cracks located completely in the paste can be caused by the same issues as those that run throughout, but if you look closely at them and there are small, round holes lining the crack, then contamination bacteria, or perhaps yeast, have caused the flaw. These types of crevices are usually accompanied by eye formation in other parts of the cheese.

HOLES AND SOFT SPOTS

Rinds can have a variety of issues with soft spots and holes. Brine or salting issues are common causes for softening and so-called "rind rot" problems later in aging. If the salt content of the brine is inadequate or dry salting is not thorough and even, that portion of the rind is susceptible to early invasion by molds and other organisms that will erode the rind. The rind is also vulnerable to moisture from condensation, touching another cheese, or inadequate air circulation. Even some spilled salt on a rind will cause moisture to be drawn to the salt and cause a soft spot.

Soft spot (rind rot) likely due to uneven salting or trapped and leaking whey

UNINVITED FUNGI AND BACTERIA

There is almost no limit to the number of fungi (that includes molds and yeasts) and airborne or moistureborne bacteria that can inhabit an aging room. Fortunately, most are aesthetic issues, not food safety concerns. While it is not practical, or perhaps even possible, to cover all the possibilities, here are some of the more common types you might encounter, along with the ways to identify and deal with them.

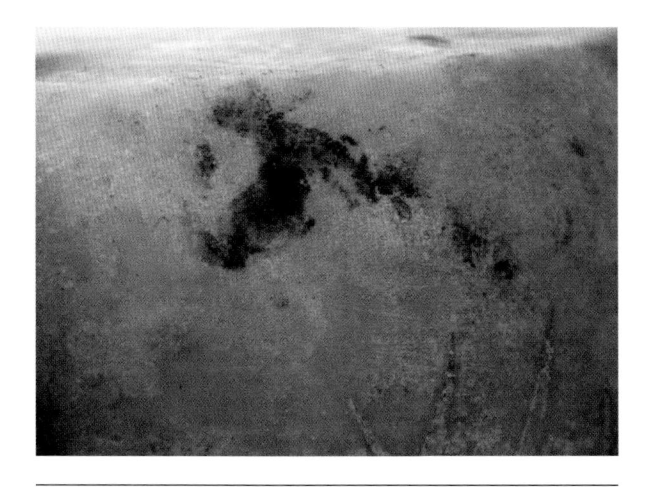

Mildew stains that occur early in aging can fade or be diminished during longer aging.

Mildew

A flat (nonfuzzy) black fungus, mildew is usually only a problem when the humidity in the environment is too high and the air exchange and movement too low. When this appears on your cheese, double-check the above causes and use a light brine– or vinegar-moistened cloth to help kill the mold. The cheese will remain stained, but the mold should cease its growth.

Furry Black/Brown Mold

Furry black or brown mold (mucor) is probably one of the most common "problem" molds for cheesemakers. It's part of the Mucoraceae family and is also known as *poil de chat* (French for "cat's fur"). Mucor isn't a big problem for rinded, aged cheeses and can be controlled with brushing or brine wiping. On white surface-ripened cheeses, however, it's a bigger problem. It's important to make sure the cheeses have dried sufficiently and that the conditions encourage the early growth of the desired white mold.

If you see a haze of tan, use a gloved, sanitized hand to gently press it down before mold spores are spread. I have heard of using a chlorine-dampened cloth (the solution should be at 50 to 100 parts per million) and lightly touching the mold. Instead of chlorine, you can try dabbing on hydrogen peroxide at 3 to 8 percent. Either way, you can rescue the cheese. Try to remember that many traditional French-surface ripened cheeses are covered with a multitude of molds, not just white! (More on maintaining the bloomy rinded cheeses in chapter 10.)

White Flaking/Tan Pitting

Initially, you will see a white, flaky growth and later a beige/tan growth that leaves craters and pits. The most likely culprit is *Scopulariopsis brevicaulis*. When observed early, it can be vigorously brushed and limited. Once it begins pitting, however, you cannot remove it. It isn't harmful, but you will lose quite a bit of rind. If you have mites, they will take advantage of the pitting and move more deeply into the cheese.

Fuzzy Blue Mold

Blue molds, both wild and those introduced during cheesemaking, are very aggressive. They need oxygen so aren't usually a problem inside cheeses with well-closed rinds and a tight paste. But if you have any seams or weak areas in a cheese, they will penetrate and follow every mechanical or bacterial opening in the cheese. Even if you don't make blue cheese, you can expect some wild invaders (unless you have a completely protected aging room, usually accomplished by using positive pressure). Surface-growing blues can be limited early by washing or brushing. In general, blue molds like a certain amount of salt and are fine with the acidity of vinegar, so brushing or wiping might work better than vinegar or brine washing.

Red, Orange, Pink, and White Smooth to Slimy
Any number of surface bacteria can change the color, texture, and aroma of a cheese. These are usually in the same family as those mentioned above under washed rinds (coryneform bacteria). At certain times of the year, you are more likely to see these changes than at others. If conditions are just right (or maybe I should say just wrong!), you will end up with an entirely different cheese from what you intended to make.

Miscellaneous and Unusual
Cheeses that are allowed to develop some surface molds can show other interesting splotches of color, including yellow, scarlet, and purple. None of these colors is any indicator of "bad" molds or contamination, but if your cheese glows in the dark or in the prescense of a black light, *Pseudomonas fluorescens* is the culprit. *Pseudomonas* is a cold-loving (psychrotrophic) bacteria that contaminates milk and can live in brine. Fortunately for most

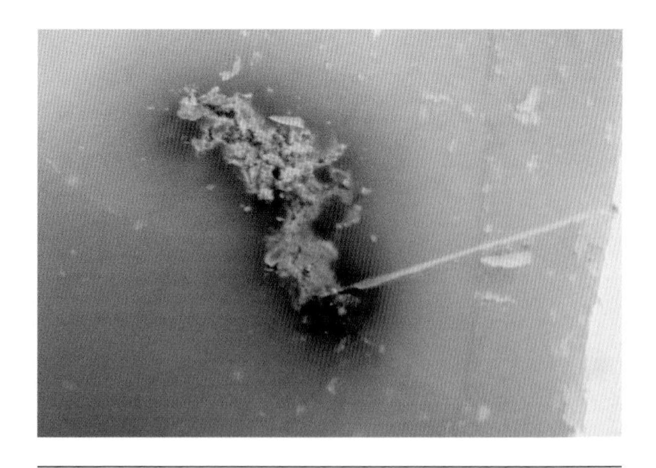

A mysterious internal flaw

cheeses, psychrotrophes like oxygen, so they usually only appear on the outside of cheeses. They do create a bigger problem, however, for soft cheeses. They can be avoided with good environmental sanitation and rapid chilling of milk to below 40°F (4°C) and good brine maintenance.

Curd/Paste Discoloration and Other Flaws
Yeasts and molds that have contaminated the cheese milk or curd can cause a variety of problems and aesthetic flaws during aging. Dark areas, bright red spots, off-flavors, bitterness, rancidity, and odd aromas are just a few of the problems that can often be traced to fungal contamination. The bottom line is, if you are having consistent issues with flaws that you can't figure out, go back to the initial milk cleanliness and the environment in which you are processing it into cheese.

5: SPICING IT UP: ADDING FLAVORS TO CHEESE

Cheese flavored with caraway and fennel seeds (Pholia Farm Spring Brook Seedy)

For the purist "flavored" cheese is something of an anathema. You rarely see a flavor-added cheese taking best in show or being the top pick of a celebrity cheesemonger. I don't quite get this, as to me, when it is done properly, adding flavor to a cheese is simply anticipating a perfect pairing. Not that all pairings are perfect—for example, I remember one cheese competition where flavors such as bacon-pineapple chevre and a hard wheel with a condensed, granular layer of hot chilis and cocoa powder running through the center really put the judges' palates and digestive systems to the test.

Personally, I love a well-flavored cheese, but since I don't like the taste of plain milk—or milky, young cheeses—this makes sense. So my goal with this chapter is to help you learn how to add flavors not only for the best pairing but also for the best visual and textural results.

ADDING SEEDS AND SPICES

One of the cheeses I make with seeds was another happenstance product. On that particular cheesemaking day, I was all set to make one cheese when suddenly I decided I wanted to make something different. I decided to try adding caraway seeds. But unfortunately, we only had a couple of tablespoons in the cupboard. After ransacking the cabinets I found a jar half full of fennel seeds. The two combined would provide enough to flavor the volume of milk that I needed to process that day. So I used both and fortunately loved the results.

CHOOSING SEEDS AND SPICES

Your first thought when adding seeds and spices should be about *quality*, both the quality of the flavor and then the safety of the product. Don't use out-of-date, old, or possibly suspect-sourced spices. For the commercial producer it is important to be able to trace the source of the ingredient and know the expiration date. This should also be important for the home producer, of course, but documentation needn't be done.

When choosing flavors, you can, of course, do whatever you personally like! But if you are selling commercially, think about what would pair well with

TABLE 5-1. POPULAR HERBS AND SPICES USED IN CHEESES

Seeds and Spices	Cheeses
Caraway seed	Gouda and Kruidkaas (Netherlands), Muenster (Germany)
Chili (hot pepper, jalapeño) flakes	Pecorino Peperoncino (Italy), hot pepper Jack (USA)
Cloves	Nagelkaas (Netherlands), Nökkelost (Norway/Sweden), Svecia (Sweden)
Cumin seed	Komijnen Kaas, Gouda, Leyden (Netherlands), Gouda types
Fennel seed	Spring Brook Seedy (USA)
Fenugreek seed	Fenacho (USA), Gouda (Netherlands)
Mustard Seed	Gouda (Netherlands)
Peppercorns	Gouda (Netherlands), Ser Korycinski (Poland), Siletz River Stones (USA)

the cheese if it were unflavored. While it is always fun to create a cheese that no one else has thought of, sometimes there is a good reason that it isn't already being made. Fad flavors (such as the bacon-pineapple disaster earlier) are just that. Still, if *you* love it, go ahead. If you are trying to come up with a marketable product, however, and feel that your sense of taste is a bit challenged in that regard, stick with some known combinations.

PREPARING AND ADDING SEEDS AND SPICES

There are two goals when adding seeds and spices: first, using the right amount of it to distribute flavor, and second, considering the right visual appeal. For strong spices, such as whole peppercorns, far less should be used than for a more delicate spice such as caraway. The size of the seed is a factor as well—you don't want to ruin the texture of the cheese with too many large seeds. A general rule of thumb is about 1 to 3 teaspoons per 2 gallons of milk. But use the rules about visual and flavor balance to help you decide.

When you are preparing seeds and spices for addition to the cheese curd, usually at the end of draining and just before pressing, they should first be boiled for 5 to 10 minutes to kill any unwanted contaminants, as well as to soften them so that they more readily imbed in the cheese. Add enough water to keep the spice covered during its time on the stove. After the spice has cooked, the simmer water can be strained off and added to the cheese milk before adding the rennet; this will help incorporate the flavor of the spice throughout the paste of the cheese.

The whole seeds are added after the whey has been drained from the vat (otherwise, much of the spice will be lost in the whey). Stirred-curd cheeses are especially easy to incorporate whole spices and seeds into, as the stirring process gives you time to spread the ingredient evenly. Seeds can also be added by sprinkling on the layers of curd as it is put in the forms, but there is the risk of creating a layer that will not knit back together and could have too much flavor.

USING HERBS AND INFUSIONS

Herbs are often mixed in soft cheeses and spreads or appear as a coating on the outside of a cheese rather than being included inside an aged variety, the simple reason being that their delicate nature makes them prone to breakdown and color change over long aging. Infusions (or teas) are a way of incorporating flavor from a visually or texturally unpalatable ingredient, such as mushrooms. "Infusion" is also a fancy word that can make a cheese sound really special.

The sky is pretty much the limit for coating a cheese with herbs (and spices, too). But as with the recommendations I made for choosing seeds and spices, sometimes less is more. There are some gorgeous cheeses out there that are almost like eating something that you feel on the forest floor—so much texture and coating that you can't appreciate the cheese. Table 5–2 lists some better known herb-coated cheeses.

Infusions can be done similarly to the steps for preparing seeds: The ingredient to be infused is simmered in water to make a tea, then strained. The solids are

BOTULISM CONCERNS

Botulism is a sometimes fatal condition caused by the bacterium *Clostridium botulinum*. Botulism is rarely a concern in cheese production and consumption, but whenever plant matter is added to a product, the risk is somewhat increased. *C. botulinum* is a spore-forming bacterium naturally present in soil and the environment (meaning it's hard to kill using the usual methods, such as pasteurization and boiling). If you know anything about canning food, you know that botulism is prevented in one of two ways (for canned goods): by ensuring that the acid level in the food is high (as in pickles, fruit jams, and tomatoes) or by pressure canning the food (where the temperature reached is higher than could be attained by boiling).

Fortunately, cheese is usually a quite inhospitable environment for *C. botulinum*. The acid level, salt content, moisture level, and storage temperature all help prevent spores from germinating (basically, sprouting). The higher the moisture of the cheese, however, the greater the risk that spores will germinate. In fact, documented cases of botulism from cheese implicate high-moisture varieties, or cheeses placed in a high-moisture situation, such as in another product.

When adding herbs, spices, or other plant ingredients to cheese, it won't hurt to keep *C. botulinum* in mind. The greater the proportion of plant matter and the higher the moisture content, the greater the risk. This especially comes into play when marinating cheeses in oil or other anaerobic solutions (botulinum spores germinate only in an oxygen-free environment). For this reason such combinations should usually be stored at refrigeration temperatures of less than 38°F (3°C).

Chevre logs rolled in various herbs and spices (Black Mesa Ranch, Arizona)

TABLE 5-2. POPULAR HERB BLENDS USED IN CHEESES

Herbs	Cheese Example
Cedar, sumac, and juniper berries	Big Holmes (USA)
Herbes de Provence	Julianna (USA), Banon (France)
Rosemary, savory, juniper berries, and chilis	Brin d'Amour (France)
Thyme and sesame seeds	Shanklish (Middle East)

discarded, and the tea is added to the cheesemaking milk. So if you have flavor you want to experiment with, but don't know how to distribute the flavor evenly—or the ingredient is too expensive or chunky to include in the paste—you can try an infusion. Mushrooms, especially rare and expensive truffles, are a good example of infusion candidates. The trick with this method is to have the tea strong enough that the flavor is distinctive; you don't want to simply make the cheese taste odd! Again, I would suggest choosing a cheese type that you would enjoy with the flavor if they were eaten separately. One side effect of using an infusion is a change in the moisture content of the final cheese, thanks to the added liquid of the infusion. While not exactly a washed curd technique, the additional moisture can have a beneficial effect on the texture of certain cheeses, such as bloomy rinds, which are also good candidates for working well with mushroom flavors, by the way.

WORKING WITH ALE, WINE, AND SPIRITS

This is an intriguing and often surprisingly complementary category of flavors. I suppose it should come as no surprise that something that is pleasing to drink with cheese would also taste good as a part of the cheese itself. At first it may feel a bit unusual to pour beer onto warm cheese curds, but it sure smells grand! While ales and beer incorporate well into the paste of cheese, wine and spirits seem to work better when used on the outside (as a wash) or to macerate (soak for an extended period) leaves that then are used to wrap the cheese (more on that later).

ADDING ALE

One of our cheese recipes includes nut-brown ale from a local brewery. The cheese, called Covered Bridge, is a washed curd style. After the curd is washed and the vat is drained, I plug the vat, add the ale, then stir the curd for 5 to 10 minutes (similarly to adding seeds and spices, you don't want to lose your added ingredient in the whey). Ale can be added to cheddar curds after milling, to stirred curd during the stirring phase, to washed curd after draining, or to any type of curd that can hold for a bit before going into the forms. The beer will bring some yeast to the curds, but it doesn't have any negative effect on aging. We use about one bottle of beer for every 2 to 3 gallons of curd (not milk).

If you are a commercial producer, be sure to document any allergens (such as wheat) on your product labels. You may also consider pouring beer from its original glass bottles into a different, unbreakable container, such as a plastic or stainless steel pitcher, then adding that to the vat, as having glass anywhere near your product is not a good manufacturing process due to the possibility of breakage and glass shards getting into the cheese.

Ale can also be used as a wash, either as a base for making a light brine or directly applied to the cheese. When it is used as a wash, the moisture and yeast can help attract surface-ripening bacteria such as is desired for making "stinky" cheeses (more on that in chapter 11).

I remember the first time our dairy inspector found a beer bottle (half full) in our aging room; he was a bit concerned about our professional future. . . .

USING WINE, SPIRITS, AND EAU-DE-VIE

Several traditional cheeses utilize wine and wine by-products to add flavor and visual appeal to cheese. Tomme au Marc is an aged French cheese that spends several weeks packed in marc (the seeds and skins remaining after pressing grapes for wine) along with some added wine. Similarly, the Italian sheep's milk

CHEESEMAKER MOTIVATIONS: THE PROFESSIONAL OR ENTREPRENEUR

There are a few folks who enter cheesemaking with the hope of creating and growing a business. While their first love may not be the actual making of cheese, they will be the ones whose passion for the product and business will help create jobs and increase the respect for artisan cheeses. While some start out intending to build a thriving enterprise, some entrepreneur cheesemakers fall into the business thanks to serendipity rather than a master plan.

Such is the case for our good friends at the Rogue Creamery in the nearby town of Central Point, Oregon. The Rogue Creamery, in business since the 1930s, was literally at the brink of closure in 2002 when David Gremmels and Cary Bryant visited the creamery to investigate the purchase of cheese for a wine shop the pair was considering opening. They bought more than cheese; they bought the entire business.

Within a year one of Rogue's blue cheeses had won the impressive title of "best blue cheese" at the World Cheese Championships. To date they have continued to reap awards far and wide for their amazing blue cheeses, including several not produced before they purchased the business. Not only has the Rogue Creamery gone beyond plenty of its previous high-profile successes, but through the leadership and vision that David and Cary have provided, cheesemakers throughout the state of Oregon, and beyond, have benefited. In addition to hearing about several of Rogue's cheeses throughout this book, you should know that David was one of my expert readers for, you guessed it, the blue cheese chapter!

cheese Pecorino Ubriaco is soaked in red wine pressings, and Drunken Goat is a Spanish goat cheese that is soaked for several days in red wine.

If you decide to try wine washing or soaking, be prepared for a softening of the outside of the cheese. Most traditional varieties go through a drying phase following their wine baths. You can experiment with the frequency of washing and the humidity in the aging room as well as soaking times. Both the hardness of the rind and the texture of the cheese will also influence the results from washing.

Eau-de-vie are clear, colorless fruit brandies that are used as aperitifs and are great for soaking leaves meant to wrap cheeses. I would say that most cheeses that have flavor added via spirits do so through the use of soaked leaves or other plant matter. Eau-de-vie and other spirits, including hard cider, can also make appealing washes.

WRAPPING A CHEESE IN LEAVES

There are many cheeses, both semisoft and semi-hard, that utilize some variation of leaf wrapping. Sometimes the leaves are dry, but more often they have been macerated in a strong alcohol, such as brandy, bourbon, or eau-de-vie. Table 5–3 shows a sampling of some of the world's better known leaf- or plant material–wrapped cheeses.

One of the first cheeses I made in 2003 was a leaf-wrapped disc of aged chèvre. We had a fig tree at the time, and I have a penchant for cream sherry. I prepped the fig leaves by washing, then soaking them in sherry in the fridge for a few weeks (this preserved the leaves until I was ready to wrap the cheese). I made a lactic/rennet curd (you'll learn about this type of make in chapter 8) in little Camembert-size forms and aged it for several weeks. Then I wrapped the cheese in the fig leaves (my

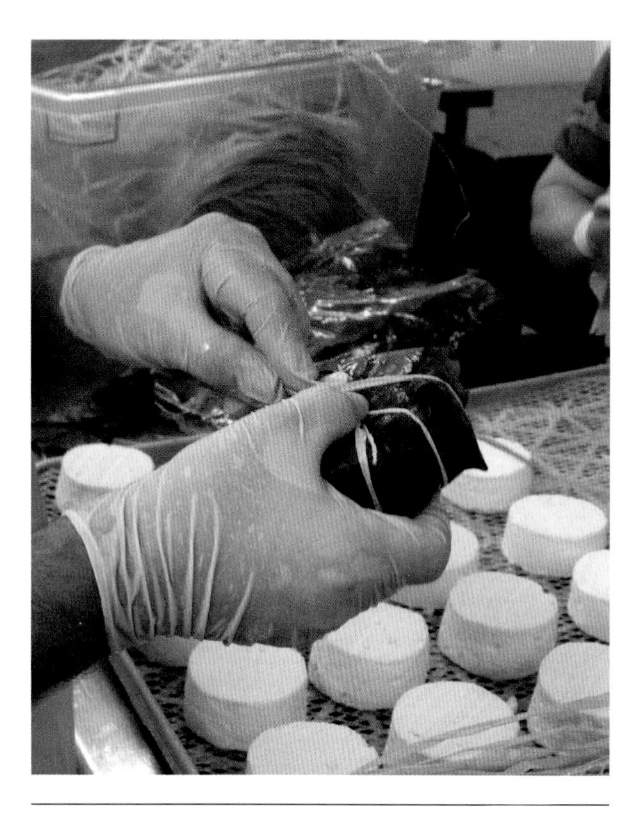

Wrapping O'Banon at Capriole Farm, Indiana

Cold smoking cheeses using a soldering iron

working name for the cheese was "Adam's Package") and let it age for a full 60 days. The cheese took on the flavor of both the leaf and the alcohol and also developed its own distinct flavor. I entered the American Dairy Goat Association amateur division cheese competition in 2004, and that little cheese won best in show. I haven't made any for years, but I still enjoy a slice of lactic-curd cheese with a glass of cream sherry.

SMOKING CHEESE

In Table 5–3, with the leaf-wrapped cheeses, is a great little disc called Up in Smoke made by my cheesemaking friend Pat Morford of Rivers Edge Chévre here in Oregon. Up in Smoke is a fantastic little cheese that combines leaf wrapping with smoking. If you ever get a chance to have this cheese (or any of Pat's, for that matter), you are in for a treat!

Smoking is an age-old method that may have come about simply by chance, as cheeses were often hung or stacked to age in the open fire huts of shepherds. The smoke would have had the consequence of preserving the cheese, keeping away pests, and adding flavor. Many cheeses today have smoke flavor added by using liquid distilled smoke but are not *actually* smoked. I am a fan of smoked cheeses—I think that perhaps because I am a vegetarian these cheeses satisfy an unfulfilled craving for something grilled and fatty.

The secret to successfully smoking cheese is the use of "cold" smoke. If you have a smoker or have seen how they work, you know that they utilize a flame or heat source to heat wood chips (from a flavorful type of wood) until they don't quite ignite but do emit smoke. Most smokers are also designed to eventually heat meat to a high enough temperature to kill bacteria and cure the product. Even if you chill the cheese and soak the wood chips in water, the temperature inside most smokers will still get too hot, and the cheese will lose oil and dry out.

I found some great tips about building a cold smoker online. We have a small meat smoker (the classic "Little Chief" brand, very affordable and easy to use) that I

TABLE 5-3. EXAMPLES OF CHEESES WRAPPED IN NATURAL PLANT MATERIAL

Leaf/Plant Material	Cheese and Country of Origin	Alcohol
Chestnut	Banon (France)	Eau-de-vie
Chestnut	O'Banon (USA, Indiana)	Bourbon
Grape	Rogue River Blue (USA, Oregon)	Eau-de-vie
Hoja Santa	Hoja Santa (USA, Texas)	None
Maple	Up in Smoke (USA, Oregon)	Bourbon (and smoke)
Nettle	Cornish Yarg (England)	None
Nettle	St. Pat (USA, California)	None
Sycamore	Valdeón (Spain)	None
Walnut	Pecorino Foglie di Noce (Italy)	None

was able to quickly convert to a cold smoker. Instead of the provided heat source, I purchased (from a hardware store) the smallest soldering iron I could find. You can also use a wood-burning tool (designed to burn designs and patterns into wood). Both of these tools have small tips that reach high enough temperatures to cause wood to smoke and burn without a flame. Next, take an empty soup or tomato can, and with a can opener cut the lid halfway around the top. Then fold the cut part of the lid in. Place wood chips inside the can (you can use any variety that is good for smoking other foods, such as alder, hickory, and apple) and lay the can on its side at the bottom of the smoker. Insert the tip of the iron into the chips and plug it in—voilà, cold smoke! The smoke isn't actually cold; there just isn't much heat being generated by the iron.

You can smoke whole cheeses when they are young and are able to absorb the smoky flavor. Alternatively, you can slice an older whole wheel and smoke the slices. How long they stay in the smoker is up to your taste, but usually 2 to 4 hours is plenty. If you can, do the smoking on a cool day so the ambient temperature doesn't overheat the cheese. For this book I made a provolone type and smoked some braids, which were amazing in flavor and appearance. We used it on pizza and pasta, and by itself!

USING ASH OR CHARCOAL

Ash is not technically a flavor, but since you might have some questions about adding and using it, I figured this chapter was probably the best place to include that information.

Ash, as used in cheesemaking, is usually vegetable charcoal, rather than actual ash, but we call it ash anyway. What's the difference, you ask? Basically, ash is what is left over after something burns down to almost nothing. Charcoal is created when a woody source is burned without the presence of oxygen, creating a chunk of carbon, plus a little ash. This carbon can be burned in the presence of oxygen to create high heat with little smoke (this is what makes it good for grilling). Food-grade charcoal is what is commonly used to ash cheeses (although currently there is some concern over its use as a food additive because there is a link between charcoal and a possible carcinogenic effect).

You can make your own cheese ash by washing and wrapping organically grown grapevines in aluminum foil and "cooking" them on a barbecue grill or even in a fireplace. The time it takes will depend on the thickness of the vines. Choose pencil-size woody vines for the best results. When you open the packet, you will have charcoal sticks. These sticks must be ground in a mortar to a fine texture, then sprinkled on the cheese with a fine-mesh sieve.

Ash is often used on the surface of white mold surface-ripened cheeses. It initially serves the purpose of raising the pH (you'll learn more about this topic in chapter 10). Later, as the white mold covers it, the ash layer disappears, but upon cutting, it is quite distinctive underneath the white layer. A couple of well-known cheeses, specifically the French Morbier and the US cheese Humboldt Fog, incorporate an ash layer in the center of the cheese. Historically, Morbier was made from two batches of curd. The first curd was drained and sprinkled with ash to protect the surface overnight. The next morning the new batch of curd was made and layered onto the first. Today ash in the center is purely ornamental. Other hard cheeses use ash as a coating, sometimes combined with leaf wrapping or layering, such as the Pecorino Foglie di Noce.

In some cases the ash is mixed with salt and applied at the same time to bloomy rinds. I find it is more effective to add ash at the end of the drying phase if using it on bloomy rinds. Otherwise, the moisture from the cheese tends to cause the ash to run. But try it both ways, and see what works for your cheeses.

Feta-type cheese marinated in oil and herbs

MARINADES FOR FLAVOR AND PRESERVATION

While the process is not quite the same as marinating a steak, cheeses can be marinated as a means of adding flavor and also to preserve both texture and quality. As you might expect, cheeses preserved in oil evolved in regions that lacked cool storage facilities but had a ready supply of food oil, such as olive oil.

We make an aged feta style we call Evans Creek Greek (ECG for short). One of the ways that sells well for us is when we marinate it in an olive oil blend—to keep the oil from gelling when refrigerated—with dried oregano, basil, garlic, and dried tomatoes. It is quite pretty, and for some reason people are willing to pay well for it, despite the fact that we keep telling them they could easily do the same thing at home, using our plain version, of course.

Oil is an effective preservative while at the same time preventing rind formation and drying of the cheese. Though, traditionally, oil-preserved cheeses did not have to be refrigerated, if other ingredients have been added, refrigeration might be called for (see the earlier sidebar on botulism on page 103). The herbs and spices added can be personalized to your own tastes, but some lovely flavored ingredients will be hindered from lending their taste to the mixture by the oil. Large spices, such as juniper berries and peppercorns, might need to be "bruised" before adding and then will need longer time in the marinade before you will be able to detect their flavor.

Try marinating balls of fresh mozzarella, feta, and even blue cheese crumbles. I remember at one tasting event, someone dropped a chunk of one of our hard cheeses into the ECG marinade, dipped a toothpick in, retrieved the sample, ate it, and was proclaiming it the best cheese there.

6: DESIGNING, EQUIPPING, AND MAINTAINING YOUR HOME CHEESEMAKING SPACE

No matter the scale of your cheesemaking dreams, having a proper space with the right equipment and knowing how to keep the space and tools properly cleaned and maintained will go a long way to making the process both fun and successful. In this chapter I focus mostly on the needs of the home-scale cheesemaker. For those considering a career in the artisan cheese world, I have dedicated an entire book to the subject in *The Farmstead Creamery Advisor.* The topic is quite vast and cannot be adequately addressed in one chapter! But the important issues are the same, whether for home creations or professional ones—you will want cleanable, easily maintained equipment and surfaces that pose no risks when used to manufacture food for human consumption. So let's take a look at all of those issues and help you get set up to make cheese!

CHOOSING EQUIPMENT

In this section we'll go over how to choose some of the many things you will need and want for making cheese. (See appendix B for some of the many sources for these items. Thanks to the popularity of home cheesemaking, you can now often find local stores carrying many cheesemaking supplies as well.)

POTS AND VATS

The primary considerations when choosing the proper receptacle to make your cheese in are *material* and *condition.* Whether you make your cheese in a 1-gallon pan on the stove or in a steam-jacketed cheese vat, your receptacle is best if made of high-quality stainless steel. You may or may not know this, but stainless steel comes in varying grades. The cheaper it is, the less likely it will hold up well over time. You may even find that it will rust in spots. That's right, stainless is not "stain-free" just "stain-less." In the "Keeping Things Clean" section later in this chapter, I'll give you some tips for maintaining stainless steel, but for now here's a general rule to go by: *If it is too cheap to be true, it is too cheap for you.*

For home use stainless steel pots with stainless or tempered glass lids—a type of glass that is fine in high-heat situations—can be used. In commercial settings, however, glass lids would not usually be acceptable, as there is a risk of breakage, which brings to light questions of food safety. By the same token, a coated or enameled surface will be fine for home use, as long as you inspect it before and, even more importantly, after use. (If you notice a chip missing from your pot when you are done making cheese, you should suspect that it is in your cheese and consider discarding the product.)

A common double-boiler–type setup for home cheesemaking

When inspecting the condition of your future cheese vat, keep in mind the code that all licensed cheesemakers must live by: *The surface must be easily cleaned.* So any deep scratches or rivets on the inside, the surface, or in nooks are undesirable—and for the commercial cheesemaker, not allowed.

The next factor in choosing a vat will be *size*. The most common at-home cheese vat is a double-boiler–type setup of a pot set in a sink of warm water or inside a larger pot of water on the stove top. Look for a large stainless steel pot and an enamel or stainless steel water bath canner sold through canning supply retailers. I have a nifty little platform at the bottom that keeps the stainless steel stockpot (which holds the milk) from sitting directly on the bottom of the water bath canner. This allows water to circulate underneath your pot of milk. I have been unable to find the same platform for sale, but you can make one using a round wire cooling rack—or even just set a few mason jar lid rings at the bottom of the canner. They will rust a bit, but if the water bath pot is not also used for cooking, it isn't a problem.

For small batches some cheesemakers use and are very happy with slow cookers (commonly known by the brand name Crock-Pot). If you use this type of appliance, choose one with a removable crock/liner if possible. I have not used one myself, but I can see the advantages, especially if you don't have room on your stove for a water canner and are only making small batches.

A couple of simple, rather inexpensive options for slickly processing 3 to 5 gallons of milk are the use of an electric countertop roaster oven or an electric countertop food warmer (used by caterers to keep stainless steel trays of food warm). The roaster ovens are inexpensive and can be easily purchased at department stores in the kitchen section. They have an enamel-coated deep pan (usually in an 18- to 22-quart size) that sets down into an enamel-coated well that can hold your water bath. They can easily be set to a low enough heat to hold milk at ripening temperatures, then take it up to the goal cook temperature later in the process. The only drawback with these vessels is that the enamel coating can chip, so if you choose this type of home vat, be sure to inspect it carefully before making cheese (or roasting your Thanksgiving turkey or Cinco de Mayo tamales, for that matter).

WHAT ABOUT COPPER?

If you plan on making alpine-type cheeses, such as Gruyére and Emmental, a copper vat might be just the thing to help duplicate the conditions that lead to the uniqueness of the traditional varieties. Also, the copper truly does interact with the cheese milk at a molecular level that influences the flavor of the cheese. If you decide to shop for a copper vessel, look for what is called a "jam pan." These are large copper pots that will work well as a cheese vat for high-temperature cook cheeses, where the acidity stays low while the curd is in the vat.

Copper should not be used to make cheeses that develop a lot of acid in the vat, as the copper will react to the acid in a harmful way, not the same way it will react to help create the traditional alpine cheeses. Copper jam pans can be purchased in a size that will hold several gallons of milk. They range in price from under $100 to several hundred dollars. Oh, this is a great item to put on your cheesemaker's wish list!

I really love the all-stainless-steel food warmers available through restaurant supply stores or online. They come closer to meeting the standards for actual licensed cheesemaking, but on a much smaller scale. It can be a little confusing when you first start looking at the choices. The lower-priced ones, for one thing, don't have the same heating power as the more expensive versions (the latter are usually above $200). But that can be good, since you don't need to attain high temperatures during cheesemaking. They don't, however, come with a liner pan (as the roaster ovens do), so you will need to buy one separately. Look for the full-size deep pans, which are usually about 6 inches deep. A lid must also be purchased separately. I suggest choosing

An electric food warmer (designed for keeping catering dishes warm) is easy to "pitch" the curd in by using a section of cheese mat.

one with a slot so you can leave your stirring tool in the milk and still close the lid.

There are a couple of really nice things about using either one of these two choices. First, the rectangular shape and shallow depth make it possible to cut better-shaped curds, and you can tilt the liner pan and keep the curds warm for easy cheddaring. About the only drawback I have experienced with ours is that if you don't anticipate the temperature and the vat begins to get too warm, it is pretty awkward to try to lift the liner pan out of the hot water when it is full of milk. After working with it for long enough, though, you should be able to anticipate how to keep the temperature adjusted so you don't need to pull the pan out.

If you end up needing a vat that will hold over 5 gallons of milk, you may need to move up to either an "official" cheese vat or a soup kettle that's made for restaurants to produce large batches of soup. Soup kettles, also known as steam kettles when they are heated by steam rather than circulating hot water, range in size from 5 to 60 gallons. They can have their own heat source or, on larger units, be plumbed to a hot-water or steam supply. Some kettles are stationary, and others tilt to pour. I use a 30-gallon unit, and I find the tilting kind very handy for pouring whey off the top of the curds.

When new, these units are quite expensive, but fortunately they can often be purchased used for a fraction of the new cost. The larger units are very durable and have few parts that wear out or need replacing, so they are a fairly low-risk purchase when buying used. Models designed to be heated by steam are readily adapted to hot-water heating without any additional cost. The hot water can be circulated via a closed system (meaning the same water is reheated and circulated in the vat's water jacket) or filled with hot water, then emptied out of the bottom.

You may wonder about using a food-grade plastic tub or bucket for making cheese. This is frequently a good choice when making low-temperature, acid-coagulated cheeses that need to ripen for several hours. Some commercial small vats are even made of such plastic. I am personally not a fan of using any plastics for food preparation when warm temperatures are involved, as well as possibly abrasive chemicals and mechanical scrubbing that could erode the surface and make future cleaning attempts inadequate.

MILK STORAGE CONTAINERS

Whether you are collecting your own milk or buying milk from others, chances are good you will be storing it for a day or more before making cheese. The same consideration you paid to choosing a vat should be given when choosing milk storage containers. Glass has a wonderful reputation and feels so traditional, but it can chip easily and also allows the loss of some vitamins from the milk when exposed to light. Still, it is the most common choice for the small home dairy. If you feel like upgrading a bit, there are a variety of small (4 to 12 quart) stainless steel milk totes available from home dairy supply companies. Remember, though stainless steel is expensive, it should last for a lifetime.

FORMS, HOOPS, AND MOLDS

Let me tell you from the beginning that you are probably going to spend more money on cheese forms than you need to. It is pretty much impossible to know early on what types of cheese you will end up making the most, what size wheels you will be happiest with, and how many forms you really need. So try not to get too stressed about it! The main thing to remember when trying to make or choose cheese forms—besides that the terms "hoop," "form," and "mold" are pretty much interchangeable—is that they must be made of some kind of material that is easily cleaned and not negatively affected by the acid produced by the draining cheese

Various homemade and purchased cheese forms

(this rules out materials such as aluminum and copper). Most cheese hoops were made in the past from stainless steel and even earlier in history from woven reed baskets and wood. Today most forms are made from food-grade plastic.

As long as they are easy to clean and made of food-grade material, you can come up with some pretty creative options, such as plastic colanders, salad-spinner bowls, or even food-storage bowls with drainage holes drilled in the sides. Some people are even using large-diameter PVC pipe (white pipe used in plumbing cold potable-water lines) as cheese forms.

Keep in mind the ruggedness of the form if you will be applying weight during draining. The less pressure that is used, the more options you have when selecting forms. If you'll be making a cheese that requires a great deal of pressure, such as cheddar, you will need a very sturdy, straight-sided form or a curved-sided form (such as Kadova forms, described in a moment) designed to take the pressure exerted by a mechanical press.

When shopping for forms made specifically for draining cheeses, you will have many choices. You will likely end up with some that you only use once or twice, and maybe you can even give those away to another cheesemaker who is in the early experimentation phase (but for some of us, it never ends). The plastic forms come in four basic types: those meant to be used with cheesecloth and pressure; basket-type forms for ladled, unpressed curd; microperforated forms with many tiny holes to simulate the effect of cheesecloth; and Kadova-type forms that have a built-in mesh lining. Kadova and microperforated forms are designed to be used with a mechanical press. However, both of these types are notorious for being difficult to clean, especially the microperforated types. But they also eliminate the need for cheesecloth and the cleaning involved with its use.

Most places selling cheese forms will include some guidelines as to which cheese types the forms are best suited for. Many of the forms will have dimensions in millimeters and volume in kilograms, so be prepared to do some conversion if you live in a country that for some reason is sticking with the imperial system of measurements. . . .

CHEESECLOTH, DRAINING BAGS, MATS

Cheesecloth serves two purposes during draining: First, it helps keep the loose curd in a shape while it knits back together, and second, it helps wick whey away from the cheese and toward drainage holes in the form. It is important to choose the right fabric to properly drain the type of cheese you are making. In general, you can apply the following policy: The finer and softer the curd, the tighter the weave of the cloth should be. The fabric sold in most kitchen and department stores that is labeled as cheesecloth has a very open, gauzelike mesh and is not suitable for draining cheeses. Instead, you will want to buy real cheesecloth from a cheesemaking supply company.

When buying cheesecloth, you may encounter some labeling confusion. Some companies sell cheesecloth labeled with two designations: "butter muslin" with a 90-thread-per-square-inch count (a tighter weave) and "cheesecloth" with a looser 60-thread count. Some companies call the 90-thread count simply "finely woven cheesecloth." Others may give no indication of thread count. In general, the finer-weave cloth will serve well for all applications. But sometimes traditional cheeses call for a looser, more open cloth or a coarser, heavier cloth called hessian or burlap.

> The fabric sold in most kitchen and department stores that is labeled as "cheesecloth" has a very open mesh and is not suitable for draining cheeses.

Curd pressed in microperforated forms

Plyban is a type of disposable plastic sheeting that can be used in place of traditional linen muslin. It is used mostly for the manufacture of blocks of cheddar and is designed to drape in square forms, leaving no crease marks. Since large manufacturers would have logistical issues trying to adequately launder linen cloth, this disposable type offers them an option. Some home and small artisan cheesemakers love this plastic cloth (it can be washed and reused), but I have never liked it myself. Give it a try, and draw your own conclusions.

Draining bags are designed specifically for making soft, spreadable cheeses. They have a very fine weave and are sturdily manufactured. A great substitute is a white pillowcase, which you will, of course, clean and sanitize before using. If you have any sewing skills and the right equipment, you can sew your own draining bags.

Draining mats can be used directly under cheeses such as Brie and Camembert during draining and aging, under forms during draining (to help keep the form itself up out of the draining whey), and under hard cheese during aging (to promote airflow around the cheese). Plastic mats come in several weaves—from coarse, with five squares per inch, to fine, with thirteen squares per inch. Which one you use will depend on the amount of airflow you want and the texture of the imprint. As you might imagine, the finer weave will not allow for as quick a drainage but will leave less of an imprint on the surface of the cheese.

If you are using the mats only for airflow around aging hard cheeses, definitely choose the more open weave matting. Many people use plastic needlepoint matting as a substitute. It looks almost the same and works as well. It is low in cost and is usually available for purchase locally (rather than having to be ordered, like the "official" cheese matting). Reed mats, such as those designed to make sushi rolls, can also be a good choice for the home cheesemaker. Many commercial cheesemakers are not allowed to use these, as they are considered "uncleanable." Reed mats, however, leave a lovely imprint on the surface of bloomy rinded cheeses.

WEIGHTS AND PRESSES

In the recipe section of this book, you will learn that there are many pressed cheeses that can be made without the investment of a mechanical press. I find that many cheesemakers don't try to make hard and semihard cheeses because they think they must fork out a large sum for one or two mechanical presses—either that or build a lever press. So if you aren't ready for this sort of commitment, just choose the many types of pressed cheeses that don't require extreme pressure.

> You only need as much pressure as it takes to accomplish the goal!

The options are almost limitless when rigging up a way to lightly press cheeses—from stackable cheese forms and jugs of water or sand to barbell weights. Just keep the following things in mind when designing your pressing system: stability—will the weights shift and come crashing down when you are not observing the pressing? And cleanability—are the weights cleanable and contained so they don't leak or leach any nonfood substances or chemicals into your cheese?

Pablo Battro's small, stackable cheese press (Argentina)

If you find you must move up to a mechanical press, you have two basic options, then multiple choices within these options; a single-wheel screw-type press or a lever press (larger commercial creameries usually use pneumatic, hydraulic, or large lever presses). Your choice will likely revolve around the number of wheels you plan on pressing at the same time. Small single-wheel presses that can press 3- to 4-pound wheels can be expensive, and it will be quite time consuming to have more than one or two wheels that you need to remove from the press, redress, and turn.

When choosing a single screw-type press, look for durability, cleanability, and a pressure scale. The most expensive are made from all-stainless-steel parts—and of course, these are the most cleanable and durable. Some single presses are made from wood and laminate parts. While wood can be cleaned and maintained for home use, the laminate parts have the potential to separate over time. The screw mechanisms on single-wheel presses are not created equal; some are not able to provide enough pressure to knit cheddar curds into a smooth mass without any openings in the paste or in the rind. If you are buying from a reputable supplier, you should have no problem researching their models and making an informed choice, but if you are buying from one of the growing number of small, online manufacturers, it's not a bad idea to get a recommendation. (The cheesemakers' forum that I recommend in appendix B is a great way to research some of these purchases.)

It is possible to build or buy a single-cheese pneumatic or hydraulic cheese press. These types of presses are not often seen in home/hobby situations and can be a bit expensive, but they do offer precision, when proper calculations for friction are factored, and pressure, as well as a bit of a thrill for those who love fancy equipment! If you are into mathematical calculations and mechanics,

A table top lever press at La Suerte Creamery, Argentina

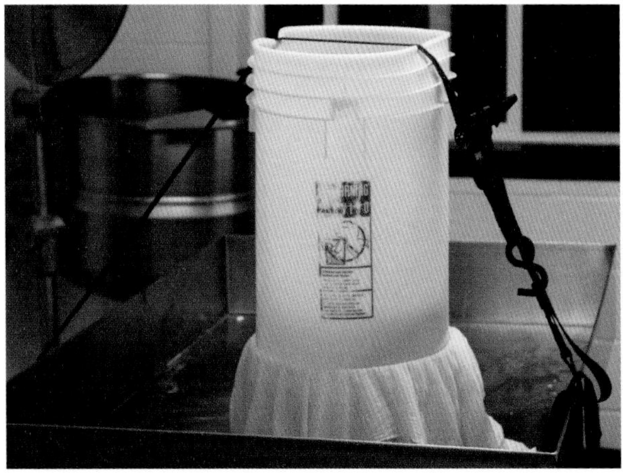

Creativity can lead to some simple, inexpensive solutions: using a ratcheting strap to apply high pressure to a 22-pound (10 kg) wheel of cheddar.

A typical single-wheel home-size screw press

you can find ways to do the math regarding calculating pounds per square inch (or the equivalent) online.

Finally, we get to the most common type of mechanical press for pressing more than a single wheel: the lever press, also called a Dutch or Holland lever press, because of its long history of manufacture and use in that country. Lever presses designed for home use are usually made of a hardwood such as maple. It is important that they not be made of a softer wood, as the lever arm from which weights will be hung can crack if it is not sturdy enough. Also, hardwoods will hold up to cleaning and the high moisture to which they will be exposed. Lever presses can be freestanding or attached to a wall. You can purchase a premade kit for single-wheel versions of these presses and also find plans online (see appendix B). Or you can attempt to make one on your own.

One day I was making a large wheel of cheddar. The only mechanical press we have is a simple screw type that can handle the curd from about a 4-gallon batch of cheese. But I had made 23 gallons of milk into lovely cheddar curds. After scratching my head a bit, I went out to our truck, grabbed a medium-duty strap and brought it to the creamery. I scrubbed and sanitized it, then strapped it around the drain table and over the top of the bucket follower on top of the pressing curd. By tightening the ratchet on the strap, I was able to create perfect pressure to form the 22-pound (10 kg) wheel and create a tight rind.

This system would work even better if I had a flat follower that would fit into the form (then the strap wouldn't have malformed the bucket follower I was using). It would also be a simple matter to line up several forms filled to the same level and place a bar or board across them and strap press them all. So keep options such as this in mind if you don't have room for a large lever press but occasionally need the extra force required to form certain types of cheeses.

MISCELLANEOUS TOOLS

You will need quite an assortment of miscellaneous tools, such as a measuring stick, ladles, curd-cutting knife, measuring cups, measuring syringe, colander, thermometer, timer or clock, gram scale if measuring culture, a scale for weighing curd, and brine "tank." If you are also waxing cheeses, that will require equipment as well.

If you are making cheese in a larger vat or a pot without any measurement markings, then figuring out how much milk is in the vat can be a challenge. It is time consuming to measure the milk before pouring it into the vat. Professional vats come with a stainless steel measuring "stick" that is marked in increments. You insert the stick in the milk and read the level. For small scale producers, a clear plastic measuring stick—marked in inches or centimeters—can be used. To determine what measurement correlates with a certain amount of milk, simply add water to the empty vessel in set amounts (say a quart or a liter), insert the measuring stick, and write down the measurement. Continue to do this, recording the measurements and volumes until the vat is full. For future cheesemaking batches, you can fill the vat, insert the stick, and measure the amount.

A perforated cheese ladle can be either purchased or modified from a long-handled, slotted stainless steel spoon or skimmer. The ladle is used to gently stir the curds, and the perforations allow for the whey to flow through the spoon so you are moving more curds than you are whey, as well as to help disperse coagulant and calcium chloride when being added to the cheese milk.

A curd knife is used to cut the coagulated milk into curds. Again, you can purchase one or use a long, narrow spatula; a frosting spreader; or even a thin piece of

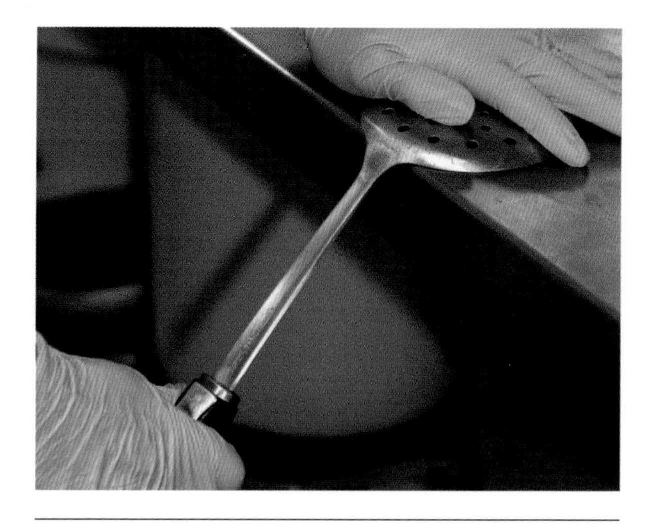

A skimmer spoon can be bent to create a cheese ladle.

stainless steel. We had a long-bladed curd knife fabricated from thin-gauge stainless with a handle at one end. It has the flexibility to help make nice horizontal cuts and is not so sharp that it scratches the sides of the vat (as a real knife would).

There are some small-scale cheese "harps" on the market, but beware of blades that are too thick or spaced too widely—once you make large cubes, it is a bit more difficult to cut them smaller evenly, as they are now moving in the vat as you try to cut them. A small-scale harp can be fabricated by creating a stainless frame that is strung with nylon fishing line. While it is not necessary to invest in such a harp, if you are making decent-size batches in a rectangular container, it will give you the advantage of nearly perfect cubes of curd. We had one made for our circular steam kettle vat, and while it works well, you have to know that it is close to impossible to get perfect cubes in a round vat. When a harp is turned in a round vat, it cuts concentric circles that must then be cut vertically, leaving you with more of a pie-shaped curd than a square one.

If you are working with sheep's milk or Nigerian Dwarf goat's milk, it is possible to cut curd with a thin wire whisk, because of the heavy-bodied nature of the coagulum from these types of milk. I originally used a long whisk from which I had removed two wires. It worked well for us, but it definitely makes more uneven sizes of curd.

Different tools for cutting curds

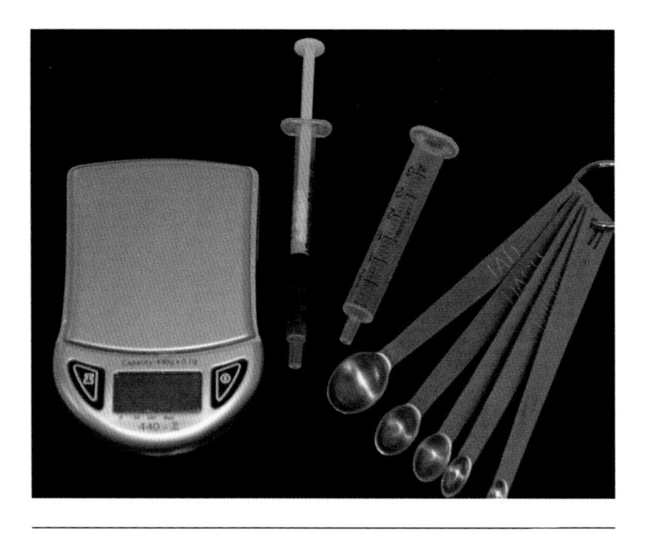

Syringes in varying sizes and a set of tiny measuring spoons can help ensure accurate measuring of cheesemaking ingredients.

TINY MEASURING SPOONS:
SORTING OUT THE SIZES

When I first bought my set of adorably cute little measuring spoons on Amazon.com, I was a bit perplexed about just how to decide what a "smidge" and a "tad" really meant. While I can't guarantee that all sets of small spoons will be the same, here is an approximate translation of these vague, old-fashioned terms:

Tad = scant ¼ tsp = 1 ml
Dash = scant ⅛ tsp = 0.5 ml
Pinch = 1/16 tsp = 0.25 ml
Smidge = 1/32 tsp = 0.12 ml
Drop = 1/60 tsp = 0.05 ml
Skewer tip = 1/120 tsp = 0.02 ml

I knew that I could never remember the equivalent measurements for these tiny spoons, so I used a Dremel tool to engrave the comparable measurement right onto the handle. Incidentally, the "drop" measuring spoon does hold literally a drop when you are using a typical dropper bottle such as holds small amounts of calcium chloride and rennet.

For measuring liquids you will need a variety of measuring cups (when possible, choose plastic over glass for safety), measuring spoons, and a couple of plastic syringes in 5 cubic centimeter, 3 cc, and 1 cc sizes (remember, a cubic centimeter [cc] is the same as a milliliter). If you have a nice little set of syringes, you will be able to make superaccurate measurements of coagulant and calcium chloride, pretty much eliminating the need to do the somewhat complicated dilutions that are taught in some courses and books.

For measuring dry powders, such as cultures, accurately weighing and dividing into unit doses (see chapter 2 for more) is always the best choice but is often not practical for the small-scale home producer. In the recipe section I have given culture measurements for 2-gallon (8 L) batches of milk in teaspoons. For measuring really tiny doses of mold and ripening cultures, I recommend using the tip of a metal meat skewer (there is a photo and more details on measuring mold cultures in chapter 10). The tip of a skewer holds roughly 1/120 of a teaspoon (0.02 ml), so a really small amount. You can also purchase (usually online) a set of tiny measuring spoons that will help measure amounts from just under ¼ teaspoon (1 ml) down to 1/60 teaspoon (0.05 ml). See the sidebar on page 117, this chapter, and appendix C for more on using tiny teaspoons.

In the home or commercial creamery, it is a good idea to have a plastic or stainless steel colander for holding draining bags while filling, draining curd, and other surprisingly handy uses. Don't use aluminum strainers, as the metal will react with the acid in the whey and cheese.

You don't need a fancy thermometer for making cheese, but you do need to make certain it is accurate. I like using the simple, metal-probe thermometers you can get for a few dollars at the grocery store. Using a piece of stainless steel wire, you can fashion a nifty little hanger so the thermometer stays suspended and you can still close the lid on your vat. Periodically, you can do an amateur, but pretty accurate, calibration of any thermometer by filling a glass with ice, adding a little water, agitating the ice and water, then submerging the probe of the thermometer in the cold mixture. It should drop to just a hair over 32°F (0°C). If you want to attempt a double-check calibration, bring water to a boil and see if the thermometer reaches 212°F (100°C). Of course, boiling point varies slightly based on elevation and atmospheric pressure, but as I said, this is an amateur calibration! It will tell you if your thermometer is off by enough to matter.

A clock or timer should be available for monitoring times during your cheese makes. I know cheesemakers who use their cellular phone timer to help remind them, for example, of when to check the cheese pH and when it is time to take the wheels out of the brine. Now if only we could get those smart phones to clean up the pots when we are done. . . .

I just recently invested (and it wasn't much money) in a tiny little scale for weighing out bags of culture. It is really quite shameful that I waited this long to get more precise with the measurements, but there you have it! For more on choosing a scale and weighing culture, refer back to chapter 2.

If you are brining cheeses—and you will be if you do a lot of the recipes in this book—you will need a container large enough to float all the wheels in a batch (or you can use multiple containers). I suggest using a transparent container, for two reasons: First, you can easily see if the brine is clear of sludge (see chapter 2), and equally important, you can see your cheese, so you reduce the chances of forgetting it in the brine and ending up with a dehydrated, oversalted cheese. Brine should be stored at either aging-room temperatures of 50 to 55°F (10–13°C) or in the refrigerator between use. Chapter 2 covered all the details of brining cheeses.

If you will be waxing cheeses for aging, plan on a double boiler, pan, or dedicated Crock-Pot that is used only for this purpose. The wax pretty much sticks forever on all utensils. A variety of natural bristle brushes will also be needed for waxing or applying other rind treatments, such as so-called "cream wax."

CREATING YOUR CHEESEMAKING SPACE

While most of us don't have too many options when choosing our cheesemaking space—usually, whatever open counter space we can temporarily lay claim to in the busy, cluttered family kitchen—whenever possible there are some things to take into consideration. Here are the main points to think about for your cheesemaking area:

- Storage space for pots, forms, press
- Adequate counter space
- A hot-water source for warming milk and for cleanup
- A place to hang or set draining cheeses
- An area away from pets, dust sources, stored chemicals, and cleaning products
- Proper ambient room temperature
- A place to store cultures and coagulants properly
- An aging fridge located where it is convenient to check daily

STORAGE SPACE

You will need a good-size cupboard for several large stainless steel pots, your plethora of cheese forms (molds), and your miscellaneous equipment, such as ladles, spoons, and probably at least one countertop cheese press. Choose a cupboard that does not share space with any cleaning products, chemicals, pet or animal products (including brushes and medications), human medications, compost or trash bins, or any other product or equipment that could dirty or contaminate your equipment.

ADEQUATE COUNTER SPACE

This seems like an easy one, but unless you happen to have an oversize and underused kitchen, counter space is probably at a premium in your household. You may think that it will be easy to clear a space on the days you make cheese, and this may be your only option, but remember that you may be occupying that space for up to 24 hours, so you might want to make sure that someone else doesn't have a major project planned—such as a large family dinner—at the same time.

> The more accessible your equipment is, the more likely it is that you will make cheese!

HOT-WATER SOURCE

You will most likely be warming your cheese vat using hot water (versus direct heat, such as setting the pot right on the burner), usually in a double-boiler–type setup or in a sink. If using a double boiler, you will need a stove top or hot plate. If you have access to a sink for the period of time you will be "cooking" the cheese, that will work just as well. (Remember, if you are making cheeses that require temperatures over 100°F (37.8°C), the hot water from the sink may not reach adequate temperatures.) Also, if you are using the same sink for cleaning equipment, you could run into some problems when trying to keep wash water out of your cheese pot. If you are using a countertop appliance, such as a food warmer or soup kettle, you may want to dedicate a space on the counter to full-time residence. Just keep in mind that the more accessible your equipment is, the more likely it is that you will make cheese!

DRAINING SPACE

You'll need a place to hang draining curd (such as fromage frais, quark, and ricotta) and a place to set cheeses that drain in forms or in a press. Small amounts of curd can be bag-drained by suspending the bag from a utensil that is placed across the top of a tall pot. Larger volumes, though, will need something such as a ceiling hook mounted under a cupboard.

For draining cheeses in forms, you will need a surface with either a slight slope that drains to a sink or container or a level perforated or grooved surface to collect and divert draining whey. If your cheeses don't need any weights for pressing, a sloped surface, such as a dish rack drain board, works great—but if you will be stacking forms or adding weights to the top, a surface with too much slope will cause the stacked forms to tip and most likely topple over. A simple flat, grooved drain board can be made using a food-grade cutting board into which you (someone will have to be handy with a table or circular saw and a Dremel-type tool) cut shallow grooves that direct the whey to one edge of the board. This drain board can then sit at the side of a sink. Be warned: This type of board can only handle so much whey at a time.

Another option is placing baking racks over shallow baking trays or in a larger plastic tub. This works fine for lighter-weight forms but will not support too much weight without collapsing the racks. To use the same system, but with more weight, place a large plastic cutting board over the racks, then cheese matting on top of that (or use the grooved drain board that I described above).

PETS, DUST SOURCES, STORED CHEMICALS, AND CLEANING PRODUCTS

I mentioned before to be sure that you store your equipment away from hazards such as cleaning products and medications, but you will also want to keep (or at least try to limit) access to your working space by critters (both the pet variety and the uninvited types). This is not only for cleanliness's sake, but also because a draining cheese is often quite tasty to your dog, and you often don't find out until you see the lick and bite marks on the wheel of Havarti you just spent 12 hours making.

Think about things like windows that open to animal pens or dusty driveways. If these are in your work space, do your best to keep them closed during cheesemaking time. Even a window that opens to a lovely forest will allow mold spores to enter the milk, and while they may not cause health issues, they will cause flavor flaws and more.

Since your work space will likely be in the family kitchen, become aware of the natural hazards that will exist when a space is shared with products such as drain opener, oven cleaner, and so on. Let other household members know that when cheesemaking is occurring, these types of products should not be used in the same space. Even if these cleaners are completely organic, you still want them out of sight during cheesemaking.

ROOM TEMPERATURE

As you learned in chapter 2, room temperature is important when making cheese. The usual ideal temperature during the make and during draining is 70 to 72°F (21–22°C). Ideally, your space will be climate controlled. If not, you will need to take this into consideration when making cheese; some things can be adjusted in the cheesemaking process to accommodate less than ideal conditions (often that something is simply your expectations!).

STORAGE FOR CULTURES AND COAGULANTS

I have recommended earlier in this book that you choose freeze-dried direct-set cultures for your cheesemaking. These are the most convenient and reliable. These types of cultures will be best stored in the freezer (you don't need a fancy freezer for this, but definitely one that has a manual defroster). Coagulants (rennet and its substitutes) will be stored in the refrigerator. There is no concern over storing this alongside your bottles of ketchup and mayo, so sharing the family fridge is just fine. Although, since coagulants should be protected from light, if you are the parents of teens who regularly stand at the open refrigerator door awaiting an inspiration, you might want to wrap your bottle in foil. I am only halfway kidding.

To help keep air and moisture from deactivating the cultures, we use several vacuum containers meant to keep deli meats and cheeses fresh. A small, handheld vacuum pump removes air (and along with air, moisture) from the container. This helps keep the cultures fresh for much longer.

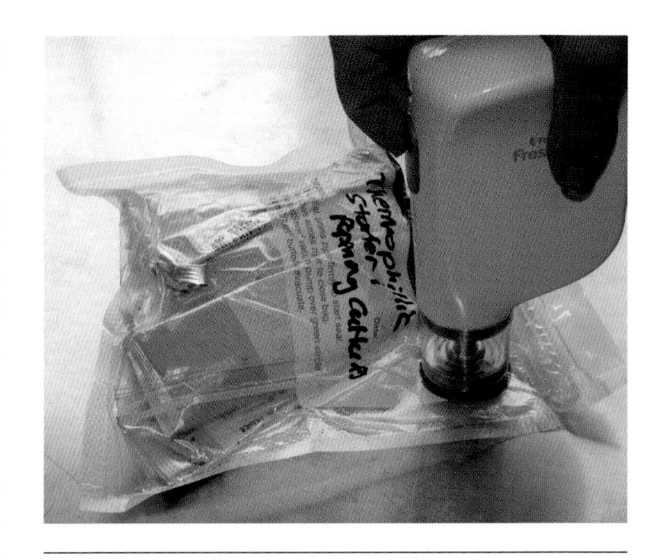

Reusable vacuum-sealed bags make great storage containers for freeze-dried culture.

CHEESE AGING-UNIT LOCATION

If you will be aging cheeses (and almost every cheesemaker will eventually give it a try), you will have an aging unit, often a wine storage fridge or other type of refrigerator. Try to locate it in a place that you pass by daily. I once tried keeping mine in a basement, which sounds great, but I didn't go to the basement daily, and the cheese suffered the consequences. Trust me, even with the best of intentions, out of sight is out of mind.

KEEPING THINGS CLEAN

When you are making cheeses for yourself or to share, you will take extra pride in creating the best product possible. No matter how well you can make a recipe, if your equipment isn't kept impeccably clean, your cheese will be tainted as well. So let's take a look at how to clean and sanitize your cheesemaking equipment and space properly (in appendix B I list more sources for cleaning and sanitizing products).

CHEMICALS AND THEIR PROPER USE

While you might associate the term *chemical* with something manmade and harmful, let's remember that everything in life is made up of chemical compounds. Even so-called *natural* cleaners are composed

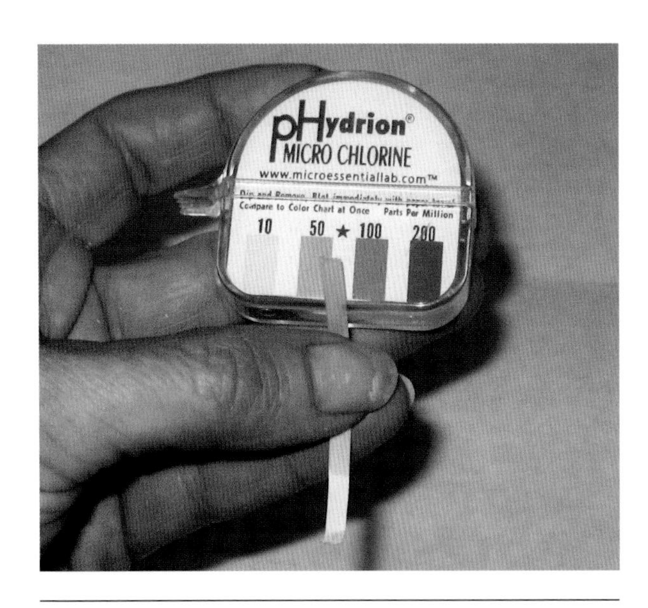

Inexpensive chlorine test strips can be used to measure the dilution of chlorinated sanitizers precisely.

of chemicals, but more than likely they are naturally occurring compounds. Remember that naturally occurring chemicals can still be harmful, though, so let's start by emphasizing safety when using cleaning and sanitizing chemicals—of all kinds.

There is a reason that cleaning and sanitizing products work well at removing residues from surfaces: They are harsh, caustic, and not something you want on your skin, in your eyes, or in your lungs. Always start by reading the warnings on the labels, and make sure your space is well ventilated. Gloves and goggles should be on hand; use them any time exposure is possible, such as when pouring chemicals from a large container into a smaller one, scrubbing a particularly troublesome area, and using strong dilutions of chemicals—when splatters and splashes of chemicals are likely.

There are basically three categories of chemicals that we will be covering: detergents for cleaning, sanitizers for sanitizing, and acids for removing calcium deposits and sanitizing. These three basic categories have a lot of overlap in their usage, however. For example, chlorine, a commonly used and readily available sanitizer, is also combined with detergent, as it has the ability to help with removal of proteins during cleaning. And acids can also be used as sanitizers. Brace yourself for a little confusion at times!

DETERGENTS

When it comes to cleaning, detergents are quite dependent on water temperature, pH, and mechanical action. In other words, you will need hot water and physical exertion to do the job. Detergents by nature are alkaline (that means above 7.0 pH, as you might remember from chapter 3). You can buy fancy "dairy detergent" that has chlorine in it, but for most home situations, Dawn brand (or an equivalent) dish detergent works just fine; in fact, it's what we use in our creamery. Don't get any fancy scented versions, though; unscented is best.

We have recently switched to a more environmentally friendly and effective cleaner that relies completely on the activity of enzymes to remove protein and fat deposits. The product we use is manufactured by Ecolab and is available only in containers larger than most home cheesemakers could use. But do consider purchasing large pails and splitting with a friend! Not surprisingly, these types of products are more expensive than their more widely used cousins. For the small producer and home cheesemaker who might not need to worry as much about covering these costs or whose product retail price can absorb the difference, these more expensive choices can be sustainable.

SANITIZERS

Sanitizers are used to eliminate any bacteria that scrubbing and washing might not have removed. There is an old saying: "You can't sanitize something that isn't clean." So don't try to use sanitizers in place of cleaning—they are meant to follow thorough cleaning. Sanitizing can be done with chemicals, both those that break down into very environmentally friendly components and those that don't, or by using heat.

The most readily available sanitizer to use at home and in small creameries is often chlorine. Chlorine, in the form of grocery store bleach, is very effective, easy to find, and inexpensive. Quite often, however, people use too much, leading to sanitizer residue on equipment (which can harm your product and produce undesirable flavors). This is a waste of money, results in erosion of the surface of stainless steel and other metals, and can harm septic and wastewater systems.

TABLE 6-1. GUIDE FOR CLEANING AND SANITIZING PRODUCTS FOR THE HOME CHEESEMAKER

Category of Product	Common Source or Brands	How Much and How to Use	Notes
Cleaning: grocery store purchase	Unscented dishwashing detergent; Dawn brand is popular with cheesemakers as its degreasing ability works well in removing milk fat.	Using the hottest water you can hand-wash in, add enough detergent to create ample suds. Soak equipment for a few minutes, then hand-scrub.	Hand-washing allows for mechanical scrubbing that lessens the need for stronger cleaners.
Eco-friendly cleaning: order online or purchase from a specialty store	Straight A Cleaner	1 tbsp in 1 gal (4 L) warm water	Environmentally friendly cleaners with quite a bit of sanitizing effects; available from brewing and winemaking suppliers
Sanitizer: grocery store purchase	Household bleach, unscented	1 tsp (5 ml) per gal (4 L) to ideal of 50–100 ppm (water should have a "pool-like" odor)	Many instructions call for using up to 2 tbsp (30 ml) per gallon of water, which is usually far too strong but is better than a solution that is too weak. For best results use chlorine test strips.
Eco-friendly sanitizing: no purchase	Heat	**Oven:** Place heatproof equipment in a 220°F (104°C) oven for a minimum of 20 minutes **Hot water:** Immerse heatproof equipment in 170°F (77°C) hot water for at least 5 minutes	The FDA accepts hot water as an approved sanitizing method for dairy equipment. Remember, you can use hotter water than listed here.
Eco-friendly sanitizing: grocery store purchase	Hydrogen peroxide 3%	Use undiluted and allow 10 minutes of exposure time when using as a sanitizer. Can be heated to 130°F (54°C) to decrease effective time to 1 minute. Rinse after use, as grocery store versions may have additional ingredients.	Readily available, but in a much more diluted form than if purchased for commercial use or in a form such as listed in the next table
Eco-friendly sanitizing: order online or from specialty store	Easy Clean No-Rinse	1 tbsp per gal (4 L) warm water	Sanitizes by producing hydrogen peroxide during rinsing; marketed as a cleaner but has ample sanitizing effect; more economical than using straight hydrogen peroxide from grocery store

While many books call for adding 2 tablespoons of bleach (60 ml) to 1 gallon (4 L) of water, you may need as little as ½ teaspoon (2.5 ml). If you are using chlorine, I recommend buying inexpensive chlorine dilution test strips. These are usually available at stores that sell bulk food and supplies to restaurants and the public. They

TABLE 6-2. GUIDE FOR CLEANING AND SANITIZING PRODUCTS FOR THE MARKET PRODUCER

Name	Common Chemical	Purpose	Water Temp	Approx. pH or Concentration Goal	Cycle Run Time
Detergent, traditional	Potassium or sodium hydroxide (lye)	Dissolving of fat	120–160°F (49–71°C) (dependent upon detergent and water)	pH of 12.0	Wash cycle 8–10 minutes
Enzymatic detergent	Enzymes	Removing fats and proteins	Follow instructions	N/a	Follow product instructions
Chlorine added to wash cycle (sometimes included in detergent)	Chlorine	Dissolving of protein	Same	50–100 ppm	Part of traditional, detergent cycle
Acid rinse/ sanitizer	Phosphoric acid, phosphoric-sulfuric blend, phosphoric–citric acid blend, hydrogen peroxide blend (peracetic acid)	Prevention of calcium deposits (milkstone); neutralizing of alkaline cleaners, acidic residue, for between milkings	70–110°F (21–43°C)	pH of 2.0–3.0	Acid rinse cycle 2–5 minutes
Acid wash	Same as for rinse but with the addition of surfactants for additional mineral removal power	Used periodically after alkaline cleaner cycle	110–120°F (43–49°C)	pH of 2.0–2.5	Acid wash 5 minutes
Sanitizing prerinse	Chlorine, iodine, hydrogen peroxide (or peracetic acid)	Presanitizing of equipment	100°F (38°C)	50–100 ppm (chlorine); 12.5–25 ppm (iodine); follow instructions (hydrogen peroxide and blends)	Presanitizing 30 seconds (no potable rinse needed)

can also be ordered online. Only by using these strips periodically will you be able to guarantee that the proper amount of sanitizer is being used. Chlorine can lose its effectiveness over time, or you might be using a more concentrated solution, so a one-time measurement might not be consistent in its results.

KEEPING STAINLESS STEEL RUST FREE

Here is a little trick that will help you keep rust off any stainless steel implement (and off other metals, too). If you see an area of rust developing, polish it with steel wool or another lightly abrasive substance. Polish it as smooth as possible, then treat the area with lemon juice. Other acids will work, too. The acid will help prevent the rust from returning. It works really well!

Use a sanitizer solution on equipment just before use; with cheese brushes soak them and then air-dry before use. When it is mixed properly, you do not need to rinse a chlorine solution with plain water. A cloth dipped in the mixed solution can be used to wipe down surfaces and other areas that come into contact with your equipment.

ACID RINSES

Acid, at the right strength, plays two roles in the cheesemaker's studio, first as a solvent of mineral deposits and second as a residual sanitizer. For the home cheesemaker it need only be used periodically to prevent the buildup of what is commonly called milkstone. Milkstone builds up slowly as the minerals in milk are steadily deposited on surfaces. While most are rinsed away during cleaning, they are not all dissolved by the alkaline detergents and will eventually form a residue on all surfaces, including plastic and stainless steel. The goal is to remove the minerals before you see the buildup by rinsing regularly with a strong acid solution. (If you are a coffee drinker, you might have periodically run a vinegar solution through your coffeemaker for the same purpose.)

The strength of the acid and the frequency of the rinse will depend on the amount of use your equipment receives, as well as the hardness of your water. Hard water has a higher mineral content and will contribute to the buildup. With softer water and minimal use, you may be able to use white vinegar for your rinse. If this is not sufficient, you will want to use an acid cleaner approved for use on stainless steel and any other material that you are cleaning (such as those listed in the usage guide for dairy chemicals table on page 123).

In addition to the enzymatic cleaner that I mentioned earlier, we have recently begun using an Ecolab product called Vortexx (formerly called Matrixx) that is made from acetic acid (vinegar), hydrogen peroxide, and caprylic acid. While more expensive than phosphoric acid–based rinses, it breaks down quickly into completely harmless components (again, see appendix B for more information on finding products).

BRUSHES AND SCRUBBERS

You can use pretty much any kind of scrub brush and scrubber to tackle cleanup, but I don't recommend using sponges. They simply don't dry out between uses and are therefore great habitats for bacteria. If you are using a green scrub pad, watch for it to leave little green "hairs" on forms and equipment. This isn't a food-safety issue, but it isn't pleasant to find them in your cheese or teeth. . . . You can purchase color-coded brushes (from a company such as Nelson Jameson—see appendix B) that will help keep them segregated by usage.

THE SIX STEPS TO SPARKLING CLEAN

A good cleaning regimen consists of at least four steps: rinse, wash, acid rinse, and presanitize. While these steps need not be as laborious for the home cheesemaker, they are still important for the production of the best possible cheese. Please remember that, depending on your water type (hard or soft) and the products you are using, the following steps may vary. Ideally, you should work with a sales representative from the company that manufactures or markets the products you are using to determine the best set of steps. The procedures I have included here are fairly typical for most situations.

STEP 1: PRERINSE

Immediately after using, rinse all equipment with lukewarm water, about 100°F (38°C), to remove visible milk and curd residues. This step is important to do before washing so the heat of the wash water doesn't "cook" proteins onto the surfaces.

STEP 2: WASH

Fill the sink with very hot water, and use detergent product from the "cleaning" category of products from the charts earlier to wash all surfaces. Use a clean bristle brush and scrub pads to manually scour the surfaces of all equipment.

STEP 3: RINSE

Rinse with clean water. Some people like to use a sanitizing acid rinse at this stage (see step 6).

STEP 4: AIR-DRY

Allow all equipment to air-dry between uses.

STEP 5: SANITIZE

Just prior to use, sanitize all equipment by dipping in a food-surface-approved sanitizer, such as from the product Tables 6–1 and 6–2 on pages 122–23. Most chemical sanitizers need 30 seconds of exposure to ensure proper killing of any residual germs. Alternatively, you can use hot water at 170°F (77°C) and immerse equipment for 5 minutes.

STEP 6: ACID WASH/RINSE

An acid wash is done on a periodic basis to remove mineral deposits that are not completely removed during the daily cleaning process. Some acid-wash products include surfactants (cleaners) to help with this step. An acid rinse (without cleaners) can be done on a daily basis instead of the stronger, periodic acid wash. If you choose to do a daily acid rinse, which we prefer here on our farm, you can perform it either just following or in place of Step 3 (rinse). When you are doing periodic acid washes, the frequency will depend greatly on the amount of calcium and other minerals in your water, as well as the frequency of use for cheesemaking. Try to observe your equipment, especially when it is dry, for hazes and colors that might indicate the need for a stronger cleaning (both through scrubbing by hand and with chemicals).

AUTOMATIC DISHWASHERS

As an alternate to hand-washing, you can effectively clean equipment by using an automatic dishwasher. Many small creameries use short-cycle commercial units that can completely clean and sanitize equipment (using extremely high heat) in just a few minutes. Needless to say, these commercial units are quite expensive and use far more energy than a domestic unit.

Milkstone (a buildup of calcium and other minerals) on a plastic form

PART II:

RECIPES FOR SUCCESS

7: GETTING TO KNOW THE FAMILY

Now that you have an intimate knowledge of all of the things that go into making cheese, both ingredients and techniques, it is time to step back and look at the big picture: the family tree of cheese. Not unlike many families, some of the branches of this particular tree are intertwined and difficult to name definitively and precisely. But it is these overlaps and blurred relationships that are rather exciting—once you get a handle on the basic steps in making these different families of cheese, you will start to see how you, the director of the show, can tweak and nudge *any* recipe and create a distinctive cheese.

THE MYRIAD WAYS TO CATEGORIZE CHEESE

As in any family, there are some members that we would all like to believe were adopted. Some cheese types are very easy to put into a strict category, while others cross boundaries, break rules, and act as hybrids. Don't get too caught up in the "need" to put cheese into a category; instead, use the family tree as a way to grow comfortable with understanding how your skills as a cheesemaker can nudge a cheese recipe onto different branches of the family tree.

Academic cheesemaking and dairy science books use several ways to divide cheese into groups. One of the most common methods defines a cheese by its finished texture based on moisture content: extra-hard, hard, semihard and semisoft (sometimes used interchangeably), and soft. Yet another approach groups cheeses based on the method of coagulation—acid, heat and acid, and rennet. For this book, and for my own understanding, I have divided the family tree by both of these sets of criteria and added a third—the method of ripening: unripened, surface mold, surface bacteria, internal mold, and internal bacteria. Let's take a more detailed look at these three ways of categorizing cheese.

CLASSIFYING CHEESE BASED ON FINISHED TEXTURE

Even if you don't know how a cheese was made, you can easily determine its texture just by feeling it, tasting it, or scooping or cutting it. Extra-hard cheeses are perfect for grating. Hard cheeses (they might fall into the extra-hard category someday, if aged longer) are nice to slice. Semihard cheeses bulge and slump a bit when sliced, while semisoft cheeses can barely be sliced. And soft cheeses, you guessed it, are spreadable and spoonable.

TABLE 7-1. MOISTURE CONTENT OF THE MAJOR CHEESE TYPES

Cheese Type	Example	Moisture % Max.
Extra-hard	Parmesan	34%
Hard (with eyes)	Emmental	41%
Hard	Cheddar	39%
Semihard	Gouda	45%
Semihard	Roquefort	45%
Semisoft	Limburger	52%
Semisoft	Monterey Jack	44%
Soft	Brie	56%
Soft	Cream	55%
Soft	Ricotta	72%

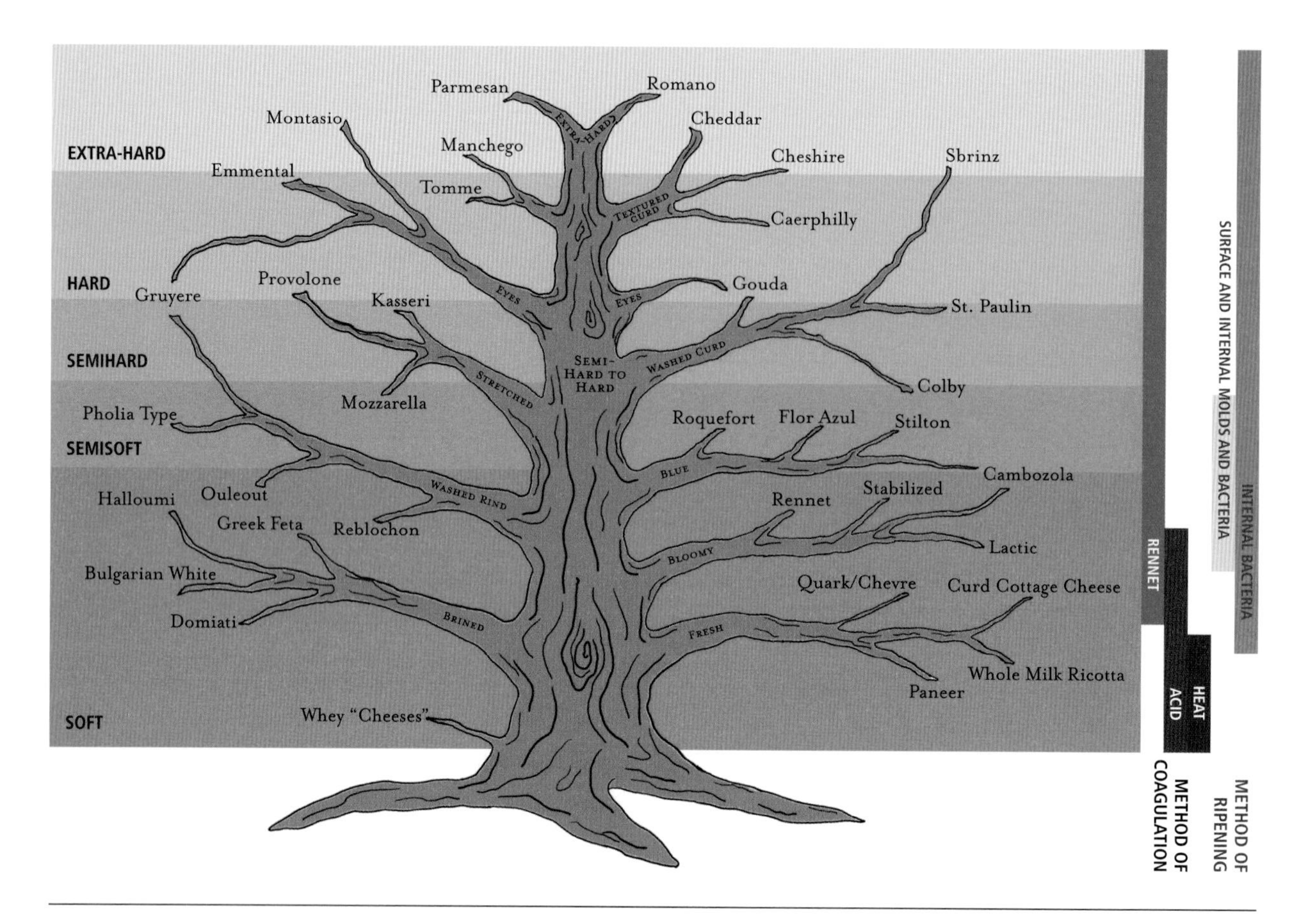

The family tree of cheeses

The pros (meaning the cheese judges, university dairy science teachers, and other technical experts) look beyond the visible texture to compare the actual moisture content of a cheese to better determine its category. But even with a specific moisture content defined, you will see in Table 7.1 on page 129 that there is not a strict curve, as other things will also determine texture (such as eyes in Swiss types, "caves" [the open spaces in an unpressed cheese] in blue cheese, and the fat content in "cream-added" cheeses such as double and triple créme bloomy rinds).

CLASSIFYING CHEESE BASED ON THE METHOD OF COAGULATION

For the cheesemaker dividing cheese types by their method of coagulation is an easy division to understand; it is something totally under the cheesemaker's control very early in the cheesemaking process. You can add rennet (or another enzyme that coagulates milk); acidify the milk to the point of curdling (by adding or developing acid); or combine high heat with acid, either developed acid or added acid. Pretty straightforward, right? It would seem so, but dividing by method of coagulation alone might mean grouping cheeses such as blue and cheddar or feta and Camembert in the same category, so using this method all by itself is a bit unsatisfactory.

CLASSIFYING CHEESE BASED ON THE METHOD OF RIPENING

Since a cheese is greatly altered by the method of ripening and aging, sorting cheeses into categories based on the ripening techniques that were employed can be a great way to sort them. From fresh, unripened cheeses (such as fromage blanc and cream cheese) to those ripened by molds and bacteria (such as blues and stinky washed rind cheeses), the environment and conditions surrounding aging have a profound influence on the final appearance and characteristics of the cheese. Therefore, this category of classification is a very important one.

You can divide methods of ripening into four categories: *unripened*, in which no aging occurs and no organisms are added that would improve or change flavor if aging did occur; *internal bacteria ripened*, in which bacteria are used to produce specific flavors and textures over an aging period; *surface ripened*, in which bacteria and mold are used on the surface of a cheese to produce changes; and, finally, *internal mold ripened*, in which molds are used to change the interior of a cheese. Obviously, even with this method of grouping cheeses, you cannot draw distinct lines between each category. Consider Gruyère, an aged hard cheese ripened by internal bacteria, with a rind that is ripened by washing with surface bacteria.

WORKING WITH THE RECIPES

The recipes that follow are for small home-size batches ranging from 1 to 4 gallons of milk. Because of the different amount of cheese that each type will yield, as well as varying goal sizes for the cheese itself, I tried to size the batches to produce a reasonable and satisfying amount of cheese. You can increase these batch sizes without altering time and temperature suggestions. Do take into consideration, though, how a larger batch will affect the amount of time it might take to do such things as fill molds with curd or turn cheeses during pressing. If the time to accomplish these types of steps increases too much, it could affect the outcome of the batch. Sometimes a helper might be needed at this stage.

For the small market producer, I have provided several options that will make the recipes more accessible to you. In appendix D you will find a couple of sample make sheets that you can duplicate for use in the production of larger batches (feel free to create and customize your own). Beginning with chapter 9 I have included "larger batch" guidelines sized for 12 gallons (46 L/100 lb). (Remember that the actual weight of 1 gallon or any liquid measurement can vary based on the components in the milk.) For the producer measuring milk in pounds, this should make changing the size of the batch as easy as multiplying or dividing in increments of 10.

Where did I find the recipes for this section of the book? My first priority was to make sure I worked with guidelines from more than one source. As you probably know by now, it is my goal to have you learn how to make cheese *intuitively*, not just follow a recipe. To that end my aim was to include "recipes" that would show as well as tell; in other words, recipes that would teach a distinctive process that you will be able to compare to other recipes in the same family and, in turn, expand your vision of the various ways each cheese can be made—and consequently develop your own signature style and specialties.

One of my main resources (and much-referenced book throughout my research) is *Cheesemaking Practice* by Scott, Robinson, and Wilbey (Springer, 1998). Though it is an academic textbook, many of the directions are translatable into home- and small-artisan instructions. Margaret Morris's fantastic book *The Cheesemaker's Manual* (Glengarry, 2003) provided an opportunity to compare some of her recipes to the ones I was working on for this book, as did Peter Dixon's handouts (from a class I attended years ago) and his website at www.dairyfoodsconsulting.com. Another great source for comparison was volume 2 of *Cheese: Chemistry, Physics, and Microbiology* (Elsevier Ltd, 2004); though it does not contain make sheets or recipes, it does have many process descriptions, finished pH levels, salt content, and so on for some of the world's traditional cheeses. Frank Kosikowski and Vikram Mistry's classic two-volume set, *Cheese and Fermented Milk Foods* (F. V. Kosikowski, LLC, 1997), has many outlined procedures for cheese makes. I wasn't clever enough to borrow these great books from my friend Carly Payne (see her profile later in chapter 15) until the manuscript was almost done, but they still provided a great fact-check resource. To round out my sources, I pulled together every recipe that I have learned from various workshops and classes over the years.

After comparing, combining, and recreating each recipe, I made sure to follow the instructions and make the cheese it described. For styles that I could not make, for styles that were out of the area of my expertise, or if I knew someone who was already creating such a cheese, I have relied on some talented and gifted cheesemakers

MEASURING UP

I'm one of those US citizens that longs for the metric system. It is such a logical, predictable way to measure and calculate. Until the time that my country makes the switch, however, recipes for multination audiences will need to be "bilingual." To that end I have included options for US and metric measurements in several fashions. For milk I have included the equivalent in gallons, liters, and pounds. For measuring culture and lipase for small batches, I provide it in teaspoons and the grams measurement (rounded to tenths). Please remember that culture powders vary greatly in density; for example, ¼ teaspoon of a culture such as Flora Danica weighs under half as much as ¼ teaspoon of Thermo B. In addition, the way the dose is poured from the packet will affect volume. The only truly accurate way to dose culture is by knowing how many doses are in a packet, then measuring by weight.

The guidelines given here for small batches are meant to get you started without being too restrictive. But I hope one day you will become a bit more keen on "doing the numbers." For very tiny measurements of cultures and mold powders I recommend using the tip of a meat skewer as a measuring device. (See more on measuring and measurement equivalents in chapter 6 and appendix C.) No matter the size of the batch, I highly recommend measuring liquids, such as coagulant and calcium chloride, in milliliters, as this method is much more accurate, but I have included as well the approximate teaspoon equivalent. (Think about your physician: Would you expect her to give you an injection in a syringe marked with ccs [same thing as milliliters] or teaspoons?) For salt I have included teaspoons and grams, for the greatest accuracy dry salt is best measured by weight.

When large recommendations are given, the culture amount is offered in units or grams, depending upon the way that the culture is labeled by the manufacturer. For liquid measurements of rennet, I am only providing milliliters as this type of measurement is far more accurate and will no doubt be preferred by the serious home cheesemaker as well as the commercial producer. For large cheese batches I often recommend a salt measurement based on the goal salt content (by weight).

from several corners of the globe to provide guidance. You will read about these wonderful contributors throughout this section. I have also included most of the recipes for the very same cheeses that we make from Nigerian Dwarf milk here at Pholia Farm Creamery.

To make sure that I could speak to the unique qualities of other milks, I worked with several sources—including, for some amazing Jersey and Guernsey cow's milk, fellow southern Oregon farmers Ken Muller from Rogue Valley Brambles and Art and Teri White of Runnymeade Farm. For some recipes I used LaMancha (a standard-sized breed of dairy goat) milk (which behaves very differently from Nigerian goat's milk), and at other times I used store-purchased cow's milk. Unfortunately, it was not very realistic to try to make each recipe from the many different milk types that might be available to you readers; for that I apologize and hope that what you learned in part I will help you adapt each cheese to your milk of choice.

EXERCISES FOR BUILDING INTUITION—"THINKING OUTSIDE THE VAT"

At the end of each recipe section, you will find a little exercise designed to help you start thinking on your

ACRONYMS OF HONOR: DECIPHERING A CHEESE'S "LETTERS"

We've all seen them, those acronyms that are often on the label of some of the world's greatest cheeses. We all know they mean something special, right? Well, in case you want to know a bit more, here is a simple summary of some of the most common letters of recognition that you might see accompanying some prestigious, historical cheeses. The most important thing to remember, no matter the acronym, is that each designation is meant to protect a product that has evolved in a specific location. The designation is meant to recognize the importance of both the history of the cheese and the importance of its production to that region.

- AOC (Appellation d'origine contrôlée): France
- DOC (Denominazione di origine controllata): Italy—mostly used for wine but also for cheese
- DOP (Denominazione di origine protetta): Italy
- DO (Denominación de origen): Spain
- DOC (Denominação de Origem Controlada): Portugal
- PDO (Protected Designation of Origin): European Union and other participating nations

Designated cheeses must conform to specific regulations that govern such things as the region where milk is produced and the cheese is made; the species and even the breed of the animal that produces the milk; the cheesemaking process itself; and the composition and characteristics of the cheese. Keep in mind that the newer PDO designation is often concurrent with the older, national designation (such as AOC or DOC). While they are not precisely synonymous, they are close.

There are several other designations that have to do with name and region protection. For this book, though, I am listing only those certifications or designations that are more comprehensive.

In recent history several historical (but previously without certification) cheeses have applied for PDO status. One of the most famous cheese types to receive PDO status recently is feta. Though feta had been previously made in many European countries, Greece received the PDO rights to use the name "feta" in 2002. While the United States does not recognize these rights, it would be wise for cheesemakers in all nations to pay attention to trends in this arena.

feet and outside the box (or vat, as it were). I have tried to make them pertinent to the cheese type that has been covered, but you will find that many of them will apply to other types of cheeses—and that is a good thing! Start trusting your instincts, and think back to the science behind the recipes. Before long you may even find yourself a master cheesemaker.

As you study the cheese family tree and read more about each of the main branches and their members, pay attention to the overlaps, and appreciate the ability of the cheesemaker and the environment to move cheeses from one family to the next. Be ready in your cheesemaking for some surprises when your cheese decides to move to a new category; it might be the best cheese you have ever created! My best piece of advice to you before you make your first recipe is to take extensive notes for each batch of cheese you make. Only through documentation will you begin to learn to adapt to the many seasonal affects on milk and cheesemaking. So get a log book and let's make cheese!

Soft, fresh acid-coagulated cheeses are tasty and relatively easy to make.

8: FRESH, ACID-COAGULATED CHEESE

FINISHED TEXTURE: Soft
METHOD OF COAGULATION: Acid, acid and heat, and occasionally a small amount of rennet
METHOD OF RIPENING: None or brief

Soft, fresh cheeses are the simplest to make and therefore also some of the most fulfilling to consume, and not only do they provide immediate gratification, but their simplicity as a cheese makes them very versatile as a product. But don't think that just because they are simple to make that they don't offer important lessons to cheesemakers at all levels of learning! It is possible to make these cheeses without knowing the science behind the process, but by learning more about this group of cheeses, you will take an important step toward a complete cheesemaking education. So enjoy the simplicity of fresh cheeses while you also learn the fundamental knowledge that mastering them can bring.

THE FUNDAMENTALS OF ACID COAGULATION

Milk will curdle, or coagulate, when it's at just the right temperature and acid level—think tangy—which is usually read using pH values. *The warmer the temperature of the milk, the less acid is needed to cause the milk to curdle.* At room temperature, 72 to 80°F (22–27°C), milk will coagulate when it also has a pH of 4.6 (which is about as tangy as a fresh tomato). We'll go over a few more examples of the relationship between temperature and pH in a bit. First let's talk about the three basic approaches to curdling milk and some basic information that applies to all fresh cheeses.

The first approach is *spontaneously*, in which raw milk is left sitting at room temperature for the right period of time and will sour and curdle spontaneously, thanks to acid-producing bacteria that have "contaminated" the milk naturally from inside the teat of the animal or from the environment. Remember, there are good and bad bacteria that can enter the milk while it is leaving the udder and after collection. Both categories are called *adventitious*, but the good ones are also *advantageous*.

The second method is by *added acid*, in which a food acid (such as lemon juice or vinegar) is added directly to the milk to lower the pH. This form of coagulation is often used in conjunction with heating the milk to a temperature at which it will coagulate more easily.

Finally, the third approach is with a *starter culture*, in which bacteria are added to the milk, then left to ferment (another word for souring) at room temperature.

In chapter 1 we covered some historical cheeses and one contemporary cheese that use spontaneous coagulation. In this chapter we'll focus on the more common and predictable methods of coagulating cheese using acid.

When producing cheeses that are acidified by the addition of a food acid (meaning no starter culture added) or by added acid plus heat, it is important to know that the resulting cheese will have a fairly short shelf life. If the coagulation occurs thanks to the production of lactic acid by starter bacteria, the shelf life is usually a bit longer. (Remember, a cheese that is heavily populated with good bacteria will be less likely to provide an attractive growth medium for any unwanted and possibly dangerous microorganisms.) But by the same token some high-heat-processed cheeses, such as paneer, are initially protected by the high-heat treatment of the milk, which kills all the bacteria present in the milk. No matter the initial processing, any high-moisture cheese will have a shorter shelf life than a drier variety. Remember, bacterial activity (including ripening and rotting) can only occur in the presence of water.

The flavor of the milk is readily apparent in unripened, fresh cheeses. So any defects in the milk will be magnified once it is condensed through the cheesemaking process. High-heat-processed cheeses will have a different flavor thanks to the cooking of the milk. While most of us tout the beauty of raw milk, the truth is that cooked milk can have a lovely flavor as well, kind of a caramel-custard, flanlike flavor. Also, the high heat changes some of the proteins so they can be captured in the curd, making these types of cheeses a bit higher in protein and total yield.

A small amount of rennet is often included in soft, fresh cheese recipes, such as quark and American-style cottage cheese. This will increase curd firmness and minimize some protein loss. Some culture packets from home cheesemaking supply companies include a tiny amount of rennet, making it simple for the beginning cheesemaker to get started without a lot of measuring.

A DEEPER LOOK AT THE SCIENCE OF CURDLING MILK

Now you know that milk coagulates naturally when it reaches a certain pH. Let's take a more in-depth look at why this happens and how it influences the resulting quality of your cheese. Take a quick look at some benchmarks for acid-induced coagulation in Table 8-1.

Can you see how by increasing the temperature you can coagulate the milk at higher pH readings? (Remember, that means less acidic; in other words, sweeter.) This will have a marked effect on the final flavor and texture of your cheese.

Processing at higher heat also has the bonus effect of capturing more protein, because much of the whey proteins—those that would have otherwise been lost in the whey—are denatured, or changed from their original way of behaving, and stick to the other type of proteins (casein).

The curd thickness of acid-coagulated cheese is always softer than that produced by rennet coagulation and, as you might have guessed, more tangy. In fact, cheeses not coagulated by acid alone are often called "sweet curd" cheeses, and their whey is called "sweet whey." Another moniker you might hear when the technical guys are talking about cheeses produced by acidification is the term "lactic technology," as opposed to "rennet technology."

So just what happens when milk coagulates thanks to the right acid level and temperature? It all comes down to chemistry. You might remember from chapter 1 that milk protein (caseins) exist in clumps suspended in the milk. These clumps, otherwise known as *micelles*, carry an overall negative electrical charge. This causes them to repel each other like two magnets pushing away from each other. They are also considered *hydrophilic*, or water loving. This allows them to remain suspended in the milk—they don't clump because they repel each other, and they float, intact for the most part, in the solution of the milk.

As the pH of a solution changes, it has an effect on the electrical charge of any molecules within the solution. When the pH and temperature both reach the point where the electrical charge is essentially neutralized, we refer to it as the *isoelectric point* of that molecule. The coagulation table reflects the different isoelectric points for casein. The warmer the temperature, the more

TABLE 8-1. HOW TIME, TEMPERATURE, AND PH INFLUENCE COAGULATION TIMES

Method	Time	Temperature	Coagulation pH
Slow coagulation	12–16 hours	72–86°F (22–30°C)	4.6
Short coagulation	4–5 hours	86–90°F (30–32°C)	5.1
Instant coagulation	The time it takes to heat the milk to goal temperature	176°F (80°C) or >	5.9–6.0 or >

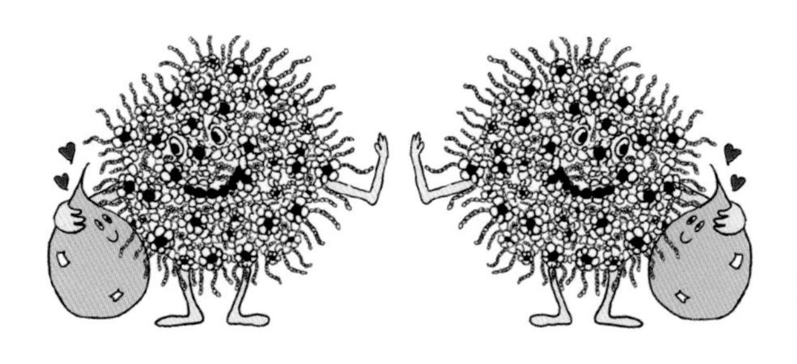

Author's illustration of casein micelles in their negatively charged and hydrophilic state—repulsing each other and binding with water

Author's illustration of casein micelles after the kappa casein has been cleaved they lose their negative charge—drawn to other micelles and repulsing water

hydrophobic the casein micelles become (this influences and helps speed draining during the manufacture of all cheese types by helping the curd expel whey).

Fresh, soft cheeses can suffer from texture defects and lack of proper acid development, such as the production of too much acid or too little. So while they seem like simple cheeses (and process-wise, they are indeed among the simplest), don't take the process of making them for granted.

TIPS FOR MAKING FRESH, ACID-COAGULATED CHEESE

In preparation for cheesemaking, all equipment should be thoroughly clean and sanitized just before use. If using chlorine or other chemical sanitizers (see chapter 6 for sanitizer mixing rates and options), be sure to rinse any tools you will use to measure rennet or culture with clean water first. Rinse sanitized cheesecloth and draining bags with warm whey just before filling with curd to remove any sanitizer residue, as well as to prepare the cloth to drain the curd.

CHOOSING MILK

The texture of the cheese will vary greatly based on the fat content of the milk. Some recipes for ricotta types, the "Quick and Easy—An Added-Acid and Heat-Coagulated Cheese" recipe, for example, call for adding butter to the finished cheese. This is simply to improve texture and is optional. The same holds true for adding cream to American-style cottage cheese: It is optional, based on your preference.

If you want to create a cream cheese style, you can replace a portion of milk in the recipe with cream. Two of the better-known cream-added soft cheeses are Neufchatel and cream cheese. Both recipes have different final goal fat and protein contents that will be difficult to duplicate exactly without standardization techniques (see chapter 2 for more on standardizing), but you can come close. To make cream cheese the milk/cream mixture should be about 11.5 percent fat, and to make Neufchatel it should be closer to 5 percent. If you know or can estimate the starting fat content of the milk, you can calculate how much cream to add.

When choosing cream that is sold commercially in the United States, you will have a few choices:

- Heavy whipping cream: 36 to 40 percent fat
- Light cream: about 18 percent fat
- Half-and-half: 10 to 12 percent fat

Most commercially produced cream sold in the United States is ultrapasteurized. This does cause flavor changes, but not the same damage that ultrapasteurization does to milk. It will damage the fat globule, however, making the fat more susceptible to breakdown by lipases. Since fresh cheeses aren't aged, this is usually not an issue.

If you only have access to store-bought cow's milk, you can still make some great cheese, but plan on adding calcium chloride—up to ¼ teaspoon (1.25 ml) per gallon of the most common strength solution available from cheesemaking supply companies—to help make up for the loss of calcium from inside the protein structures during cold storage, which leads to a weaker curd

YOGURT, CRÈME FRAÎCHE, AND KIN

I really struggled over whether to include yogurt, crème fraîche, sour cream, and cultured butter recipes in this book. While these products are cultured by the addition of starter bacteria, they are not (usually) concentrated after production. Yogurt, crème fraîche, and sour cream have 100 percent yield (unless the yogurt is drained), and butter (which is often cultured and ripened a bit) is concentrated, but in a much different fashion from any cheese. I finally decided that I had to draw the line somewhere. I do want to talk a bit about yogurt, however, because it offers some insights into cheese production. So let's touch on some basic principles about making yogurt.

Making great yogurt is relatively simple and very rewarding. If the milk you are working with makes a yogurt that is too thin, you can do what we Greeks have done for centuries—simply drain it in muslin for an hour or more until you achieve the thickness you desire. When it's drained to a spreadable consistency, you can rightfully call it cheese (*labneh*, for example, or simply *yogurt cheese*). Most commercial yogurts have nonfat dry milk powder added, or other thickeners, but you don't need to do that yourself.

Yogurt recipes call for the milk to be scalded at 180°F (82°C), then cooled to the ripening temperature. I used to think this step was unnecessary (you know how I love raw milk), but the high-heat treatment does two wonderful things for yogurt: It clears it from all bacteria (normal pasteurization doesn't), leaving a clean slate for the yogurt cultures to work on and provide their probiotic qualities, and it helps create a yogurt with a better texture and body by capturing the whey proteins in the coagulum. You might not even need to drain it.

Many yogurt cultures that you buy from cheesemaking supply centers have some bacteria in them that produce what is not-so-palatably called "slime." This so-called slime is actually made up of compounds called *exopolysaccharides* that are produced by certain mesophilic and thermophilic bacteria. In some circumstances, such as in brine and cheese, these are not desirable, but in yogurt they can help prevent the yogurt from separating when stirred and help to create a creamy texture. You might have wondered why some yogurts separate as soon as the first spoonful of yogurt is removed, and others do not. Now you know!

set. You can also try using skim milk and replacing 1 cup of the milk with 1 cup of light or heavy cream. Because whole milk is usually sold homogenized, the cream in the whole milk will be more damaged than that in the separated, pasteurized cream. If you remember from chapter 2, homogenized fat globules can coat the milk proteins, which will also negatively affect cheese yield. Of course, if you have access to store-bought cream-top (also called cream line) milk, you are ahead of the game!

For commercial sale in the United States, all fresh cheeses, which current laws define as any cheese that has not been properly aged for a minimum of 60 days, must be made from legally pasteurized milk. If you are a home cheesemaker making the products for only your own consumption and you are confident of the quality and safety of your milk source, you can make the decision without involving regulations. I personally suggest gently pasteurizing the milk (see chapter 2) for any fresh cheese that involves a long ripening process to ensure the best flavor, if not safety. Fresh cheeses that involve high heat will be essentially—but not legally—pasteurized.

Crème fraîche is a cultured cream product that can't truly be called a "cheese," but it sure can be called delicious!

CHOOSING CULTURE

When choosing culture for a fresh cheese for which acid is developed by starter culture as opposed to those with added acid, such as vinegar or citric acid, choose one that will help develop flavor over a shorter period of time, since the cheeses will not be aged. Popular choices often include a little bit of *Lactococcus lactis diacetylactis* (LLD), which will help create a buttery flavor early in the process. For fresh cheeses made with higher heat, a thermophilic culture is usually used. Keep in mind that at the temperature at which high-heat varieties are coagulated all starter bacteria will be destroyed.

The following chart begins with two types of prepackaged blends often available from various home cheese supply companies. The first listed (which includes several strains of culture) also includes a small amount of rennet in the packet, just right for making long-set, long-draining varieties such as quark, chèvre, and fromage blanc. Following that is a culture that is also available from larger companies, but is also sold repackaged for smaller batches. Next are two blends available from larger supply companies that are good choices. Following that are two thermophilic blends for high-heat varieties, one prepackaged for home-size

TABLE 8-2. CULTURE CHOICES FOR FRESH CHEESES

Name	Contains	Manufacturer	Small-Batch Dosage:* 3–4 Gal (12–15 L/26–34 lb)	Larger-Batch Dosage: 12 Gal (46 L/100 lb)	Characteristics and Application
Chèvre, Fromage Blanc, and Crème Fraîche	LL, LLC, LLD	Packaged by home cheese supply co.	1 packet	N/a	Starter and aroma cultures packaged for small batches; includes rennet
Flora Danica	LL, LLC, LLD, LMC	Packaged by home cheese supply co.	1 packet	N/a	Starter and aroma cultures packaged for small batches; does not include rennet
Flora Danica or CHN 11, 19	LL, LLC, LLD, LMC	Chr. Hansen	¼ tsp (0.4 g)	10U	Starter and aroma cultures
Meso Aromatic B	LL, LC, LD, LM	Abiasa	¼–½ tsp (0.6–1.2 g)	10 g	Starter and aroma cultures
Thermophilic	ST, LB, LH	Packaged by home cheese supply co.	1 packet	N/a	All-inclusive blend for several higher-temperature-ripened and cooked cheeses
Thermo B	ST, LB	Abiasa	⅛ tsp plus (0.4 g)	5–10 g	Blend for several higher-temperature-ripened and cooked fresh cheeses
TA 60 Series (Choozit) (includes TA 60, 61)	ST (fast-acid producer)	Danisco	⅛–¼ tsp (0.4–0.8 g)	2–5U	Use to make your own blends by combining with LB 340
LB 340 (Choozit)	LB	Danisco	1–2 skewer tips	0.05–0.15U	Use with TA 60

*Remember that culture is always best measured by weighing and dividing into unit doses based on the strength of the culture, instead of using volume or set weight.

batches and the other in multiple-dose packets. Last are two single strains of thermophiles that you can purchase separately and blend yourself.

CHOOSING COAGULANT

Very little coagulant is used in the making of most fresh cheeses. In fact, acid-coagulated versions don't use any, depending instead on the added acid alone to coagulate the milk.

Soft, fresh varieties, such as quark and chévre, can be made without rennet, but the addition of a small amount will help yield and allow for better coagulation at a slightly higher pH and therefore less acidity. Most recipes call for a certain number of drops of rennet diluted in water or a measurement, such as ⅛ teaspoon, diluted in water to create a solution from which you measure out a certain volume.

People often ask me just how to define a "drop," which seems like such an arbitrary measurement. A drop's volume will change depending on the composition of the liquid and the size of the opening of the dropper bottle. But if the same liquid is dispensed by an eyedropper or typical dropper bottle, the volume will be essentially equal each time. By my measurements a drop is roughly equal to 0.05 milliliters. If you have a single milliliter (1 cc) syringe, you can measure

TABLE 8-3. DROPS AND DILUTIONS: DOSING SMALL AMOUNTS OF RENNET

Cheese Type	Drops per Gal (4 L)	Milliliters per Gal (4 L)	Amount of Diluted Solution of ¼ Cup Water (60 ml) + ¼ Tsp (1.25 ml) Rennet for 1 Gal (4 L)
Long set, long drain	2–4	0.08–0.12	1¼ tsp (6 ml)
Curd-type fresh cheese	8–10	0.16–0.24	2½ tsp (12 ml)

this small amount fairly accurately. A set of "tiny" measuring spoons usually includes one that will measure a single drop (see chapter 6 for more on measuring).

Here are some equivalent measurements using drops, milliliters, and measurement of a diluted solution in Table 8-3.

Coagulant should be added by sprinkling on top of the milk and stirring with an up-and-down motion, as with other cheese types. However, since so much less rennet is used in these cheeses, it is not as critical, nor is it necessary that the milk be immediately stilled. Coagulation will take place over a much longer time, so immediate stillness is not as critical.

CHOOSING ACID

When making a recipe in which the curd is coagulated by adding acid, you can choose from several food acids, with equal success. Each will add a different flavor profile to the cheese, so this should be kept in mind. Sometimes you choose what you have available. I have even combined several different types of acid when I didn't have enough of a single type of acid to properly coagulate the amount of milk I was working with. The acidity level of vinegar is usually steady, but that of citrus fruits such as lemons and limes can vary, so it's not possible to give exact measurements.

Fortunately, you can easily judge by the results when enough acid has been added, by observing at what point the whey becomes translucent to clear. Plan on between ¼ and ⅓ cup per gallon of milk for when either vinegar or lemon juice is needed. If using citric or tartaric acid powder to coagulate the milk, dissolve it in water first, then add the mixture to the milk. For powdered acids plan on 1 to 2 teaspoons dissolved in ¼ cup cool water per gallon (4 L) of milk being coagulated.

TIPS WHILE "IN THE VAT"

High-heat varieties of fresh cheese take the least amount of time to make and spend very little time in the vat. If stirred constantly, they can be heated using direct heat. If using a water bath, it is difficult to reach the goal temperatures, as milk tends to be slow to boil thanks to its components. Some paneer recipes call for the milk to be brought to a boil just before you add the acid. This isn't wrong, just another variation.

For long-set, long-drain types of fresh cheese, a cooler ripening temperature can be used, along with a longer set time (but still working with the same goal pH). This can yield a slightly sweeter cheese. Keep in mind the finished goals of pH and consistency at the end of ripening. To keep a long-set variety warm for an extended period, try using an ice chest. You can set the pot down in the chest and pour warm water around

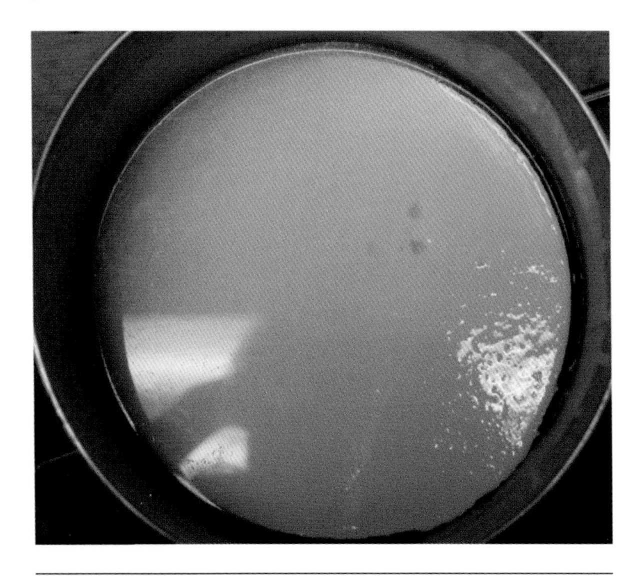

Long-set, long-drain curds are usually ready when the curd mass pulls away from the edges of the pot and whey covers the top.

the pot. Make the water a tiny bit warmer than the goal ripening temperature, as the chest will initially absorb some of the heat. Depending upon the ambient room temperature, the ice chest may hold the heat steady for the entire ripening phase or it may need more heat added. Alternatively, you can use hot-water jugs set in the chest instead of hot water. This makes it a bit easier to change them out as they cool.

If you aren't able to check pH, watch for the curd to begin pulling away from the sides of the pot and a layer of whey to cover the curds by about ½ inch (1.2 cm). This is usually indicative of proper acid development for long-set varieties.

DRAINING

Most fresh cheeses are drained in fine-weave muslin cheesecloth or bags made from tightly woven cotton sheeting. I am a fan of using a sanitized pillowcase for draining chévre types at home. Some people even get fancy and sew their own, often including a drawstring to make things very easy. For more on cheesecloth options and draining bags, see chapter 6.

Cloth and draining bags can be hand- or machine-washed, dried, then presanitized with boiling water or a properly diluted chlorine solution. Just before use rinse the cloth with some whey from the pot or vat. If you are using a colander, be sure it is made from stainless steel or plastic, as the acidic whey from most fresh cheeses will be reactive with aluminum.

During draining, the room should be kept close to 72°F (22°C) for optimal loss of whey and development of acid. If it's too cool, the drainage will be too slow; too warm, and drainage will be too rapid. The same pattern is true of acid development—the colder, the slower. You can use this knowledge to help your timing of when you will need to remove the cheese from draining and incorporate the salt. For example, if you know you will be unavailable in 12 hours but available in 10, you can drain the cheese at a bit higher temperature. You get the idea.

If you find that the curd forms a firm layer around the outside of the draining cheese, the room may be a bit too cool. You can try stirring the curd a couple of times during draining. Or place the draining cheese in an ice chest or warmer area.

Remember, when the cheese is refrigerated, it will thicken a bit more, so you may want to end the draining time when the texture seems a bit softer than your final goal.

ADVICE ON SALTING

Because fresh cheeses are not aged, salt does not play the same role in assisting with long-term flavor and texture development. It does, however, provide immediate flavor enhancement and for all but high-heat varieties—where the starter culture is destroyed and acid production halted by the high temperature—brings a halt to acid development. It is generally added at the end of draining, but you can use it to assist with whey removal for soft, fresh varieties by adding some to the draining curds if the pH has reached its goal but the curd is still too wet. For example, if you notice that the curd flavor is just right, but the texture is too moist, go ahead and stir the salt into the curd and hang it to drain a bit longer.

CREATING SHAPES

Soft, fresh cheeses can be shaped in several ways. They can be directly ladled and drained in forms; partially drained in bags, then ladled into forms to gravity drain if a shaped cheese is desired; or they can be drained in the bag to a dry texture, then formed into logs or shapes and packaged plain or encrusted with herbs. Shapes may need to drain for up to 24 hours before they can be unmolded and will hold their shape.

A NOTE ON THE RECIPES

I have included four basic fresh cheese recipes in this chapter. You will see, I hope, that with just a few variations, you can make many different cheeses. Some of these go by different names in other cheesemaking books or in different cultures but are in reality virtually identical. My goal is to help you understand the underlying processes and be able to customize them into products that you enjoy, no matter their names. Remember, all the detailed instructions can be found in the tips for making each kind of cheese in chapters 9 through 15. Be sure to refer back to these if you have any questions or are uncertain about a process.

QUICK AND EASY—AN ADDED-ACID AND HEAT-COAGULATED CHEESE

A great place to begin your cheesemaking journey is with the creation of simple, acid- and heat-coagulated cheeses. These types of cheese go by many names throughout the world, including *queso blanco* ("white cheese") from Mexico and US border states; whole milk or sweet ricotta from the United States and Italy; vinegar or lemon cheese from the United States; sheep's and goat's milk *brousse* in France; *mizithra* from Greece (sometimes curdled with sour [acidified] whey); and *paneer* or *panir* from India and other Middle Eastern countries. The curd can be drained for a soft, crumbly cheese or lightly pressed for a firm slicing, grilling cheese. By changing the cream content of the milk, the type of acid, and the method of draining, you can make a large variety of cheeses. With only slight variation, the same recipe can be used to make Italian style *mascarpone* (useful for desserts and pastries), whey ricotta (also called *ricottone*), and *anari* (a Greek version).

Paneer is a quick and easy fresh cheese that is high in protein and extra tasty when fried.

INGREDIENTS

Milk: 1 gal (4 L/8.6 lbs) whole or partly skimmed milk
Acid: About ¾ cup (177 ml) lemon juice or vinegar (white or cider)
Salt: ¼ tsp (1.5 g), or to taste, pure salt

STEPS

Prepare Equipment: Make sure all equipment is cleaned and sanitized and that your cheesemaking space is free from possible contaminants. Refer to chapter 6 for tips on proper equipment preparation.

Prepare Milk: Place milk in pot, and using direct heat, and stirring constantly, bring milk to 195 to 200°F (91–93°C). Remove the pot from the heat source, and allow to cool to 190°F (88°C). Stir to help the milk cool a bit more quickly.

> TIPS: When using direct heat, it is very easy to scorch the milk. Use a heavy-bottomed pot, and stir constantly but gently. The milk may foam and look as if it is ready to boil. As the temperature nears the goal, the rate of increase will slow.

Curdle: Add acid 1 tablespoon (15 ml) at a time, stirring gently after each addition, until curd separates, leaving clearish whey. Let set uncovered for 5 to 20 minutes.

> TIPS: Curd should float to the top and consolidate into a large mass.

Drain: Ladle curds into cheesecloth-lined colander. You can briefly rinse the curd with cool water to assist with cooling, but this is not mandatory. Allow the curd to drain in the colander for 20 minutes if you are proceeding to pressing. If not, drain for 60 minutes or more until you like the texture.

> TIPS: If a lot of curd remains in the pot after you have ladled the floating portion, carefully pour the remaining hot whey through a sieve and place the collected curd in the lined colander. I do not recommend pouring the entire mass through the cloth, as the curd mass can block the flow of the hot whey and splash back and burn you. Also, you want the curd to be cooling, and the hot whey will simply slow this process.

Salt: Using a spoon, gently stir and work curd until it is smooth and even in texture. Then stir in ¼ teaspoon salt or to taste. If cheese is to be used soft, it is done at this stage. Simply use or transfer to a sealable container, and store in the fridge for up to about a week. If it is being used as a slicing or grilling cheese, proceed to the next step.

Pressed Variation: With the curd still in the cheesecloth, push and form the mass into the center of the cloth. Then close the cloth by gathering the corners together and gently twisting them to form the curd into a compressed mass. Place the cloth-wrapped ball on a smooth surface, and press gently down to form a disc about 1.5 inches (4 cm) thick and as even in circumference as possible. Open the cloth, and pat the curd ball a bit to smooth the top. Then fold the cloth one corner at a time over the curd, making it as smooth as possible. Wrap any excess so it is tucked under the mass, forming a tightly closed packet.

Place the packet on an inverted plate in a larger bowl or container. Place another upside-down plate on top of the packet, and set a bowl or pan on top of the plate. I like to use a heavy cast iron skillet. The total weight at this point should be about 3 pounds (1.5 kg). After 10 minutes add 3 more pounds of weight. Press for a total of 1 hour.

Finishing and Using: After an hour or more of pressing, the curd packet should feel firm. If you press on it with your fingertips and it gives or whey leaks out, press longer.

When your goal texture is reached, the cheese can be used or wrapped tightly and chilled for use later.

Keep for up to one week in the refrigerator. The texture of the cheese will become smoother and more sliceable after it chills for a day or more. But it is ready to use immediately and will hold its shape fried, grilled, or in stews. Superversatile!

To press a paneer-type cheese, form a packet of curd wrapped tightly in cheesecloth, then press on an inverted plate in a bowl.

VARIATIONS

MASCARPONE TYPE:

Use heavy or light cream. Remember, don't use ultra-pasteurized cream, which is very common in grocery stores. Also look for brands with the least number of additives; for example, carrageen (a thickener) and dextrose (a sugar). If you have access to farm-fresh cream, all the better. In a double boiler heat cream to 185°F (85°C). For each quart (950 ml) dissolve ¼ teaspoon tartaric acid in 1 to 2 tablespoons cool water. Turn off the heat, and stir the solution into the cream. The curd will be very fine and soft—not like the whole-milk variation, which is clumpy and coarse. Let set for 5 to 10 minutes, then pour into a colander lined with a double layer of cheesecloth or fine organdy. You will need to allow it more time to drain, usually about 12 hours. It is normally not salted. Mascarpone is traditionally used for desserts, such as the classic Italian treat tiramisu.

WHEY RICOTTA:

The best whey ricotta is made from "sweet" whey obtained during the making of a pressed cheese. The whey is called sweet because the pH is usually just above 6, so it's not too tart. To make 100 percent whey ricotta, replace the milk in the above recipe with whey and follow the same instructions. When the whey reaches the goal temperature, the whey proteins will begin to precipitate out of the solution. At this point you can add some acid to increase the yield, but it isn't necessary and will make the resulting curd more tangy than if the acid is omitted.

WHEY AND MILK RICOTTA:

For a higher yield and milkier ricotta, add milk at one-quarter to one-half the volume of the whey, then follow the instructions in the main recipe.

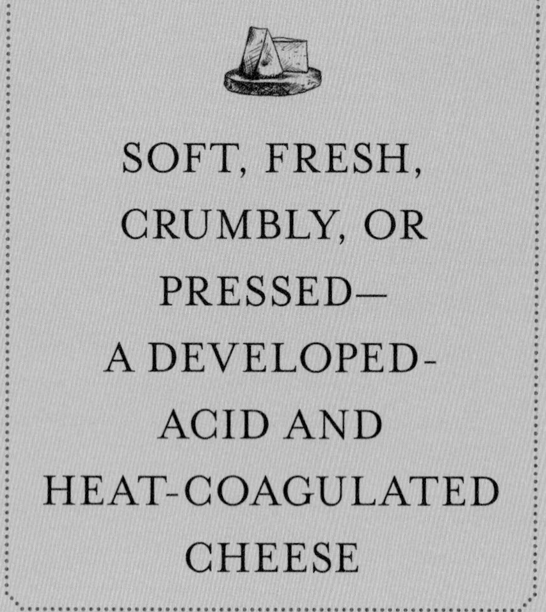

SOFT, FRESH, CRUMBLY, OR PRESSED— A DEVELOPED-ACID AND HEAT-COAGULATED CHEESE

The next recipe is quite similar to the previous one but uses acid developed by added starter culture bacteria followed by high heat. The resulting curd can be managed much like the first recipe: bag-drained or pressed. While this method takes a bit longer, the flavor is more complex thanks to the starter culture. The curd will be very fine when forming, not like the added-acid variety in the first recipe. Cheeses from this recipe will resemble whole milk ricotta and European-style cottage cheese. If lightly pressed and drained, it will resemble ricotta salata. Bellweather Farms in California makes amazing sheep's milk and Jersey milk ricotta, using the developed-acid method. In chapter 1 I talked about some historical cheeses being made by allowing cream or milk to sour naturally, through the activity of starter bacteria that had populated the milk from the environment, then using heat to consolidate the curd. Traditional cheeses such as Schmierkase and Devonshire clotted cream are made using these techniques.

INGREDIENTS

Milk: 1 gal (4 L/8.6 lbs) whole or partly to fully skimmed milk
Culture: ⅛ tsp (0.4 g) Thermo B culture (or equivalent blend)
Salt: ¼ tsp (1.5 g), or to taste, pure salt

An aged, pressed, acid-coagulated cheese, similar to ricotta salata

STEPS

Prepare Equipment: Make sure all equipment is cleaned and sanitized and that your cheesemaking space is free from possible contaminants. Refer to chapter 6 for tips on proper equipment preparation.

Prepare Milk: In a water bath warm milk to 95°F (35°C).

Culture: Sprinkle culture on top of milk. Let set for 3 to 5 minutes, then stir gently for 2 to 5 minutes.

Ripen: Cover, and maintain at 95°F (35°C) to ripen to pH 5.9 to 6.0, usually 2 to 4 hours.

TIPS: If you have trouble maintaining the temperature, place the water bath and cultured milk on a counter and wrap with towels or a blanket. Alternatively, you can set the pot in an ice chest with a few jugs of 98°F (37°C) water.

Salt: Add salt, and stir for 1 to 2 minutes.

Heat: Place on direct heat, and stir continuously to 176 to 185°F (80–85°C) until curd clumps and floats.

TIPS: If curd does not form and separate, this means the pH did not reach the goal and there is not enough acid to coagulate the milk. You can add some vinegar or dissolved citric acid to complete the coagulation.

Set: Turn off heat, and allow to set 5 to 15 minutes to allow curd to consolidate and float.

Drain: Ladle curds into cheesecloth-lined colander. You can briefly rinse the curd with cool water to assist with cooling, but this is not mandatory. Allow the curd to drain in the colander for 20 minutes if you are proceeding to pressing. If not, drain for 60 minutes or more until you like the texture.

TIPS: If a lot of curd remains in the pot after you have ladled the floating portion, carefully pour the remaining hot whey through a sieve and place the collected curd in the lined colander. I do not recommend pouring the entire mass through the cloth, as the curd mass can block the flow of the hot whey and splash back and burn you. Also, you want the curd to be cooling, and the hot whey will simply slow this process.

Pressed Variation: While the curd is still warm, spoon into basket-style forms, making sure they have many perforations. Pack with a spoon. If the curd is already drained, you may need to add a small amount of weight to form the cheese into a more solid shape. This can be done by stacking the forms or placing an empty form on top of the curd and setting a cup of water (of about equal weight to the curd) on top. You can judge if the weight is enough by watching the color of the whey that drips from the cheese—if it is clearish, then you are not using too much weight. Press for up to 12 hours, depending on how firm you want the curd. There are no rules for this type of cheese; it's all about your goals. Follow the simple rule of "the higher the moisture, the shorter the shelf life."

Finishing and Using: For soft, spreadable variations, salt the drained curd to taste and use immediately, or place in a sealable container and keep in the refrigerator for 1 to 2 weeks. Formed varieties can be sprinkled with herbs, more salt, spices, and so on. If you want to make an aged version (such as ricotta salata), sprinkle the well-drained, low-moisture cheese with a bit of salt; air-dry in a low-humidity, clean environment if you desire a grating version. Depending on your goals, you can adjust the humidity and length of aging time.

RHONDA GOTHBERG, *GOTHBERG FARMS*, WASHINGTON STATE

I first met Rhonda in 2005 while doing research on building our own farmstead creamery. Rhonda's small goat cheese dairy just happened to be located minutes from where my father and his wife were living in the tiny, artsy hamlet of Bow, Washington. Rhonda graciously not only let me visit her farm and tour her dairy but was generous with information and tidbits of wisdom that I have never forgotten—including the quote I still repeat regarding our chosen professions, "It is an income, not a living." Despite this perspective, or perhaps thanks to Rhonda's ability to face reality, Gothberg Farms has grown and prospered since my first visit those several years ago.

Rhonda milks a small herd of pasture-fed, registered LaMancha dairy goats (the breed with such tiny ears that uninformed folks will often ask if their ears have been removed) that she cares for and tends with the love and attention so common among small herdsman. Her fresh cheeses, made in the fashion of the "developed-acid, long-set, long-drain" recipe that follows, are drained in forms, then packaged and sold at various retail locations and farmers' markets throughout the Puget Sound area of Washington state. The Gothberg Farms label also appears on several semisoft and hard aged cheeses that Rhonda and her team of part-time cheesemakers and assistants produce in the on-farm creamery.

Rhonda Gothberg at her farmers' market booth. Photo courtesy of M. E. Bowlin

Rhonda's business reflects a perfect balance of fantastic location; available workforce; wonderful farmland; great business skills; healthy, happy goats; and, of course, gifted cheesemaking.

Soft fresh cheeses, such as chevre, fromage blanc, and ricotta, are versatile for both savory and sweet dishes. Pictured here encrusted with toasted nuts and spices

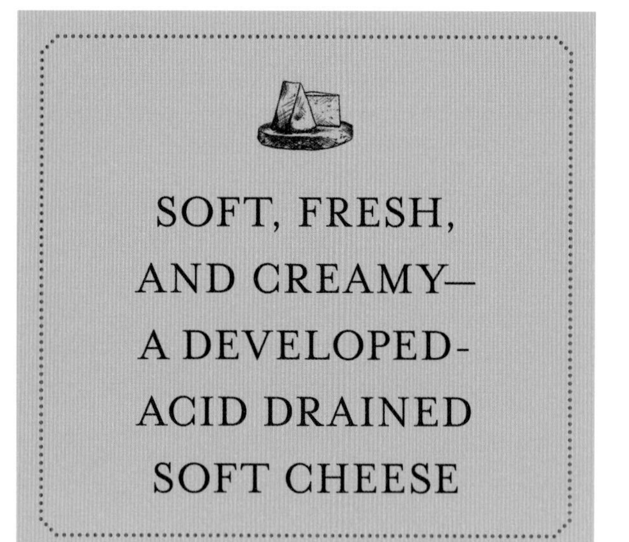

SOFT, FRESH, AND CREAMY— A DEVELOPED-ACID DRAINED SOFT CHEESE

Slowly ripened, long-drained fresh cheeses comprise a huge segment of dairy products consumed throughout the world. These tangy, soft cheeses go by many names, come in many shapes and textures, and are flavored with an almost limitless variety of herbs, spices, and flavorings. In France you will find *faisselle* (named for the tiny basketlike drainer in which it is formed); *fromage blanc* and *fromage frais*, usually made with lighter-fat milks; in Germany *quark* (sometimes known as "baker's cheese") is a staple; and in the United States we have dubbed soft, fresh goat cheese chevre (shortened from the French *fromage de chèvre* or goat's cheese). If you use high-fat or cream-added milk, you will end up with cream cheese. You can think of this category as an "anything goes" with regard to type of milk, texture at the end of draining, and possible uses. While one of the easiest to make, soft fresh cheese from this category is also one of the most practical and versatile.

INGREDIENTS

Milk: 1 gal (4 L/8.6 lbs) whole milk; for *quark* style skim milk is often used, and for *cream cheese* varying amounts of cream can replace part or all of the milk
Culture: scant ⅛ tsp (0.2 g) culture blend such as Flora Danica (or equivalent blend). If you are using a prepackaged culture from a home cheesemaking supply company, it may include rennet. If so, do not add any more rennet to the recipe.
Calcium chloride (optional): maximum ¼ tsp (1.25 ml) diluted in ⅛ cup (30 ml) cool, nonchlorinated water
Rennet: 2–4 drops (or equivalent from Table 8–3) single-strength rennet diluted in ⅛ cup (30 ml) cool, nonchlorinated water
Salt: ¼–½ tsp (1.5–3 g) pure salt

STEPS

Prepare Equipment: Make sure all equipment is cleaned and sanitized and your cheesemaking space is free from possible contaminants. Refer to chapter 6 for tips on proper equipment preparation.

Prepare Milk: Warm milk in double boiler to 86°F (30°C).

> TIPS: If the cheese is for personal consumption, it can be made from raw milk but is more likely to suffer from the growth of other bacteria during the long, slow ripening period.

Culture: Sprinkle culture on top of the milk. Let set for 3 to 5 minutes, then stir gently for 2 to 5 minutes.

Additions: Stir in calcium chloride (if using). Stir in rennet if not included in premixed culture.

Ripen and Coagulate: Ripen at 72°F (22°C) for 12 to 24 hours.

> TIPS: If you have trouble maintaining the temperature, place the water bath and cultured milk on a counter and wrap with towels or a blanket. Alternatively, you can set the pot in an ice chest with a few jugs of 75°F (24°C) water.

Drain: When the pH of curd is 4.6 to 4.7 (about ½ inch of whey will cover the top of the mass, and the curd will be pulling away slightly from the sides of the pot), cut the mass into ½ inch (1.3 cm) vertical columns, then ladle into a draining bag or fine-

Chevre and long-drain types made in small batches can be drained over a pot using a rod or sturdy utensil.

cheesecloth-lined colander. Tie corners together or tighten bag, and hang to drain. Room temperature should be about 72°F (22°C) until desired texture is achieved, usually 4 to 6 hours (longer in cooler room, shorter in warmer room).

TIPS: Check the curd partway through to make sure that it is draining evenly. If the room temperature is not optimal, the outside layer of the cheese may drain too quickly and trap soft, poorly drained curd inside. If you see this, you can stir the curd to encourage more even draining.

Salt: Empty into bowl, add salt to taste (start with ¼ tsp), and mix gently but thoroughly, allowing time for salt to dissolve and be dispersed. Goal salt content is 0.8 percent of curd weight.

Finishing: Store in the refrigerator for up to 3 weeks. The cheese will develop more flavor after a few days, so a period of ripening is helpful. These types of cheeses freeze well.

VARIATIONS

FORMED:

To form shapes the curd can be partly drained in a bag, salted, then spooned into well-perforated basket forms and drained overnight. Alternatively, the shapes can be salted after draining. Formed cheeses can be sprinkled with herbs, spices, and so on. They can also be formed in layers, with seasonings and other ingredients alternating with layers of cheese. These are often marketed as "tortas" here in the United States.

AMERICAN-STYLE COTTAGE CHEESE

The last recipe in this chapter is for American-style curd cottage cheese. This cheese, in some ways, is similar to semihard cheeses, since a curd is formed that is cooked. But because it remains soft and is used fresh, it fits best in this chapter. I developed this recipe after trying many whose results were quite disappointing. Nothing resembled what I think of as cottage cheese (like you get in the grocery store). The process is fairly lengthy, but the results are quite good. You can vary it a bit by making it more or less acidic at the time of cutting the curd. Also, the curd size can be varied according to your tastes.

INGREDIENTS

Milk: 1 gal (4 L/8.6 lbs) whole to partly skimmed pasteurized milk
Culture: ¼ tsp (0.2 g) mesophilic blend such as Flora Danica (or equivalent blend)
Calcium chloride (optional): maximum ¼ tsp (1.25 ml) diluted in ⅛ cup (30 ml) cool, nonchlorinated water
Rennet: 8–10 drops single-strength rennet diluted in ⅛ cup (30 ml) cool, nonchlorinated water
Salt: ¼ tsp (1.5 g) pure salt
Cream for dressing: ⅛ cup (30 ml) heavy cream

STEPS

Prepare Equipment: Make sure all equipment is cleaned and sanitized and your cheesemaking space is free from possible contaminants. Refer to chapter 6 for tips on proper equipment preparation.

Prepare Milk: Warm milk to 86°F (30°C).

Culture: Sprinkle culture on top of milk. Let set 3 to 5 minutes, then stir gently for 2 to 5 minutes.

Additions: Stir in diluted calcium chloride. Wait 5 minutes.

Rennet: Stir in diluted rennet solution, using an up-and-down motion for 1 full minute. Still milk.

Ripen and Coagulate: Cover pot, and maintain temperature at 86°F (30°C) for 2 to 4 hours or until pH of whey is 6.0 or curd is just pulling away from the sides of the pot.

> TIPS: If you have trouble maintaining the temperature, place the water bath and cultured milk on a counter and wrap with towels or a blanket. Alternatively, you can set the pot in an ice chest with a few jugs of 88°F (31°C) water.

Cut: Cut curd into ½ inch (1.3 cm) cubes. Rest at 86°F (30°C) for 10 minutes.

Cook: Begin heating very slowly and stirring very gently. Heat to 115°F (46°C) over 60 to 90 minutes; increase the temperature a bit more slowly in the beginning.

> TIPS: The curd is very delicate in the beginning. If it seems as if stirring is causing the curds to shatter, for the first few minutes of heating you can gently shake the pot slightly to agitate the curds and move the whey. Once the curds have firmed up a bit, stir frequently or continuously to prevent uneven heating, ensure even shrinkage, and prevent matting.
>
> As the temperature increases, the pH will continue to develop until the temperature reaches 104°F (40°C), when the mesophilic bacteria will reach the limits of their temperature range. The goal pH of curd at this time is 4.9 to 5.0 (or more acidic, if you prefer it that way). The final cook stage to 115°F (46°C) is to shrink the curds to the desired texture and prevent any further acid development.

Drain: When the curd texture is springy but still tender, remove from heat and let the curds settle in the pot for 5 minutes. Pour off excess whey, and add cool water to the pot. Rinse and drain the curds four times with increasingly cold water.

Salt: Place curds in colander, and mix in half the salt. Let drain for 30 minutes. Mix in the rest of the salt or to taste.

Finishing: Dress with heavy cream if desired. Refrigerate. Use within 2 weeks.

THINKING OUTSIDE THE VAT

SCENARIO: You are making fresh *chèvre*-type cheese. You add the culture and coagulant to the milk just before bedtime, then place the pot in your oven with the pilot light on and the door ajar, as you have found the oven will maintain just the right temperature for ripening the cheese. But in the morning there is much more whey covering the cheese than usual and the curd is shrunken in the pot, with many fissures running through it. You do which of the following?

A. Check the starting pH of the milk before the next batch to see if the milk is starting out more acidic than usual.
B. Don't use any starter culture in the next vat; you can rightly assume that the milk has plenty of natural, good bacteria to properly acidify the vat.
C. Consider using less starter culture, and verify that the oven is not too warm.
D. Add less coagulant the next time.

Answer: A, C, and D. Answer B could be correct, but you shouldn't assume that it is.

Feta and its many cousins are popular brined cheeses throughout the world.

9: BRINED CHEESES, FRESH AND AGED

FINISHED TEXTURE: Soft to semisoft
METHOD OF COAGULATION: Rennet
METHOD OF RIPENING: Internal bacteria or none

Salt has served to preserve and protect food for many centuries, especially in seacoast areas where high rehydration temperatures make storing in other manners difficult to impossible and salt is a ready ingredient. It is not surprising, then, that Mediterranean countries would evolve as the leaders in producing and consuming brine-aged cheeses. There are many derivations of these cheese types, both in countries located along the Mediterranean sea and in neighboring regions. These cheeses go by many names, and a few require some unique procedures for their manufacture.

In the United States the most common brined cheese is without a doubt feta. Greece now holds the PDO rights to produce and label cheese with the name "Feta" (Greek Feta must be made from not less than 70 percent sheep's milk, with the balance being composed of goat's milk). The popularity of this cheese, however, and a pre-PDO history of manufacturing similar cheeses in many countries, means that you can buy cheeses called feta (or made in the same style) all over the world.

Feta-type cheeses aging in brine

While the making of a cheese in the style of feta is very similar in most countries, there are a few other brined cheeses whose manufacture departs greatly from that of feta. They are usually put in the same family, though, because of the distinction that aging in brine gives them (i.e., producing a white, rindless, very salty cheese). Because the steps in making these other types are unique, I have offered special tips for each cheese style.

TIPS FOR AGING, STORING, AND USING BRINED CHEESES

There are several methods for preserving these cheese types. If you are using the cheese fresh, follow the dry-salting instructions for each recipe.

For brine aging I prefer the method taught by one of my first teachers, Peter Dixon. The wheels are first submerged in a fully saturated brine for 8 hours per pound (16 hours per kg) of cheese in a single wheel, then moved to an 8 percent brine for long aging. We rarely have anyone "complain" that our feta is too salty, and we can age it for 6 to 8 months for a really complex end product. (For more on making and maintaining brine, see chapter 2.)

When aging in brine, it is best to fill the container to the top, leaving no airspace. This will help prevent molds and yeasts, both of which need oxygen, from growing. I also like to lay a sheet of plastic film across the top before securing the lid. Translucent food-grade containers are best, as you can inspect the cheese for problems without opening them. The brine, if made from scratch instead of using whey, should change from clear to a pale yellow-green during aging. It should not look opaque and white; if that occurs you should open the container, remove the cheese (which is probably going soft), mix a new brine with a bit more calcium chloride than the original recipe, and rebrine the cheese. Also watch for gas buildup and expansion of the container, although a little bit is acceptable. You may need to "burp" the container to release some of the gas.

When eating brined cheeses, remember that they are quite salty, so they work well in dishes where they can replace some or all of the salt. Brined cheeses can also be briefly soaked in a low-salt or salt-free solution to leach out part of the saltiness. If you choose to do this, remember that the calcium content of the solution should be adjusted so you don't soften, or possibly dissolve, the cheese. For this reason, traditionally Greeks have been known to soak extra-salty feta in fresh milk to remove the unwanted salt.

FIRM AND CRUMBLY GREEK FETA-STYLE BRINED CHEESE

Fortunately, feta-type cheeses are relatively simple to make and usually problem-free. For the beginner cheesemaker they are a great transition cheese to take you to the next skill level. That being said, there is a lot of feta being made that could be better!

Feta can be made from pasteurized milk or raw milk if the cheese will be aged (or used only for home use). As with other cheeses, raw milk brings many nonstarter bacteria into the process along with native enzymes (e.g., lipase) that can lead to greater flavor development. Still, choosing the right culture will help even raw-milk varieties reach their full potential. Because feta types are usually not heated to over 90°F (32°C) in the vat and most acid development occurs during draining, it is important to choose a starter that can develop the right amount of acid under these conditions. If proper acid levels are not reached over a certain amount of time, the cheese can become vulnerable to flaws such as coliform blowing.

In the following table you will notice some lovely blends that will help create acid and add flavor and aroma during aging. The three blends highlighted in brown are composed of the same bacteria types, but in undefined ratios. Below those are two blends that include some thermophilic bacteria—the heat lovers. While this may leave you scratching your head (didn't I just emphasize the low temperature at which this curd is processed?), these thermophiles will be fairly inactive in the vat and during draining but will provide some really helpful enzymes for the breakdown of proteins during aging—if you age it—leading to

TABLE 9-1. CULTURE CHOICES FOR FETA TYPES

Name	Contains	Manufacturer	Small-Batch Dosage:* 3–4 Gal (12–15 L/26–34 lb)	Larger-Batch Dosage: 12 Gal (46 L/100 lb)	Characteristics and Application
Flora Danica	LL, LLC, LLD, LMC	Packaged by home cheese supply co.	1 packet	N/a	Starter and aroma cultures packaged for small batches
Flora Danica or CHN 11, 19	LL, LLC, LLD, LMC	Chr. Hansen	¼ tsp (0.4 g)	10U	Starter and aroma cultures
Meso Aromatic B	LL, LC, LD, LM	Abiasa	¼– ½ tsp (0.6–1.2 g)	10 g	Starter and aroma cultures
Feta A and B (Choozit)	LL, LLC, ST, LB, LH	Danisco	¼ tsp (0.4 g)	10U	Starter cultures plus small amount of aroma and 3 thermophiles for texture and flavor during aging, only comes in 100U pkg
MA 4000 Series (Choozit)	LL, LLC, LLD, ST	Danisco	⅛ tsp (0.3 g)	4U	Starter and aroma, plus one thermophile for adjunct flavor during aging

*Remember that culture is always best measured by weighing and dividing into unit doses based on the strength of the culture, instead of using volume or set weight.

more flavor and improved texture. (For lots more on cultures see chapter 1.)

Rennet choices can include all three major varieties: traditional veal rennet; microbial coagulant; and fermented chymosin. Small scale, traditional feta often is made using rennet paste, which includes lipase from the young animal's stomach. Lipase lends its own piquant (spicy) flavor and is also the enzyme that will break down fat, leading to even more flavor. If you decide to add lipase to your cheese milk, add it just after the culture, dissolving it first in a bit of nonchlorinated water. When purchasing lipase you can select from three (sometimes four) different varieties: calf lipase for the mildest, kid for medium, and lamb for the strongest picante results (the fourth choice offered by some companies is a combination of lamb and kid lipase). Remember that a little bit of lipase goes a long way, so measure carefully; if you find your cheese is too strong, use less, or none, the next time.

> Keep in mind how closely related some of these cheeses are, and how you, as the cheesemaker, can turn them into something "new" with a simple choice.

When it is time to drain feta types, you can place the curd in cheesecloth and hang to drain or drain in a form. If the form has microperforations or is lined with mesh, you do not need cheesecloth. If it is a regular form, cheesecloth will help the curd drain evenly and will make it easier to turn the wheels. Remember, very little to no weight is used for most of these cheese varieties.

During draining, the pH of Greek-type feta should drop to under 4.8, while Bulgarian-style feta finishes a bit higher at 5.0. Because the final texture of cheese is greatly influenced by the amount of whey present during draining, as well as by the final pH, these goal numbers are important for the development of the typical feta texture—slightly crumbly and moist. In most cases the higher moisture content means that feta types retain a bit of lactose even after aging, about 1 percent

Ideal feta texture after draining is slightly open.

according to the data. The finished texture should be smooth and free from eyes formed by bacteria and any large openings left by curd that doesn't knit well together. Small mechanical openings are fine and can actually help transport brine more quickly throughout the cheese.

Feta comes in different ways sometimes and even can appear by accident. One day, for example, I was making blue cheese and ran out of blue mold. So I made two wheels that did not have any *roqueforti* mold added and would therefore never be truly blue. The texture and process was so similar to feta that I decided to brine them and see how they turned out. Six months later, voilà—feta! When you get to the blue cheese chapter, keep in mind how closely related some of these cheeses are and how you, as the cheesemaker, can turn them into something "new" with a simple choice.

INGREDIENTS

Milk: 1 gal (4 L/8.6 lbs) whole milk
Culture: scant ⅛ tsp (0.2 g) Flora Danica (or equivalent blend)
Lipase (optional): 2 skewer tips lipase dissolved in 1 tbsp (15 ml) cool, nonchlorinated water
Calcium chloride (optional): maximum ¼ tsp (1.25 ml) diluted in ⅛ cup (30 ml) cool, nonchlorinated water
Rennet: ⅛ tsp (0.7 ml) single-strength rennet diluted in ⅛ cup (30 ml) cool, nonchlorinated water

Salt: 1–2 tbsp (15–30 g) pure salt *or* follow brining instructions

STEPS

Prepare Equipment: Make sure all equipment is cleaned and sanitized and that your cheesemaking space is free from possible contaminants. Refer to chapter 6 for tips on proper equipment preparation.

Prepare Milk: Warm milk to 88 to 90°F (31–32°C).

Culture: Sprinkle culture on top of milk, and let set 3 to 5 minutes. Stir gently for 2 to 5 minutes.

Additions: Add lipase solution if using, and stir for 30 seconds. Stir in diluted calcium chloride, if using.

> TIPS: Lipase will impart a stronger, more pungent flavor to the cheese. It is often added to feta types made from cow's milk to help mimic the naturally stronger flavor of traditional sheep's and goat's milk feta. The source of the added lipase will greatly influence the end flavor. The amount added can be adjusted to your personal taste as well.

Ripen: Maintain temperature at 88 to 90°F (31–32°C) for 45 to 60 minutes.

Rennet and Coagulate: Stir in diluted rennet solution, using an up-and-down motion for 1 full minute. Still milk. Maintain 88°F (31°C) for 45 to 60 minutes or until curd is at the clean-break stage.

Cut: Cut curd mass into ¾- to 1-inch (2–3 cm) cubes, and rest 10 to 15 minutes.

Stir: Stir at 88 to 90°F (31–32°C) for 20 minutes. Settle for 5 minutes.

> TIPS: Feta curds will be soft and high in moisture compared to most cheeses (other than soft and fresh). They need to drain slowly so the residual whey can help develop the high acid content desired by the end of draining.

Drain: Remove excess whey down to the level of curds. Scoop curds into a well-perforated form or pour into a cheesecloth-lined colander. Drain in form, without pressure, for 6 to 12 hours. Turn after 1 hour and again in 2 hours. If draining in bags, tie cheesecloth in a knot and hang for 6 to 12 hours. The goal cheese pH at the end of draining is 4.6 to 4.8. Flavor should be tangy but not sour.

Salt: For aged feta place whole cheese in heavy brine for 8 hours per each pound (approx. 16 hours/kg) in one wheel. Then transfer to an 8 to 10 percent brine at 50 to 55°F (10–13°C) solution for aging at 50 to 55°F (10–13°C). For fresh consumption lay blocks on a draining mat, and sprinkle all sides with 1 to 2 tablespoons (15–30 g) dry salt. Let drain at room temperature (under 70°F [21°C]) for 12 hours, and salt again with same amount.

LARGE-BATCH GUIDELINES

Milk: 12 gal (46 L/100 lb) whole milk (for larger batches increase ingredients proportionately)
Culture: 5U Flora Danica or other from chart
Lipase (optional): ⅛–¼ tsp (0.7–1.4 g)
Calcium chloride (optional): 10–15 ml (about 1 tbsp)
Rennet: 7–9 ml single-strength rennet (3.5–4.5 ml double strength)
Salt: Heavy brine and 8–10% brine

SMOOTH AND CREAMY BULGARIAN-STYLE BRINED CHEESE

If you prefer a creamier texture in your feta, this recipe will help you make that style of cheese. A larger curd cut allows for more moisture retention that helps create a different texture in the finished cheese. In addition, a pressing stage helps close the texture and remove whey before acid development removes calcium, which also contributes to a smoother, more mineralized texture. However you slice it, there is a feta style for everyone! If your tastes lie in between, this is your chance to create your own hybrid recipe by adapting this version with the previous, more crumbly type.

INGREDIENTS

Milk: 1 gal (4 L/8.6 lbs) whole milk
Culture: 1/8 tsp (0.25 g) MA 4000 or Feta B (or equivalent blend)
Lipase (optional): 2 skewer tips lipase dissolved in 1 tbsp (15 ml) cool water
Calcium chloride (optional): maximum 1/4 tsp (1.25 ml) diluted in 1/8 cup (30 ml) cool, nonchlorinated water
Rennet: Scant 1/8 tsp (0.6 ml) single-strength rennet diluted in 1/8 cup (30 ml) cool, nonchlorinated water
Salt: 1–2 tbsp (15–30 g) pure salt *or* follow brining instructions

STEPS

Prepare Equipment: Make sure all equipment is cleaned and sanitized and that your cheesemaking space is free from possible contaminants. Refer to chapter 6 for tips on proper equipment preparation.

Prepare Milk: Warm milk to 87 to 89°F (31–32°C).

Culture: Sprinkle culture on top of milk, and let set 3 to 5 minutes. Stir gently for 2 to 5 minutes.

Additions: Add lipase solution if using, and stir for 30 seconds. Stir in diluted calcium chloride, if using. Wait 5 minutes.

> TIPS: Lipase will impart a stronger, more pungent flavor to the cheese. It is often added to feta types made from cow's milk to help mimic the naturally stronger flavor of traditional sheep's and goat's milk feta. The source of the added lipase will greatly influence the end flavor. The amount added can be adjusted to your personal taste as well.

Ripen: Maintain temperature at 87 to 89°F (30–31°C) for 30 minutes.

Rennet and Coagulate: Stir in diluted rennet solution, using an up-and-down motion for 1 full minute. Still milk. Maintain 88°F (31°C) for 45 to 60 minutes or until curd is at the clean-break stage.

Cut: Cut into 3-inch (7.6 cm) vertical columns. Rest 10 minutes, maintaining temperature.

Drain: Ladle or scoop curd layers into cloth-lined colander or form. Hang cloth to drain for 4 to 5 hours. Place in form, and apply weight equal to that of curd. Room temperature should be about 72°F (22°C) for acid development during draining. Press to cheese pH 5.0 to 5.1 (usually 3 to 5 hours). Curd will taste sweet with a slight tang.

Salt: Cut into 4 x 4-inch (10 x 10-cm) blocks or slabs. For aged cheese immerse slabs in heavy brine for 8 to 12 hours, then transfer to an 8 to 10 percent storage brine for the duration of aging.

For use when fresh lay blocks on a draining mat and sprinkle all sides with 1 to 2 tablespoons dry salt. Let drain at room temperature of about 70°F (21°C) for 12 hours, and salt again with same amount. Transfer to storage container for use.]

LARGE-BATCH GUIDELINES

Milk: 12 gal (46 L/100 lb) whole milk (for larger batches increase ingredients proportionately)
Culture: 2U MA 4000 or Feta B (or equivalent blend)
Lipase (optional): ⅛–¼ tsp (0.7–1.4 g)
Calcium chloride (optional): 10–15 ml (about 1 tbsp)
Rennet: 7–9 ml single-strength rennet (3.5–4.5 ml double strength)
Salt: Heavy brine and 8–10% brine

Filling forms with feta-style curds

Feta cheeses work well in many dishes and can replace added salt. Pictured here on a Mediterranean-style pizza

BEAU SCHOCH, *SCHOCH FAMILY FARMSTEAD*, CALIFORNIA

When we visited the Schoch Family Farmstead in Monterey County, California, they were in the midst of building an on-farm cheesemaking facility. The family dairy, a small Holstein cow operation, has been in the Schoch family since the early 1940s, when Beau's grandfather and great-uncle, both Swiss immigrants, established the business. While many companies and farms adorn their names with the word "family," this business is true to its title. Most of the milking is done by Beau's father, John, who also manages the herd—a herd that is rather famous in Holstein genetic circles—and his brother Seth, while Ty, the oldest Schoch son, handles the farmers' markets. Even Beau's mother, a successful businesswoman in her own right, is fully engaged in the enterprise.

While the creamery focuses on aged raw-milk cheeses that are linked historically to the area, they also produce a lovely, smooth raw-milk feta-style cheese (similar to the Bulgarian-type feta recipe here) that they age in brine. The Schochs' expansion of production will, they hope, soon allow Beau's skillful hands to transform more of their own milk into cheese—and in the process bring new life to a farm, a family, and a way of life worthy of preserving.

Beau Schoch of Schoch Family Farmstead

SOFT AND SALTY DOMIATI-STYLE BRINED CHEESE

Domiati is an extremely popular Egyptian cheese (also made in other countries under different names, such as *gibna beyda* from Sudan) made from several types of milk to which a percentage of salt is added during the make process. The salt serves as an antimicrobial and preserves the quality of the milk, in addition to extending coagulation time and creating a softer curd set. These two results are not necessarily goals of the process, merely results. Consequently, the recipe calls for more rennet and a coagulation time of several hours. Don't be surprised by the softness of the curd! The soft curd is not cut and cooked but is ladled directly into bags or forms for drainage.

Because this is an uncultured cheese, I suggest adding a bit of lipase to improve flavor. You can also experiment with adding the ripening culture LBC 80 (*Lactobacillus casei*) directly to the whey brine during aging. LBC is a salt-tolerant bacteria that will help break down proteins during aging for additional flavor development.

This is a bit of a unique cheese, with its soft texture and saltiness that goes beyond that of feta, so I am not sure how it will fit in most people's cooking, but it is worth making at least one time just to appreciate the amazing differences in cheese types. I found the flavor somewhat bland, but also pleasant and, of course, very salty. After aging, the flavor grows more complex, but the salt level is still quite high.

INGREDIENTS

Milk: 1 gal (4 L/8.6 lb) whole milk
Lipase: ⅛ tsp (0.7 g) lipase dissolved in 1 tbsp (15 ml) cool, nonchlorinated water
Calcium chloride (optional): maximum ¼ tsp (1.25 ml) diluted in ⅛ cup (30 ml) cool, nonchlorinated water
Salt: 1 cup (230 g) pure salt (goal 7% of the weight of milk)
Rennet: Scant ¼ tsp (1.2 ml) single-strength rennet diluted in ⅛ cup (30 ml) cool, nonchlorinated water

STEPS

Prepare Equipment: Make sure all equipment is cleaned and sanitized and that your cheesemaking space is free from possible contaminants. Refer to chapter 6 for tips on proper equipment preparation.

Prepare Milk: Warm milk to 95 to 104°F (35–40°C).

Additions: Stir in lipase and calcium chloride (if using). Stir gently for 1 minute.

Salt: Stir in salt until dissolved.

Rennet and Coagulate: Add rennet solution, stir with an up-and-down motion for 1 minute. Cover, and maintain temperature of 95 to 104°F (35–40°C) for 3 to 6 hours.

> TIPS: If you have trouble maintaining the temperature, place the water bath and cultured milk on a counter and wrap with towels or a blanket. Alternatively, you can set the pot in an ice chest with a few jugs of 100°F (38°C) water.

Drain: Curd is ready to drain when a clean break is obtained. The texture is softer than most cheeses. Ladle the curd into a cheesecloth-lined colander. Set over a pot, and cover the top. Drain 12 to 24 hours, then tie

Domiati-type curd being ladled for draining

the corners in a stilton knot (images of tying a stilton knot are on page 215), and retighten several times over the next 4 to 8 hours until the curd is firm enough to cut. Maintain a room temperature of about 72°F (22°C) during the entire draining process.

Finishing: Unmold, and cut into 4-inch (10-centimeter) slabs. Use or age in 8 percent brine for 2 to 6 months. Expect a soft, salty, rather mild cheese that is meant to be used as a spread or as a base for other seasonings. It can be used to fill savory pastries, for instance.

Halloumi-style cheeses before and after scalding, adding mint leaves, and folding

FIRM AND SALTY GREEK-STYLE GRILLING CHEESE

There is just something about fried cheese. I am sure to some it sounds almost like a sacrilege, but I find it just plain delicious—especially when the cheese was made just for that purpose. Such is the case with the popular Cypriot cheese Halloumi. As I write this, Halloumi is being considered for PDO status, but as it is often made with a combination of milk from cows, goats, and sheep, it is proving difficult to agree upon exact proportions of each milk type.

Traditional Halloumi from the Greek island of Cyprus is made from primarily sheep's milk with the addition of a smaller amount of goat's milk. Dried mint leaves are folded into the still-hot discs, and the cheeses are aged in brine. Commercial production might include the use of cow's milk and the omission of both mint leaves and the traditional folded-disc shape. In the United States the Ballard Family Dairy in Idaho produces a lovely halloumi-type cheese called Golden Greek from the milk of their Jersey cows.

Authentic Halloumi is made only on the island of Cyprus, in the Mediterranean off the coast of Greece. It is a versatile cheese that doesn't melt—a unique quality—and is therefore called a "grilling" cheese. It is the high pH of the cheese that causes its nonmelting characteristic. And while high pH is commonly associated with a less safe product, the scalding step (which we will discuss in a moment) essentially pasteurizes the cheese, rendering it completely safe.

Because of the high-heat treatment during the scalding stage, any starter culture would be destroyed, along with the coagulant. While some recipes call for the addition of starter culture, which will leave enzymes that will assist flavor during aging, this cheese is traditionally not cultured. When made with raw milk, nonstarter bacteria present in the milk and environment will influence the flavor during aging. If you are working under more sterile conditions and find the resulting flavor of this cheese a bit uninteresting, try adding a bit of lipase to the milk, or you can also try adding a pinch of *Lactobacillus casei* (LBL 80 when purchased from Danisco's Choozit line) to the brine. This salt- and acid-resistant probiotic will help break down some proteins (through its production of aminopeptidase) that will enhance flavor. (If you are a commercial producer, be sure to document this purposeful "contamination" of the brine so that any microbiological testing will be weighted by this knowledge.)

The curd is traditionally drained in tall, narrow baskets, so that after draining they can be cut to form a thick disc. During draining these discs are not usually pressed but are turned several times. I do suggest hand-pressing the curd into the form first under the whey, then during draining if needed. Because the discs will spread a bit during scalding, you don't want to start with too large a diameter. I like to use a tall form, then cut the resulting curd mass into two discs about 3 inches thick. If you don't have these types of molds available, you can press in a larger shape, use a bit of weight to assist with forming the shape, then cut the shape into 4 x 6 x 2-inch (10 x 15 x 5-cm) shapes.

The whey used for scalding must first be heated to about 195°F (91°C) to bring out the whey proteins from the liquid. The whey curd is then skimmed off and drained in baskets. If you were in Cyprus, you could call this resulting whey cheese *anari*. Most of us will call it ricotta. If you don't remove the whey proteins, they will form a messy coating on the scalded discs of cheese.

During scalding it is important that the discs heat evenly. If you are using direct heat to keep the whey hot, don't let discs settle to the bottom where they can stick. For that reason I recommend using a basket or sieve to lower the curd into the whey, but don't let it sit directly on the heat. You will know the cheeses are ready when they float to the surface, usually after 30 to 60 minutes. If you cool one, you will find it has assumed the texture of cooked chicken breast meat (a metaphor you will hear again when we get to cheddar cheesemaking). The discs will enlarge and spread during scalding and again a bit more when the hot shapes are removed from the whey to drain.

Disc shapes are traditionally folded into crescents, with dried mint leaves included and the salt in the center. The mint adds flavor during aging but can be omitted. Dry salting is done at a rate of 5 percent by weight (so if your curd weighs 10 pounds, you would use 0.5 pounds of salt). After the initial dry salting and cooling, the cheeses can be used or aged following the same protocol as for feta types. The flavor is improved by aging, especially when mint is present, but the salt content is increased. As you age the cheese, it will lose its grilling properties as the protein matrix changes (you'll learn more about that in chapter 13).

INGREDIENTS

Milk: 1 gal (4 L/8.6 lbs) whole milk
Culture: ⅛ tsp (0.25 g) MA 4000 (or equivalent blend)
Lipase (optional): ⅛–¼ tsp (0.7–1.4 g) dissolved in 1 tbsp (15 ml) cool, nonchlorinated water
Calcium chloride (optional): maximum ¼ tsp (1.25 ml) diluted in ⅛ cup (30 ml) cool, nonchlorinated water
Rennet: Scant ½ tsp (1.2 ml) single-strength rennet diluted in ¼ cup (60 ml) cool, nonchlorinated water
Salt: 1–2 tbsp (15–30 g) pure salt or heavy brine
Mint (optional): Dried mint leaves

STEPS

Prepare Equipment: Make sure all equipment is cleaned and sanitized and that your cheesemaking space is free from possible contaminants. Refer to chapter 6 for tips on proper equipment preparation.

Prepare Milk: In a double boiler warm milk to 86 to 93°F (30–34°C).

TIPS: Because of the high-heat treatment that occurs later in this recipe, milk will essentially (but not legally) be pasteurized.

Culture: Sprinkle culture on top of milk, and let set 3 to 5 minutes. Stir gently for 2 to 5 minutes.

TIPS: Culture is not used in the traditional version but will provide some enzymes that are useful during aging. For fresh versions culture can be omitted.

Additions: Add lipase solution if using, and stir for 30 seconds. Stir in diluted calcium chloride, if using.

TIPS: Lipase will impart a stronger, more pungent flavor to the cheese. It is often added to feta types made from cow's milk to help mimic the naturally stronger flavor of traditional sheep's and goat's milk feta. The source of the added lipase will greatly influence the end flavor. The amount added can be adjusted to your personal taste as well.

Rennet and Coagulate: Stir in diluted rennet solution, using an up-and-down motion for 1 full minute. Still milk. Maintain 86 to 93°F (30–34°C). Goal coagulation time is 40 minutes.

Cut: Cut curd mass into ¾- to 1.5-inch (2–4 cm) cubes, and let rest 5 minutes.

Stir: Stir gently, and heat to 100 to 108°F (38–42°C) over 20 to 30 minutes. Hold for 20 minutes more, stirring gently.

Pitch: Settle for 5 minutes, then press by hand to form mass under whey. Drain whey, and reserve in a pot for use later.

Drain: Cut curd mass into quarters, set form down into warm whey, and hand-knead curd mass into form. Move form to drain board. Stack and hand-press curd to ensure knitting. By the end of draining, curd should be well knit together.

Whey Prep: Place pot of whey on direct heat, and heat to 185 to 195°F (85–91°C). Stir occasionally. Proteins will float to the top; skim whey cheese from surface (use whey cheese as you would ricotta).

Curd Heating: Unmold cheese. If needed, cut to thickness of approximately 2 inches (5 cm). Place shapes in steam basket or colander, and immerse in hot whey for 40 to 80 minutes at 190 to 195°F (88–91°C). Don't boil whey. Curds are ready when they float in the hot brine; remove colander or basket from whey.

Salt and Finishing: Spray cheeses briefly with cool water, and place on drain table. Sprinkle with ¼ pound (113 g) salt and optional dry mint leaves. Fold warm discs in half (forming crescent shape). Let cool for 2 to 3 hours. If adding mint, sprinkle on before folding. If aging, pack cheeses tightly in tubs, and cover with 8 to 12 percent brine. Age for weeks to months.

Affinage: These cheeses, while high in salt content, have a wonderful place in cooking where salt can be omitted or reduced. They are usually not cheese-plate material. As you continue to make brined cheeses, consider letting them age for longer periods of time. While most people currently use them at a fairly young age, a long-aged feta develops incredible texture and flavor when compared to a young version. So give them a chance, and you will be pleased.

LARGE-BATCH GUIDELINES

Milk: 12 gal (46 L/100 lb) whole milk (for larger batches increase ingredients proportionately)
Culture: 2U MA 4000 (or equivalent blend)
Lipase (optional): 1–2 tsp (5.6–11.2 g)
Calcium chloride (optional): 10–15 ml (about 1 tbsp)
Rennet: 12–14 ml single-strength rennet (6–7 ml double strength)
Salt: Heavy brine
Mint (optional): Dried mint leaves

THINKING OUTSIDE THE VAT

SCENARIO: You make a lot of feta-type cheese that you share with your family. Over a week's time and several batches of cheese, you notice that the coagulation time (until the curd is ready to cut) is getting steadily longer. You do which of the following?

A. Verify that the pH is on target, as you know that acidity affects coagulation time.
B. Ignore this; it is normal at this time of year and should be expected.
C. Check the age of the coagulant.
D. Consider adding more rennet the next time, after you rule out that C is the problem.
E. Consider adding calcium chloride to see if that helps, as you have heard that sometimes milk can need calcium chloride added at certain times of the year.
F. Verify that your thermometer is calibrated, as you know that temperature can affect coagulation time.

Answer: A, C, D, E, and F.

A selection of white mold, surface-ripened cheeses

10: WHITE MOLD, SURFACE-RIPENED CHEESES

FINISHED TEXTURE: Soft to semisoft
METHOD OF COAGULATION: Lactic acid and/or rennet
METHOD OF RIPENING: Internal bacteria and surface mold

When you talk about French cheeses, most of us immediately think of Brie and Camembert. These two surface-ripened cheeses are easy to love and are arguably the most imitated cheeses in the world. They seem to represent sophistication of palate and are therefore the gateway cheese for many cultures newly entering the world of artisan cheese consumption. Camembert and Brie, however, are just the edge-of-the-wedge (my cheese metaphor for "tip of the iceberg"). There are an almost uncountable number of surface-ripened cheeses—and yes, most of them are indeed French!

With surface-ripened cheeses, often called *bloomy rinds*, the place in which they are made and aged has as much influence on the end product as does the recipe. Since these cheeses depend on molds and yeasts growing on the surface of the cheese, the wild, native, and cultured fungi of that cheese's origin can be unique. For that reason it is quite a challenge to try to duplicate some of the many exotic surface-ripened cheeses that you might encounter abroad. But fortunately, the pure white bloomies, such as Brie and Camembert, are a bit easier to approach.

Before launching into the recipes, it's important to consider some basic science of surface-ripened cheese technology, as well as the different approaches to making the curd, selecting mold-ripening cultures, and making shapes. And finally, we'll cover a few extra pointers for aging and affinage, where bloomy rinds truly bloom.

THE FUNDAMENTAL SCIENCE OF WHITE MOLD–RIPENED CHEESES

The surface-mold ripened cheeses have the most complex ripening process of any cheese made. The desired surface molds will only grow well under very particular circumstances. And once they grow, another series of conditions must occur for the desired softening and development of flavor of the interior of the cheese. Let's take a look at the life cycle of a bloomy rind and the conditions the cheesemaker will have to create for successful production.

The bloomies start their lives similarly to soft, fresh cheeses. Some begin through a long, slow acid coagulation using little rennet, while others have a quick ripening and coagulation. As with soft, fresh cheeses, the curd is rarely heated and sometimes not even stirred. The process for making bloomies begins departing from a soft, fresh cheese with the addition to the milk (or sometimes later to the formed cheese) of white, or sometimes cream- or tan-colored, molds. Often, special yeasts and even surface-ripening bacteria such as *Brevibacterium linens* are added as well.

As with their fresh cousins, draining is done while the curd is still very high in moisture. This high moisture content provides the milk sugar (lactose) that the starter bacteria need to continue to acidify the curd during draining. By developing acid while in the forms, the curd will take on the typical texture of a bloomy rind cheese and reach the ideal pH level. The correct pH at salting, usually 4.6 to 4.7, accomplishes two things: It means that there is plenty of lactic acid present to "feed" the molds, and it creates a situation in which the curd pH is easy to change later (you'll hear more on that in a minute).

The perfect amount of salt is also necessary for flavor and to promote the growth of the right surface molds. The ideal amount is usually 1 to 2 percent of the weight of the cheeses. In the beginning you should weigh your unmolded cheeses, then calculate the salt amount based on that weight. (You can note in your logbook how much each shape weighs on average, then figure out future salt amounts based on these weights.)

If the milk was not inoculated with the desired molds, a hydrated mold solution is misted on after salting.

The cheeses must go through a short drying phase after salting to limit the surface moisture or else the white molds, which don't like it too wet, will not do well. This phase usually takes 1 to 2 days at 62°F (17°C) and 80 to 85 percent relative humidity (RH). Turning daily will help ensure even drying. Sometimes a fan is used to move air throughout the space to help with drying. It's important to get as even a drying as possible—you don't want the surface to dry to the point of forming a hard crust, but it should no longer be seeping whey or wet to the touch.

The first ripening phase is done at 50 to 55°F (10–13°C) at about 95 percent RH. At this point the cheeses must be turned daily. Wearing gloves or sanitizing your hands can help prevent black and brown mold "fingerprints" on the surface. If the conditions are right, you will detect a slight, slick feeling on the surface after just a few days, which means that yeasts are beginning to grow and prepare the surface for the molds. After a few more days, close inspection will reveal a growth of delicate white molds. At this point the cheeses should be continually turned until an even coat of mold covers the entire surface.

> Because of their high moisture content and high finish pH, these cheese types are at greater risk for growing pathogenic (harmful) bacteria that was either present in the milk already or from a subsequent contamination.

The surface yeasts and molds begin to change the pH of the cheese, raising it from the outside of the cheese to the center. Remember earlier when I said that the low pH at salting would mean that the pH could change later? Well, that happens during the final phase of aging. The finished cheese will have a very high pH of sometimes over 7.0 at the surface and over 6.0 in the core. Because of their high moisture content and high finish pH, these cheese types are at greater risk for growing pathogenic (harmful) bacteria that was either present in the milk already or from a subsequent contamination.

The last phase of ripening is intended to occur slowly so the flavor will develop and the paste soften (some varieties intended for longer shelf life, however, are meant to have a firmer paste). Ideally, the cheeses should be moved to a holding refrigerator with a temperature of about 38°F (3°C). The humidity must remain high, at about 95 percent, which is difficult to sustain at this temperature. Also, a refrigerator motor runs more frequently than an aging unit and will have its own drying effects on the cheese. For that reason it is normal to wrap the cheeses in ripening paper, a special cheese paper that will still allow for an exchange of oxygen but will help prevent drying. The second ripening phase can take 1 to 3 weeks, depending on the size of the cheese and the desired end texture.

A DEEPER LOOK AT THE SCIENCE OF BLOOMY RINDS

The first target that must be attained when making these cheeses is reaching the ideal pH. For long-set, lactic curd types, it should be quite low, at about 4.6 by the time it is ready for draining. Quick-set, rennet curd types should be much higher just before draining, at about 6.0 or higher, but should drop to 4.7 or so by the end of draining. The high whey retention in the curd provides the necessary lactose to ensure that the starter bacteria can continue to produce acid. This will provide enough lactic acid to provide lactate—which is basically the negatively charged portion of lactic acid—to feed the white molds.

In addition to providing lactate, the low pH does something equally important to the outcome of the cheese. During the drop in pH to this higher acid level, the curd will lose whey and calcium phosphate. Calcium phosphate in the curd acts as a buffer, absorbing hydrogen ions in an attempt to stabilize the pH. I like to use the example of Tums, the over-the-counter medication. The active ingredient in Tums is calcium carbonate. When taken to relieve heartburn (excess acid in your stomach), the carbonate absorbs and reduces that acid (a glass of milk, high in calcium phosphate, has a similar effect). By reducing the calcium phosphate level in the curd, you are creating a cheese with a low buffering capacity, meaning that its pH will be easier to change later on. You'll see why this is so important in a minute. (For a refresher on buffering, take a look back at chapter 3.)

It is critical to use salt on the surface of the cheese to promote the growth of the desired molds, which are more salt tolerant than the undesirable types. By the same token the molds that you want to encourage to grow will not do so if the cheese is too wet, but many of the unwanted ones will grow (once these unwanted molds crowd the surface, it is difficult for the white molds to find room). So proper salting followed by a drying phase are critical.

Now a series of reactions occurs that begins the process of converting the texture of the cheese. As the yeasts and molds begin to grow, they metabolize lactate into carbon dioxide and water. Oxygen is necessary for this process, so good air exchange is important; opening the door of your aging area to turn cheeses daily is usually sufficient. As the lactate is metabolized, the pH rises. Once the surface lactate has been consumed, lactate will begin to move from the core outward and also be consumed.

When the lactate is gone, the molds begin to break down proteins and produce ammonia, which moves toward the center of the cheese. Since ammonia is alkaline, this raises the pH even further. The production of ammonia also means that you need to ensure a good air exchange for the aging cheeses. (We've all smelled a bloomy that is overaged and reeks of ammonia, right?)

The other substance that moves outward is calcium phosphate. What little remains in the curd at the end of draining now attempts to buffer the increasing pH at the surface of the cheese. But it is an unsuccessful battle, as the ammonia production far outstrips the buffering capacity, so the pH continues to rise. The rate at which lactate can move out and ammonia inward depends on the permeability of the curd, which is affected by two major things—fat content and moisture content. If the fat level is high, the curd will be more stable (meaning more solid and difficult to change), and if the moisture content is too low, the movement of substances through the curd will decrease. Double and triple crème bloomies have what is considered a "stabilized" paste. It is still rich and wonderful in texture but will not be able to ooze and run.

Now the amazing part happens, when the increased pH means that the proteins can now begin to bind to water and soften. If you remember from chapter 8 we talked about the isoelectric point of casein—the point at which it loses its charge and becomes hydrophobic (water hating) and can form networks. As the pH goes back *up*, the proteins are taken farther *away* from their isoelectric point, meaning they begin to become hydrophilic again (water loving). This magical process is called *resolubilization*. There you have it, the secret of gooey soft cheese—resolubilization of casein!

Many books still credit the breakdown of proteins, *proteolysis*, by the white molds as the primary factor in creating an oozing cheese. While proteolysis and lipolysis—the breakdown of fats—are assisted by the activity of the mold and are responsible for the development of flavor and a certain amount of texture in these cheeses, it has been shown that the actual molds do not extend beyond a few millimeters into the surface of the cheese. So while these processes are critical for a superior cheese, they do not deserve the prime credit for oozing texture.

THE SECRETS OF MAKING BLOOMY RIND CHEESES

There are three basic foundation "makes" for bloomy rind cheeses: lactic curd, rennet curd, and stabilized curd. All three lead to very different end products. We'll compare the three major approaches for making bloomies, then cover the different culture and ripening agents, and finally I'll offer some pointers for choosing and using different forms to create cheeses of different shapes.

LACTIC CURD: OLD TECHNIQUES FOR MODERN MARVELS

I would say that the majority of traditional (which almost always means French) bloomy rind cheeses are from a long-set lactic curd, usually with the addition

Vegetable charcoal being sprinkled on the surface of bloomy rind cheeses in class at Pholia Farm

of a tiny bit of coagulant. The resulting texture is fine, smooth, and without openings. The flavor is tangy, but not overwhelmingly so. The tang and texture can be the perfect balance for a complex, flavor-packed rind that is made up of multiple, powerfully flavored molds.

Lactic curd is used for many of the unusual shapes created in traditional, especially French, surface-ripened cheeses, such as pyramids, buttons, and *crottins*. Often the lactic curd is predrained for a brief time to make the filling of forms and creating of shapes more successful.

In France many of these cheeses are still made from raw milk, but many other countries require cheeses aged under a certain amount of time, often 60 days, to be made from pasteurized milk. Small shapes such as pyramids and *crottins*, a small cylinder shape in which many classic French bloomy rinds are made, are not only quite difficult to successfully age this long, but also have a very brief window of perfect ripeness. Pasteurizing can also assist with allowing the added starter cultures and molds to grow unimpeded.

If you are comfortable with the quality of your raw milk, however, and are not selling the cheeses in a jurisdiction that does not allow for this, it is perfectly acceptable to try making them from unprocessed milk. You can expect some other, native wild molds to provide some competition, leading to mottled rinds, but this is far more similar to the traditional French varieties than are the pure-white cheeses.

RENNET CURD: A SHORTER ROUTE TO DELICIOUS RESULTS

Short-set, renneted curd is used for many Camembert- and Brie-type recipes. The curd is usually set over a period of 1 to 3 hours, cut in cubes or columns, then either ladled or stirred briefly without heating, then ladled some more. The curd retains a high level of whey, allowing for acidification during the draining period, as opposed to during the ripening period. This curd knits together differently from an acid curd, thanks to the retention of more minerals, creating a cheese with a bit more structure and body than its acid curd cousins.

Rennet curd bloomies are often drained in open-ended hoops and placed on mats. For large Brie forms, the hoop has two parts that stack on top of each other. When the curd drains down to the level of the bottom hoop, the top section is removed and the remaining portion is flipped. If the forms were not in two sections, the curd would drain down to about half the level of the single form. When flipped, the still-fragile wheel is at high risk for breaking if it has to fall too far. By forming the curd in a two-part mold, the wheel can be turned without damaging the structure.

Some versions of this type of cheese are still made from raw milk in France, but in most parts of the world, they are made from pasteurized milk for the same reasons as with the acid varieties. Large format shapes, such as Brie, can take a bit longer to age and can often reach the 60-day limit.

A rennet curd Brie-type cheese aging showing an even coat of soft, white mold growth

STABILIZED CURD: TECHNOLOGY FOR TEXTURE

It took me a long time to figure out just what people meant when they used the term "stabilized curd" or "stabilized paste." I knew it meant bloomy rinds that could be aged for a longer time, and therefore are often made with raw milk, but I didn't understand how they were made. I hope after reading this you will be able to completely grasp the subject.

A stabilized-paste, surface-ripened cheese is one that either never softens completely or has a delayed softening. This increases the window of time during which this cheese can be sold, provides a more durable texture for shipping and selling, and allows for raw-milk versions that can be held to the required age. For consumers who haven't acquired affection for the fully ripened, oozing bloomy, it can also be a preferable alternative.

The primary secret to creating a stable paste lies in the pH at the end of draining being higher than for the other types. The goal pH for most surface-ripened cheeses at the end of draining is 4.7 to 4.8. For a stabilized type it is 5.0 to 5.25. There are several options for attaining a less acid, sweeter curd. The most common method is to replace part or all of the mesophilic culture with a thermophilic type; this will cause less acid to be produced during the make. The second method involves washing the curd to remove a portion of the lactose and therefore reduce the availability of food for the starter culture during draining. The final method is to increase the fat content of the curd, which is really less about stabilizing the paste than it is about creating a double or triple crème. Increased fat, as we learned earlier, creates a curd that is less permeable and consequently inhibits the movement of compounds during ripening, therefore impeding the softening process.

USING STARTER AND RIPENING CULTURES FOR WHITE MOLD–RIPENED CHEESES

Since you now know that these cheese types have the most complicated ripening system, you can probably imagine that they also have some complicated culture blends and options. These blends can be especially important for short-ripened, pasteurized versions that are aged in fairly sterile aging units. The traditional

A BOUNTY OF CULTURE CHOICES

There are actually more starters and ripening cultures to choose from than I cover in this section. Should you get your hands on a catalogue for commercial cheesemakers, you will likely be overwhelmed by the variety. I have only included the more commonly available varieties for this book, and especially those that are available in smaller packages. Just knowing how many choices exist might help you better appreciate the amazing variety and complexity of these cheese types and the lengths to which we will go to try to replicate what used to be provided by the natural environment in which these cheeses were produced.

French cheeses counted on inoculation of the milk by nonstarter bacteria that provided ripening enzymes, as well as environmental molds and yeasts present in a long-established, nonsterile aging environment to provide unique and complex flora. Let's review some of the many starter cultures and ripening mold and yeast cultures, and how to inoculate the milk directly or how to spray the molds on the cheeses after salting.

STARTER CULTURES FOR BLOOMY RIND CHEESES

Table 10–1 might look a little intimidating at first. But once we go over some of the things you learned from chapter 2, I hope all will become clear.

A starter culture's primary function is to produce acid. After the cultures die (which, as you now know, they are eventually supposed to do), they release enzymes that then break down other compounds in the cheese, thereby creating aroma and flavor. Some of these aroma producers act quickly, and others work over time. Raw milk contains nonstarter bacteria that will provide many subtle aromas and flavors. If you are working with pasteurized milk and finding your cheeses a little bland, try choosing one of the blends (or make your own) that has more aroma producers. (For more on these cultures and their individual qualities, refer back to chapter 2.)

The cultures are color coded based on their similarity. The brown highlighted blends contain the same types of bacteria, but the proportions of each type may be different—this is proprietary information that culture manufacturers don't really want us to know! I have included blends from the three major culture manufacturers. In the table are three main types of blends, in decreasing order of aroma production. There is one thermophilic strain (for a stabilized paste) that can be used alone, along with one of the first three blends, or with aroma producers alone, in varying ratios depending on your goal. I have included two individually packaged aroma cultures that can be handy to have on hand when creating your own blends. In the recipes, however, we will rely upon premixed blends.

MOLD AND RIPENING CULTURES FOR BLOOMY RIND CHEESES

When making bloomy rinds you will be selecting from several strains of *Penicillium candidum* (also called *Penicillium camemberti*) and *Geotrichum candidum*. Some bloomies use only *P. candidum*, but most include a little *Geotrichum* (at one-quarter the amount of *P. candidum*) too. In addition, some recipes will call for the addition of specific yeasts to help deacidify the surface of the cheese more rapidly, ensuring a good growth of white molds later. In the recipes that follow, I will be using the most popular strains of *P. candidum* and *Geotrichum*, but don't be afraid to buy some of the others and experiment with different combinations—this is how you can eventually create your own masterpieces!

Bloomies can be made using only *Geotrichum*. These are usually lactic curd types that are only ripened for a few days to a week or so. They are firm, mild, and pleasant—and usually small in format. You can experiment with the different *Geotrichum* strains that are available. If you have the right aging environment (i.e., something

TABLE 10-1. STARTER AND AROMA CULTURES FOR SURFACE-RIPENED CHEESES

Name	Contains	Manufacturer	Small Batch Dosage:* 3–4 Gal (12–15 L/26–34 lb)	Large Batch Dosage: 26 Gal (100 L/220 lb)	Characteristics and Application
Flora Danica	LL, LLC, LLD, LMC	Packaged by home cheese supply co.	¼ tsp (0.4 g)	1 packet/ 50 gal (190 L)	Starter and aroma cultures packaged for small batches
Flora Danica or CHN 11, 19	LL, LLC, LLD, LMC	Chr. Hansen	¼ tsp (0.4 g)	10U	Starter and aroma cultures
Meso Aromatic B	LL, LC, LD, LM	Abiasa	¼– ½ tsp (0.6–1.2 g)	10 g	Starter and aroma cultures
MO 030R	LL, LC, LD, LM	Sacco Clerici	1⁄16 tsp (0.3 g)	0.5–4U	Starter and aroma cultures
MM 100 Series (Choozit) (includes MM 100, 101)	LL, LLC, LLD	Danisco	⅛ tsp (0.3 g)	4U	Starter and milder aroma
MA Series (Choozit) (includes MA 11, 14, 16, 19)	LL, LLC	Danisco	⅛ tsp (0.3 g)	4U	Starter and some aroma production
Meso III	LL, LC	Abiasa	¼– ½ tsp (0.6–1.2 g)	10 g	Starter and aroma cultures
MO 030	LL, LC	Sacco Clerici	1⁄16 tsp (0.3 g)	0.5–10U	Starter and aroma cultures
TA 50 Series (Choozit) (includes TA 52, 54)	ST (slow-acid producer)	Danisco	Varying ratios with other starters	Varying ratios with other starters	Use alone or with other starters and/or aroma producers for stabilized paste
LM 57 (Choozit)	LM (LMC)	Danisco	Add 1 part for every 2 parts starter	Add 1 part for every 2 parts starter	Aroma culture, use with starter cultures in yellow and red
MD Series (Choozit) (includes MD 88, 89)	LLD	Danisco	Add 1 part for every 2 parts starter	Add 1 part for every 2 parts starter	Aroma culture (with some starter activity), use with starter cultures in red

* Remember that culture is always best measured by weighing and dividing into unit doses based on the strength of the culture, instead of using volume or set weight.

with an absorbent surface such as wood boards or stone walls), it will eventually become inoculated with spores not only from the introduced molds and yeasts but also from wild versions.

I have included mold and yeast strains from two major culture houses, along with the repackaged, convenient small packets from companies such as New England Cheese Supply. These versions do not offer

THE CHEESE NUN

For a fun lesson in the diversity of white molds and yeasts—especially *Geotrichum*—rent, buy, or borrow the PBS documentary *The Cheese Nun*. This 2006 film follows Sister Noella Marcellino, a Benedictine nun from the Abbey of Regina Laudis in Connecticut, as she follows her passion for making raw-milk cheese, and the support it affords her abbey, through a journey that includes obtaining her PhD and spending over 3 years in France (partially on a Fulbright scholarship) studying surface-ripened cheeses. She remains one of the world's foremost experts on bloomy rind, raw-milk cheeses. While still considered a cloistered member of her abbey, she continues to champion cheesemakers by serving as a judge and a consultant at such cheese events as the American Cheese Society (ACS) annual conference. In fact, Vern, my husband, was able to spend over an hour chatting with her and trying to troubleshoot photos of problem cheeses that he had brought with him to an ACS conference in Montreal, Canada. She is an amazing, educated, and passionate advocate for cheese and cheesemakers and, frankly, just the kind of person more of us should aspire to be.

variety, but they are premeasured so they are easier to dose for small batches. If you are not making this type of cheese often, it can be less expensive as well as more convenient to use these cultures.

Penicillium candidum is produced in several strains: some that grow more slowly (for stabilized paste), some with traditional growth, and others in a mucor (black, grey mold)-resistant strain. I will be using only one in

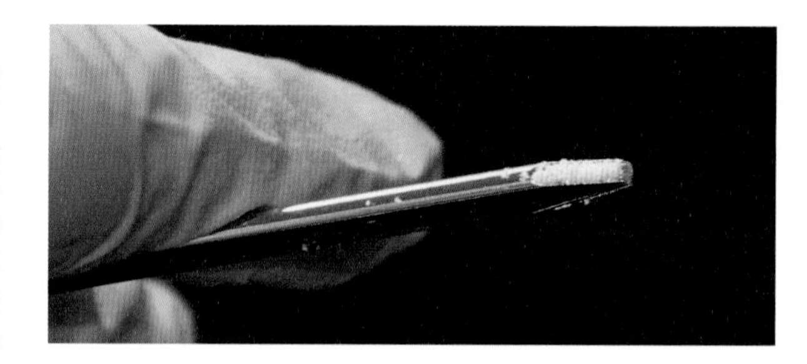

Using a metal meat skewer to measure tiny doses of mold and yeast cultures

the recipes, but you can try the others and see what happens. With all of these mold cultures, sometimes the choice will boil down to the most convenient size you can purchase, since so little is used in the cheese. The white mold cultures listed in the table are what is currently available. When shopping, keep in mind that names and strains might change and some variations might have been discontinued.

Geotrichum is not a true mold but instead is a yeast that in some forms grows in a fashion that looks and acts like a mold, creating fuzzy tufts of growth (if you can remember that both molds and yeasts are fungi, you are on your way to sorting out the necessary information). When choosing a variety of *Geotrichum*, consider the final look of the cheese. A more yeastlike strain will be flatter, slicker, and cream-colored, while a moldlike strain will be fuzzy, dry, and whiter.

Additional yeasts are helpful to add when working with pasteurized milk. Yeasts grow quicker than the white molds and prepare the surface by lowering the acid level, making it a more hospitable environment for the mold. Yeasts are present in the air and raw milk, so they don't have to be added, but again, give them a try and see what you think! The dosage rate on yeasts is very low, so for small batches the dose can be difficult to impossible to measure accurately. For small batches I recommend using the flat tip of a metal meat skewer (like the ones you use to make shish kebabs) to measure mold and yeast dosages.

INOCULATING: CHOICES FOR ADDING MOLD CULTURES

Without a doubt I prefer to inoculate the mold and ripening cultures directly into the milk, as opposed

TABLE 10-2. *PENICILLIUM CANDIDUM (CAMEMBERTI)* MOLD CULTURES FOR BLOOMY RIND CHEESES

Name/ Dosage per Package	Manufacturer	Small Batch Dosage*	Large Batch Dosage*	Characteristics and Application
Penicillium candidum **50 gal/pkg**	Packaged by home cheese supply co.	⅛ tsp/4–8 gal (15–30 L/34–69 lb)	1 packet/40–80 gal (140–300 L/310–660 lb)	Traditional and stabilized varieties
PC 33 (SAM, 3) (Choozit) 2- and 10-dose package	Danisco	1/16 tsp/2–4 gal (8–15 L/17–34 lb)	⅓–½ dose/26 gal (100 L/220 lb)	Resists growth of mucor mold
PC 43 (ABL) (Choozit) 10-dose package	Danisco	1/16 tsp/2–4 gal (8–15 L/17–34 lb)	⅓–½ dose/26 gal (100 L/220 lb)	Traditional
PC 53 (Neige) (Choozit) 2- and 10-dose package	Danisco	1/16 tsp/2–4 gal (8–15 L/17–34 lb)	⅓–½ dose/26 gal (100 L/220 lb)	Traditional and stabilized, stronger aroma and flavor
PCA3 2- and 10-dose package	Chr. Hansen	1/16 tsp/2–4 gal (8–15 L/17–34 lb)	⅓–½ dose/26 gal (100 L/220 lb)	Traditional and stabilized varieties

*When added directly to the milk before coagulation

TABLE 10-3. *GEOTRICHUM CANDIDUM* MOLD CULTURES FOR BLOOMY RIND CHEESES

Name/ Dosage per Package	Manufacturer	Dosage when used with *P. candidum**	Characteristics and Application
Geotrichum candidum **50 gallons/package**	Packaged by home cheese supply co.	Add at ¼ the rate of *P. candidum*	Traditional and stabilized
GEO 13 10-dose package	Danisco Choozit	Add at ¼ the rate of *P. candidum*	Intermediate (between yeast and moldlike growth, stronger flavor and aroma)
GEO 15 2- and 10-dose package	Danisco Choozit	Add at ¼ the rate of *P. candidum*	Yeastlike growth, mild flavor and aroma
GEO 17 10-dose package	Danisco Choozit	Add at ¼ the rate of *P. candidum*	Moldlike growth, very mild flavor and aroma
GEO CB 10-dose package	Chr. Hansen	Add at ¼ the rate of *P. candidum*	Moldlike growth

*When added directly to the milk before coagulation

to spraying them on the cheeses later. It is easier to be accurate, and you are more likely to have an even distribution on the cheese. That being said, this method uses more culture than if the cultures are misted directly onto the cheese surface. So if you are a higher-production cheesemaker, misting might be a better choice.

TABLE 10-4. YEAST CULTURES FOR BLOOMY RIND CHEESES

Name/ Dosage per package	Yeast	Manufacturer	Dosage*	Characteristics and Application
KL 71 (Choozit) **2- and 10-dose package**	*Kluyveromyces lactis*	Danisco	Innoculate outside of cheese	Can also ferment lactose (unusual for a yeast), so apply as a spray after salting
LAF 4,5 **10-dose package**	*Kluyveromyces marxianus*	Chr. Hansen	0.1 dose/26 gal (100 L/220 lb)	Assistance with acid neutralization on rind
DH (Choozit) **10-dose package**	*Debaryomyces hansenii*	Danisco	0.1 dose/26 gal (100 L/220 lb)	Neutralizes lactic acid, production of proteolytic and lipolytic enzymes, inhibits growth of undesirable fungus
LAF 3 **10-dose package**	*Debaryomyces hansenii*	Chr. Hansen	0.1 dose/26 gal (100 L/220 lb)	Neutralizes lactic acid, production of proteolytic and lipolytic enzymes, inhibits growth of undesirable fungus

*When added directly to the milk before coagulation

When adding directly to the milk, simply sprinkle the powdered cultures onto the surface of the milk at the same time as adding—or immediately after—the starter culture. The spores will rehydrate but remain dormant until the conditions are right for growth.

Spray inoculation is performed after the cheeses have been salted. If they have been dry salted, wait until the salt has been absorbed and the cheeses are no longer weeping any whey. If brine salted, let the cheeses sit on the rack for a bit until they no longer appear wet.

When spray inoculating, mold and yeast cultures must be properly rehydrated before applying to the cheese. Combine ½ teaspoon (2.5 g) of pure salt, ½ teaspoon table sugar, and 1 quart (950 ml) nonchlorinated water, add a two-dose packet of mold powder to the solution, and let it rehydrate at room temperature for about 16 hours. (Mixed inoculant can be stored in the refrigerator for about two months.)

Use the finest misting bottle you can find. I have used the olive oil sprayer called "Misto," but there is a similar bottle designed to mist vermouth onto the top of a martini. You can use a regular plastic spray bottle, too; just test it first by spraying into the air and seeing how small the droplets are. Usually, you can spray fairly high above the cheeses and let the fine droplets spread out to coat the surfaces evenly.

CREATING INTERESTING SHAPES

Many interesting shapes can be made using the lactic curd predrained method. From hearts to pyramids, buttons to logs, the limit is really only one of size and practicality during aging. Remember, these cheeses are all small of scale, as you cannot easily (or successfully) create a large-format, acid-coagulated cheese. Using the rennet curd method, however, you can make some on a bit larger scale.

Pyramids are not turned during draining, as it would negatively affect their shape. They actually don't need turning anyway: Since the wide part is at the top, the whey can drain out through the tapered sides fairly well. Pyramids are often best made from predrained curd so the forms can be filled to the limit without much shrinkage occurring during the rest of the draining. They can also be made using rennet curd and hand-ladled acid curd. When using this method, take special care to refill the forms as they settle (or you will end up with a very small pyramid!). Sometimes the bottoms (which, as described previously, are at the top during draining) will become concave, with the sides adhering to the form as it drains. To help prevent this, fill the form so it is a bit higher in the center or trim the bottom after it is unmolded.

Juniper Grove Farm "Bûche": *Geotrichum*-ripened logs beautifully presented with a wheat straw in the center for stability. Made by pioneering artisan cheesemaker (and friend) Pierre Kolisch in central Oregon. Photo by Vern Caldwell

Buttons can be hand-formed from well-drained curd. To form a button, wear gloves or use sanitized hands to roll a small shape. Place the shape on a fine-mesh cheese mat to air-dry a bit before salting.

Small forms with only one side open (as opposed to the two open ends of Brie and Camembert forms) can be flipped during drainage in two ways. First, you can carefully slide the cheese out into your sanitized or gloved hand, then quickly reverse it into the form. If it doesn't settle properly, jiggle it gently to help it seat well into the bottom of the form. The second method involves using an empty form into which to slide the cheese directly from the first form. (This is the method used in larger-scale production, with large, multiple form racks that are sandwiched together and then flipped as one.)

Layered cheeses, where you see a little line of ash or spice running through the center of the wheel, take special effort to make. The first layer is allowed to drain for a bit, then sprinkled with ash or spices; after that, curd that is not too wet is layered on top. You can accomplish a similar effect by using two forms and sprinkling the top of one with the ash or spice, then unmolding the second form onto the top of the first. This will help prevent the loss of too much of your ash layer to whey from undrained curd.

ACHIEVING SUCCESSFUL DRAINING

Keeping the room at the stated goal temperature during draining is probably more critical for this cheese type than for any other cheese. You really have to think ahead when you plan on making these cheeses, as some times of the year it might be more difficult to maintain the right temperature and humidity. For small batches you can use smaller, insulated spaces (such as an ice chest) to help control the temperature; cover the draining forms with bubble wrap and a towel to help keep them warm; or cool the room with a window air conditioner if the temperature is too warm. However you do it, try to provide the ideal conditions.

Notice if your forms have little "feet"—small bumps or protrusions on the form that elevate the bottom of

Two-part form, organdy, and mats for draining a Brie-type cheese at La Suerte Creamery in Argentina

the form slightly above the drain board; if not, you should set them on an elevated rack to help prevent whey from pooling underneath and being impeded from escaping the forms. If you are worried about the tops drying out during draining, you can lay a cheese mat over the forms, then place a damp cheesecloth on top of that. This will also help keep the temperature even.

No matter what the recipe says, if the curd looks too fragile to drain, let it rest a bit longer. If you notice that the curd is too large and whey pockets are developing, take a sanitized knife and run it gently through the curd in vertical cuts to help release the trapped whey.

> Using a piece of ultrafine weave organdy over the mat can help prevent curd from sticking and tearing when the cheeses are flipped.

Forms that have two ends open, such as Camembert and Brie forms, are drained by first placing a rack on the draining surface; a cheese mat on the rack; and a piece of organdy or other tight-weave cloth over the mat. (The cloth can be omitted if the mat is of fine enough texture to not allow curd to escape or become locked in the weave of the mat, which leads to tearing when the cheese is flipped.) The form is placed on the top layer, and a reversed layer of the same is set on top of the form. When turning the cheese, you must hold this entire stack together and turn as one.

SALTING

The most common way to salt bloomy rinds is by sprinkling them with dry, flaked noniodized salt. The ideal amount is 1 to 2 percent of the weight of the curd. But you can also submerge the cheeses into saturated brine for a brief period of time. Brining brings about some advantages, such as evenness of salt, and supports the shapes while they are still fragile. But timing is critical. Since shapes and sizes vary so much, it is difficult for me to give you accurate guidelines, so take the suggestions in the recipes with (dare I say it?) a grain of salt.

If you are dry salting the cheeses, unmold them carefully onto a rack covered by a cheese mat. Unmold with enough space between each shape to be able to salt the sides easily. Measure the amount of salt needed—remember to use a noniodized flake salt, as the flakes stick to the surface better than a grain and give the salt time to melt and absorb—and carefully sprinkle it on with your fingers or a sifter, sieve, or shaker. Wheel shapes can be rolled to coat the sides, but know that more will fall off this surface (some cheesemakers, however, omit salting the sides of very thin wheels). For pyramids I like to salt the bottoms *before* unmolding, as they are very difficult to move without damaging once they are unmolded and before they have dried.

With all shapes try to make sure that most of the salt falls on the cheeses, not the table. If you find too much not ending up where it should, use a bit more than the called-for 2 percent. Also, if you see that a lot of the salt melts off the cheeses—this will happen if they have not drained well enough by the time they are unmolded—then also use a bit more.

Brining these shapes can work very well as long as they can be handled without damaging during transition or brining. Immerse in a heavy brine for about 20 to 30 minutes for small shapes. Sprinkle dry salt on the floating surface or turn halfway through. Be sure to note the time that they were brined in your logbook, so if you find there is too much or too little salt, you can make adjustments on future batches.

If you are using ash on the outside of the rind (for more on using ash, refer back to chapter 5), you can apply it at the same time as the salt, but I prefer to add it partway through the drying phase, when the outside of the cheeses are still moist but not too wet. This helps create an even layer of ash. If the cheese is too wet, some ash may run off. If it is too dry, not enough will adhere. Use a fine sieve to apply an even coat. Ash the bottom of the shapes first (especially for pyramids), then the other sides, to help prevent fingerprints.

AGING AND STORING

Most of the secrets to the successful aging of bloomy rinds will come from understanding the science involved in making them, so here are some other tips and secrets for working with surface-ripened cheeses.

Turning daily is extremely important with these cheese types. If you don't, you will soon find that the white molds will grow around the mat, and when you flip the cheese, the entire skin will come off and stay on the mat. When this happens, the cheese is ruined, or . . . you can eat it right away! When turning, be sure to use gloved or sanitized hands, or you will be inoculating the surface of the cheeses with unwanted mold spores. Pyramids cannot be turned but should be lifted from the mat and set down in a slightly different orientation. Large-format Brie types should be flipped using a second cheese mat (sanitized and dry) placed on top of the cheese; then slide the supporting mat and the cheese onto one hand while holding the mat down on top. Flip the cheese, and remove the old mat.

If mold growth is too slow, you can respray the surface of the cheeses with a white mold and yeast mixture as described earlier. Do this very carefully so as not to wet the surface of the cheese any more than is necessary.

Monitor humidity religiously for these cheeses. It won't take much to ruin them at any stage before the final ripening. Once they are wrapped and moved into a cooler environment, they are a bit more impervious to fluctuations in the relative humidity. After being wrapped they are basically in their own life-supporting habitat.

So what about wrapping? It must be done properly to create this "biosphere." You can buy several versions of superior cheese-wrap paper. They usually consist of a two-part construction. The inner lining—the one next to the cheese—is often a sulfur-infused, greaseproof paper with a cellophane exterior. The breathable, but nonstick, sulfur paper does several things: First, it helps hold moisture near the cheese without being wet, allows for air exchange so the cheese can breathe, and helps counteract the ammonia produced by the cheese (sulfur is acidic, while ammonia is basic, so they neutralize each other). The cellophane exterior is somewhat breathable, but less so; hence, it helps keep the cheese from drying while it continues to ripen. You can substitute other papers, if you don't want to invest in "official" cheese-wrap papers (which aren't cheap and come in set sizes).

There are a few more readily available substitutes for wrapping bloomies that work decently. Some options for the inner lining are greaseproof paper (meant for wrapping hamburgers), thin parchment paper (for lining baking sheets), or even waxed paper. For the outside you can use cellophane, which is somewhat breathable (100 percent biodegradable, but its manufacture involves some pollutants), perforated cling film, or perforated aluminum foil. The cling film and foil can be purchased in sheets or rolls, but I have had good results with perforating my own using an ice pick or other perforating tool. If you are unsure of your wrap's effectiveness, simply unwrap the cheese occasionally, observe it, smell it, and feel it. You should be able to determine if your process is doing the job or not.

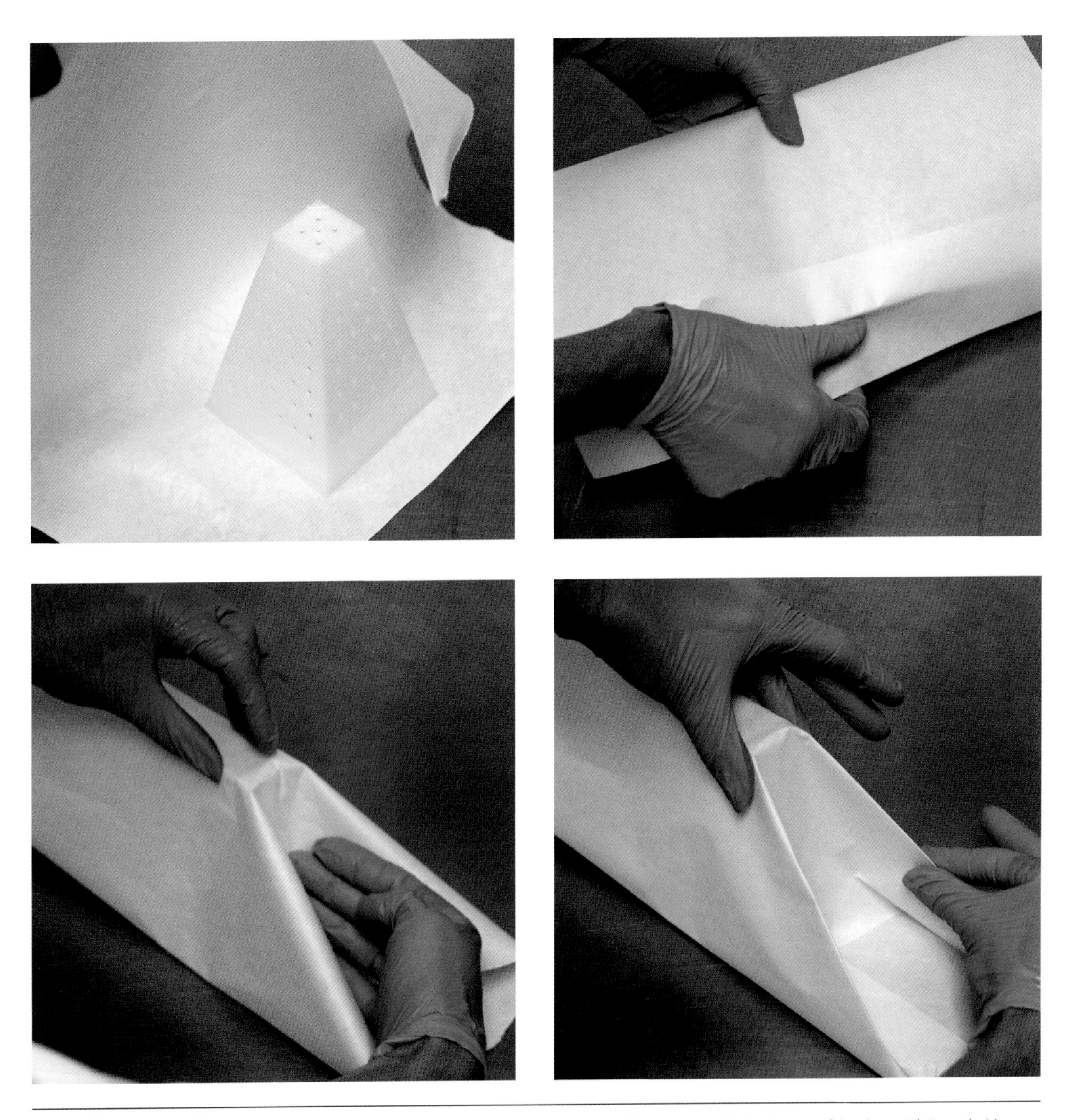

Steps in wrapping a pyramid shape: 1) Set cheese on paper and bring paper across the top; 2) Fold the paper under the bottom of the cheese; 3) On each side press the paper down to form two "wings"; 4) Crease the wings, then fold toward the cheese; tuck the remaining paper under the cheese, and apply a label or sticky tab.

MEGHAN MCKENNA, *MOUNTAIN LODGE FARM*, WASHINGTON STATE

If there ever was a cheesemaker fairy tale, it could be the story of Meghan McKenna and Mountain Lodge Farm. Meghan, a soft-spoken, lovely young woman, has no trouble fitting right into the role of cheesemaker princess. While our story is lacking an evil stepmother, there is a benevolent fairy godmother—the ever-generous, energy-filled, goat-loving owner of Mountain Lodge Farm, Sherwin Ferguson.

Meghan and Sherwin met in 2009 while both were attending the Quillisascut Farm School, a farm-based cheesemaking and food educational program founded and run by Rick and Lora Lea Misterly in Washington state. Meghan, then a cheesemonger at Metropolitan Market in Seattle, was pursuing her passion for learning more about the cheeses she already loved to sell and eat, while Sherwin, a nurse practitioner, mother, and budding farmer, was methodically steering her way toward a small goat and sheep creamery. The two women felt an immediate kinship. To top it off, Sherwin had land, capital, and the need for a cheesemaker, while Meghan had neither of the first two but did have a thirst for becoming a maker of cheese.

In the fall of 2010, Meghan and her husband Shawn spent several months in the French Pyrenees tending sheep and cows, including milking and herding, making cheese, and participating in that great agrarian event called *transhumance*, when the animals are moved from valley to highland pastures or vice versa. Upon their return to the United States, Sherwin (who by then had built up quite a herd of Nigerian Dwarf and La Mancha goats) announced that the plans were drawn and construction of their artisan dairy was about to begin—with Shawn and Meghan as official parts of the team. The two joined Sherwin and her husband, Bob, at Mountain Lodge, and Meghan began to craft the recipes for the first Mountain Lodge Farm cheeses.

With no proper facility, but knowing the importance of food safety, the team designed what they fondly referred to as "The Cheese Bubble," a plastic-sheeting-segregated area in which Meghan could experiment and hone her skills. By the summer she had taken samples of her cheeses to the American Cheese Society conference for critique, where she received glowing reviews and helpful advice from such cheese experts as Sister Noella Marcellino, a.k.a. the Cheese Nun. By the fall of 2011, she had won six firsts, including one for her ashed pyramid "Summit" and Best in Show for a crottin style called "Paradise" in the amateur category of the American Dairy Goat Association's cheese competition.

With such "wins" beneath their wings and the team working as one, we can all expect that amazing cheeses will continue to be crafted by Meghan and that this cheesemaker fairy tale is only just beginning.

Meghan McKenna, cheesemaker at Mountain Lodge Farm, Washington state. Photo courtesy of Brian Weir

LACTIC-SET BLOOMY RIND CHEESE

This large family of cheeses includes many small-production farmhouse cheeses produced originally in France. There are so many amazing variations that it is almost a sin to name just a few of the better-known varieties here. The lactic acid cheese family is home to many goat's milk varieties such as Saint-Maure de Touraine, a French AOC goat's milk cheese that is formed in a log shape with a straw down the center. Juniper Grove Farm in Redmond, Oregon, makes a beautiful *Geotrichum*-ripened goat cheese log artfully pierced with a wheat straw that cheesemaker Pierre Kolisch calls *Bûche*.

And Brie de Melun (cousin of the better-known and larger-size Brie de Meaux) is another French AOC cheese with a long history and loyal following. This raw cow's milk cheese is made in discs that age in reed mats and gain a distinctive white and red rind. California Crottin is the award-winning creation of pioneering cheesemaker Jennifer Bice from Redwood Hill Farm in California. Her delectable goat's milk crottin is made using the lactic-set, ladled method and has a *Geotrichum*-wrinkled rind and tender, earthy paste that is quite loyal to its French ancestors.

INGREDIENTS

Milk: 1 gal (4 L/8.6 lbs) whole milk
Culture: 1/16 tsp (0.1 g) Flora Danica or Aroma B, 4 skewer tips *P. candidum*, and 1 skewer tip Geo 15
Rennet: 4 drops single-strength rennet in ⅛ cup (30 ml) cool, nonchlorinated water
Salt: 1 tsp (5 g) pure salt (goal is 1.9–2% of weight of drained curd)

STEPS

Prepare Equipment: Make sure all equipment is cleaned and sanitized and that your cheesemaking space is free from possible contaminants. Refer to chapter 6 for tips on proper equipment preparation.

Prepare Milk: Warm milk to 68 to 72°F (20–22°C).
TIPS: I highly recommend pasteurizing milk (even if for home use only) as long ripening time makes milk more susceptible to growth of unwanted organisms. That being said, many of the French greats are made from raw (or thermized) milk.

Culture: Sprinkle cultures on milk surface; let set 3 to 5 minutes, then stir gently for 2 to 5 minutes.

Rennet: Stir in rennet using an up-and-down motion for about 1 minute. Still the milk.

Ripen/Coagulate: Allow to ripen for 18 to 24 hours at 68 to 70°F (20–21°C) until it reaches a pH of 4.5 to 4.6.
TIPS: Length of ripening will vary according to room temperature and amount of culture used. You can purposely shorten or extend this time according to your availability. If you have trouble maintaining the temperature, place the water bath and cultured milk on a counter and wrap with towels or a blanket. Alternatively, you can set the pot in an ice chest with a few jugs of water that are at 68 to 70°F (20–21°C).

Drain: Ladle curd in thin layers into forms set on draining mat. Drain in forms for 24 hours, turning one or two times over the draining period. Room temperature should be 68 to 72°F (20–22°C) at beginning of draining and slowly cool to 62°F (17°C) by the end.

Salt and Drying: Unmold and sprinkle with 0.01 pounds (usually about ¼ tsp per 6 oz cheese) salt. Let set on draining mats for 24 hours. Humidity in the room should be 80 to 85 percent and room temperature 62°F (17°C). Use a low-velocity fan if cheeses aren't drying within time frame.

Ripening and Affinage: Place in ripening fridge at 52 to 55°F (11–13°C) and 95 to 97 percent RH. Turn daily. When mold growth is even, you can wrap or pat growth down. Ideally, wrap the cheeses and move to a cooler fridge, 38 to 40°F (3–4°C).

LARGE-BATCH GUIDELINES

Milk: 12 gal (46 L/100 lb) whole milk (for larger batches increase ingredients proportionately)
Culture: 1.5–2U Flora Danica or Aroma B, 0.2–0.3 dose *P. candidum*, and 0.05–0.10 dose Geo 15
Rennet: 4 ml single-strength rennet (2 ml double strength)
Salt: 1.9–2.1% of weight of drained curd

RENNET CURD BLOOMY RIND CHEESE

Rennet curd bloomy rinded cheeses are probably some of the most popular and most copied cheeses in the world. Imported French Brie is what I consider a "gateway" cheese: It is often the first cheese people try when expanding their palates and budgets to include fine fromage. Brie de Meaux is considered

Goat milk, lactic curd ashed bloomy rind "Summit" from Mountain Lodge Farm. Photo courtesy of Shawn McKenna

the granddaddy of all bloomy rinded cheeses. This AOC raw cow's milk cheese has almost untold descendants, as well as imposters, made from the raw and pasteurized milk of cow, goat, and sheep.

One of my favorite US derivations is Dinah's Cheese from Kurtwood Farms in Washington state. Owner and cheesemaker Kurt Timmermeister (also the author of *Growing a Farmer* and a forthcoming book called *Growing a Feast*) makes his Camembert-size bloomy rind from the milk of his fifteen-cow Jersey herd. Another gem is Leigh Belle made by Brian Futhey of Stone Meadow Farm in Pennsylvania. His Jersey and Milking Shorthorn–cross cows graze on rotational pastures, helping create a dense, yellow milk that showcases Brian's knack for great cheesemaking.

INGREDIENTS

Milk: 3 gal (12 L/26 lbs) whole milk (this amount of milk will produce a single, traditional-size wheel of approximately 13 x 1.5 inches (33 x 4 cm)
Culture: ⅛ teaspoon (0.2 g) Flora Danica, 1/16 tsp (0.3 g) *P. candidum*, and 2 skewer tips of Geo 15 or 17
Calcium chloride (optional): maximum ¾ tsp (3.75 ml) diluted in ⅛ cup (30 ml) cool, nonchlorinated water
Rennet: Scant ⅜ tsp (1.8 ml) single-strength rennet diluted in ⅛ cup (30 ml) cool, nonchlorinated water
Salt: 3 tsp (15–16 g) pure salt (goal is 2% of weight of drained curd)

STEPS

Prepare Equipment: Make sure all equipment is cleaned and sanitized and that your cheesemaking space is free from possible contaminants. Refer to chapter 6 for tips on proper equipment preparation.

Prepare Milk: Warm milk to 94°F (34°C).

Culture: Sprinkle cultures on top. Let set for 3 to 5 minutes, then stir gently for 2 to 5 minutes.

Ripen: Allow to ripen for 35 to 45 minutes, until you see the pH drop to about 6.55 (0.05 drop).

Additions: Stir in calcium chloride for 1 minute and wait 5 minutes before adding rennet.

Rennet and Coagulate: Add rennet solution, and stir gently with an up-and-down motion, then still milk. Cover, and let set, maintaining 94°F (34°C) until clean break is achieved. Goal time until ready to cut is 30 to 35 minutes. If using flocculation factor, multiply time to flocculation by 6 for total time until cut—or watch for clean break.

Cut: Cut curd into ¾-inch (2-cm) cubes, and let rest 15 to 30 minutes. The pH at cut should be 6.45 to 6.50. Rest until ¾-inch whey covers curds.

> TIPS: During the rest phase do not increase the vat temperature or attempt to maintain it.

Stir: Stir gently to loosen the curds, rest 15 to 30 minutes, then stir again. Repeat stir and rest phases until curd texture is even throughout. The pH at the end of stirring should be 6.40 to 6.45. Do not reheat vat; room temperature should be 73 to 78°F (23–26°C).

> TIPS: If you are working in a cold room, you may need to apply a little heat to the vat. By the end of stirring, the curd should be 86 to 88°F (30–31°C).

Drain: Ladle into forms. Room temperature should be 72 to 78°F (22–26°C) at the start of hooping. When using a two-part hoop, fill to top of second part one time. For large Brie-size hoops, drain as follows:

- First flip: When to level of lower hoop
- Second flip: Two hours later
- Third flip: Two to three hours after second flip

Room temperature during turning should begin at around 70°F (21°C) and cool to 68°F (20°C) by the time the cheese is ready to unmold (usually about 12 hours later).

Salt: Unmold, usually 12 hours later, when pH of cheese is 4.7 to 4.8. Decrease room temperature to 62°F (17°C), and sprinkle top and bottom of cheese with a total of 0.03 pounds kosher salt—you don't need to salt the short vertical sides.

Drying: Let set on draining mat for 24 hours. The humidity should be 80 to 85 percent; a fan can be used. Salt should be 1.9 to 2.1 percent of the weight of curd. If applying ash, you can do it with the salt or after drying.

Ripening and Affinage: Place in ripening fridge at 52 to 55°F (11–13°C) and 95 to 97 percent RH. Turn daily. When mold growth is even, you can wrap or pat growth down. Ideally, wrap the cheese and move to a cooler fridge, 38 to 40°F (3–4°C).

TIPS: Make sure hands are sanitized and dry when turning. Pyramids cannot be turned but should be repositioned. They're ready when you like the texture! They usually start being good at 2 weeks and beyond. When checking for finished texture, remember that the cheese will be firmer when cold.

LARGE-BATCH GUIDELINES

Milk: 12 gal (46 L/100 lbs) whole milk (for larger batches increase ingredients proportionately)
Culture: 2–3U Flora Danica or Aroma B, 0.2–0.3 dose *P. candidum*, and 0.01–0.15 dose Geo 15 or Geo 17
Calcium chloride (optional): 10–15 ml (about 1 tbsp)
Rennet: 7 ml single-strength rennet (3.5 ml double strength)
Salt: 1.9–2.1% of weight of drained curd

WASHED CURD, STABILIZED PASTE BLOOMY RIND CHEESE

Don't let the term "stabilized paste" make you think that this cheese is meant only for the mass market of the large grocery store. Some of the world's most prized bloomy rinds are made using one or more of the methods that fall within this category. Take, for example, California's Cowgirl Creamery and their classic white mold cheese MT Tam. Not only is this luscious cheese made using a washed curd technique, says one of the founding Cowgirls, Peg Smith, it is also a triple crème, meaning a denser, less volatile paste and incredible decadence of texture and flavor. Another cheese course staple and cream-enriched classic is the widely available French cheese Pierre Robert.

INGREDIENTS

Milk: 1 gal (4 L/8.6 lbs) whole milk
Culture: 1/16 tsp (0.1 g) Flora Danica, Aroma B, or equivalent blend, 4 skewer tips *P. candidum*, and 1 skewer tip Geo 15 (may also benefit from a small dose—say, 1 skewer tip—of yeast, such as *Debaryomyces hansenii* contained in a blend such as LAF 3)
Calcium chloride (optional): maximum ¼ tsp (1.25 ml) diluted in ⅛ cup (30 ml) cool, nonchlorinated water
Rennet: 8 drops (0.2 ml) single-strength rennet diluted in ⅛ cup (30 ml) cool, nonchlorinated water
Salt: 1 tsp (5 g) pure salt (goal is 1.9–2.1% of weight of drained curd)

STEPS

Prepare Equipment: Make sure all equipment is cleaned and sanitized and that your cheesemaking space is free from possible contaminants. Refer to chapter 6 for tips on proper equipment preparation.

Prepare Milk: Warm milk to 95°F (35°C).

Culture: Sprinkle cultures on milk surface. Let set 3 to 5 minutes, then stir gently for 2 to 5 minutes.

Ripen: Cover, and hold at 95°F (35°C) for 3 to 4 hours to a pH of 6.2 to 6.0.

Additions: Stir in calcium chloride for 1 minute, and wait 5 minutes before adding rennet.

Rennet: Stir in rennet using an up-and-down motion for about 1 minute, then still the milk.

TIPS: Less rennet is needed in this recipe since the pH of the milk has already dropped from the starting point. Rennet coagulates easier the more acidic the milk.

Coagulate: Let the curd sit and rest until a clean break is attained. The goal coagulation time is 30 minutes. Maintain 95°F (35°C).

Cut: Cut curd into ½-inch (1.3-cm) cubes, then let rest 10 minutes. Maintain ripening temperature.

Stir: Stir every 10 minutes for three repetitions. Stir just enough to float curds and make sure they are not matting. Maintain the ripening temperature. Let settle for 5 minutes.

Whey Off and Wash: Remove 1.5 quarts (1.5 L) whey (25–30% of total volume). Add 1.5 quarts water at 95°F (35°C). Hold for 10 minutes, stir, then hold for 20 to 30 minutes without stirring.

Drain: Drain whey off to level of curds. Ladle into well-perforated forms. Flip three times over the first 5 hours. Room should be 75°F (24°C). Drain in forms for 12 hours with room temperature at 68 to 72°F (20–22°C) by end of draining. Goal pH at the end of draining is 5.0 to 5.1.

Salt and Drying: Cool room to 62°F (17°C), and sprinkle curds with 0.01 pounds kosher salt. Let set on draining mat for 24 hours. The humidity should be at 80 to 85 percent. A fan may be used to help assist the drying. Salt should be at 2% curd weight.

Ripening and Affinage: Place in ripening fridge at 52 to 55°F (11–13°C) and 95 to 97 percent RH. Turn daily. When mold growth is even, you can wrap or pat growth down. Ideally, wrap and move to a cold fridge 38 to 40°F (3–4°C).

LARGE-BATCH GUIDELINES

Milk: 12 gal (46 L/100 lbs) whole milk (for larger batches increase ingredients proportionately)
Culture: 2–3U Flora Danica or Aroma B, 0.2–0.3 dose *P. candidum*, and 0.01–0.15 dose Geo 15 (may also benefit from 0.05 dose yeast, such as LAF 3 or other from chart)
Calcium chloride (optional): 10–15 ml (about 1 tbsp)
Rennet: 2.4 ml single-strength rennet (1.2 ml double strength)
Whey Removal: 25–30% whey (by total volume)
Salt: Goal is 1.9–2.1% of weight of drained curd

THINKING OUTSIDE THE VAT

SCENARIO 1: It is time to cut the curd, according to the goal time on the make sheet. But when you check for a clean break, the curd is too soft. You do which of the following?

A. Wait until it is firm enough.
B. After waiting for the right texture, then cutting, you check the whey pH and make a note on your make sheet if it is off target.
C. Go ahead and cut the curd, or it may become too acidic.
D. Check that your thermometer is calibrated.
E. Consider making adjustments in culture and coagulant amounts, as well as adding calcium chloride before coagulating.

SCENARIO 2: When you try to make a Camembert-type cheese, it grows a lovely white coat of molds but doesn't soften during ripening. The problem *could* be which of the following?

A. The pH wasn't low enough when the cheese was salted.
B. The aging area isn't humid enough.
C. Not enough white mold was added.
D. The milk fat is too high.

SCENARIO 3: You are making cheese and everything seems fine except the pH isn't dropping in the vat. The problem *could* be which of the following?

A. Bacteriophage.
B. The starter culture is old.
C. The milk has antibiotic residue in it.
D. The milk has colostrum (first milk from animals after they have their babies) in it.

1. Answer: A, B, D, and E.
2. Answer: A, B, and D.
3. Answer: All of the above.

Rush Creek Reserve, a stellar example of a spruce bark–wrapped washed rind cheese, from Uplands Cheese Company, Wisconsin

11: WASHED RIND SURFACE-RIPENED CHEESES

FINSHED TEXTURE: Soft to semisoft
METHOD OF COAGULATION: Rennet
METHOD OF RIPENING: Surface and internal bacteria

This category of cheese is also known by the less-than-appetizing name of "stinky cheese." And boy, can they reek! Gym socks (think high school boy's locker room gym socks) are a common analogy. But fortunately, things don't always taste exactly like they smell. I *love* these cheeses. Soft and oozing with strong, meaty, baconlike flavor—and did I mention the texture?

Washed rind, surface-bacteria-ripened cheeses also include some semihard to hard varieties from the Gruyère family (more on this group in chapter 14). But in that particular style, the surface-ripening bacteria produce minimal changes and, usually, all but disappear before aging is complete. The term "washed rind" describes the affinage process of applying different liquids (such as ales, wine, and brine) to the surface of many kinds of cheeses. This type of washing has a far greater influence on the rind of the cheese than it does on the interior. So while many cheeses have a rind that is washed, they are not necessarily defined as a "washed rind" cheese. For this chapter we will focus on those where the surface bacteria profoundly influence the final texture, flavor, color, and, of course, aroma of the cheese.

These cheese types are more popular in some parts of the world than others but are of growing popularity, thanks to more adventuresome palates and an expanding array of well-produced domestic varieties and better-kept imports. In this chapter I will use "washed rind," "stinky," and "surface ripened" interchangeably to refer to those types that obtain a soft to semisoft texture, pronounced odor, and distinctive color, thanks to the activity of surface microorganisms. Let's get to know these unique, potent cheeses better.

THE ACCIDENTAL STINKER

My first stinky cheese was a product of happenstance. I'd made one of our usual hard cheeses called Elk Mountain and had a little curd left over. So I put it in a small form, pressed it, and aged it right alongside the regular wheels. It was December, so the humidity in the aging room was about 95 percent (when the aging room cooler doesn't need to run, the humidity stays higher). The rinds of Elk Mountain are washed with ale, and ale, of course, brings yeasts.

After a few weeks I noticed that the rind on the little wheel was turning a pinkish, orange color. I ignored it and kept turning the cheeses. At one point I thought of getting rid of the little one, but after about two months it started to feel soft when I turned it. After another few weeks it was very soft. About this time it just so happened that we had a special cheese celebrity visitor coming by to see our farm, so we decided to cut it when he was there so he could tell us what had happened.

Well, we cut, we tasted, we oohed, we ahhed. It was heavenly. It was stinky. It was serendipity. The point is that you actually don't have to know what you are doing to make these kinds of cheeses successfully—at least the first time. But knowing more will help you appreciate the process and increase the odds of success.

THE FUNDAMENTALS OF STINKY CHEESE

Similar to the white mold–ripened beauties from the previous chapter, washed rind cheeses have a very complex ripening process, but one that is a little less needy and sensitive to changes than that of their furry white cousins.

Washed rind cheeses are usually made with mesophilic bacteria blends, are not heated to temperatures much over 95°F (35°C), and are lightly pressed. The resulting curd is high in moisture and usually has a pH of about 5.0 at the end of draining. The higher moisture content at the beginning of draining will help lower the pH to the goal during draining, as it will provide lactose with which the starter bacteria will continue to produce acid.

During the initial stages of aging, at a high relative humidity of about 95 percent and temperatures ranging from 50 to 59°F (10–15°C), yeasts from the environment or added to the milk or washing solution begin to grow on the surface of the cheese and raise the pH. When the conditions are hospitable, surface bacteria from the washing solution or the environment will begin to colonize the cheese.

From there the surface microorganisms work their magic, feeding off the lactic acid and providing enzymes for the breakdown of fat and protein. Over time they raise the pH inward, toward the center of the cheese, and the texture softens.

During the first few weeks of aging, the cheeses must be routinely washed with a light brine solution in which you can include one of the surface-ripening

Wimer Winter cheeses aging at Pholia Farm

bacteria, such as *Brevibacterium linens*. Even if you have *B. linens* in your environment, the washing will help it grow evenly on the surface of the cheese and limit the growth of unwanted molds and organisms while the desired bacteria do their stinky work. This washing process also gives this category of cheeses their other name, "smear ripened," as the surface colonies are "smeared" across the rind.

By the end of a successful aging period of about three weeks, the cheese will have a color ranging from bright orange to almost pinkish-brown, be starting to soften, have a sticky surface, and of course have a rather strong odor. The pH at the end of aging can rise to over 6.5, making these cheeses more susceptible to postcontamination by unwanted, and perhaps dangerous, bacteria. So as always, don't take the process for granted, but don't let caution keep you from experimenting with these delicious cheeses!

LOOKING DEEPER AT THE SCIENCE OF BACTERIAL SURFACE-RIPENED CHEESES

The surface microflora of washed rind cheeses were once thought to be dominated by a specific *coryneform* bacteria, namely *B. linens*. Recent studies and rind sampling, however, have shown that even when *B. linens* is used in the brine wash, it might not be present on the finished cheese. Instead, almost innumerable adventitious bacteria, many from the same coryneform family as *B. linens*, "contaminate" the surface and perform the task of ripening the cheese. The sidebar story of my happy-accident stinky cheese is a good illustration of the availability of these adventitious "contaminants." Knowing the great variety of possible surface microorganisms will not change the way you care for the cheese, however. The desired flora will all thrive under the same care that *B. linens* would. So let's go over, in a bit more depth, the science of the stinkers.

The cheese's future texture is aided by a high moisture content at the beginning of draining that provides lactose for the starter bacteria to continue to produce lactic acid during pressing. The increased lactic acid and a pH of about 5.0 at the end of draining also ensures a loss of calcium phosphate, since calcium phosphate is coaxed out of the protein structure by hydrogen ions and is lost with the whey. This loss of minerals means a loss of overall buffering capacity in the curd. While this reduced capacity is not as pronounced as in most bloomy rind curd, where the pH at the end of draining can range from 4.5 to 5.1, it will still help create a situation in which it will be easier to raise the pH later during aging, leading to a softened texture.

The rate at which washed rind cheeses ripen is greatly influenced by the ratio of surface area to volume. So the ideal size of these cheeses is usually small, with the height being far shorter in comparison to the diameter (this is the main reason that the semihard to hard varieties that have a washed surface, such as Gruyère, do not soften and the surface flora eventually dry off).

The first organisms to grow on the surface of washed rind cheeses are a variety of salt-tolerant, or *halotolerant*, yeasts (similar to the first microflora appearing on white mold–ripened varieties), probably the most important being *Debaryomyces hansenii* and *Geotrichum candidum*. As with stinky cheese's more demure cousin, the yeasts begin to deacidify the surface of the cheese by metabolizing lactic acid (converting it

A taleggio-style cheese by Brian Futhey at Stone Meadow Farm, Pennsylvania

to carbon dioxide and water) and transforming certain amino acids to ammonia. The breakdown of lactic acid is an obvious way to raise the pH, and because ammonia is alkaline, the pH goes up. Since the surface-ripening bacteria grow better at a pH higher than 5.6, this deacidification helps create a hospitable environment for the surface bacteria.

In addition to raising the pH, yeasts perform several other key functions in the production of a softened washed rind cheese. First, they produce nutrients that stimulate the growth of surface-ripening bacteria—think of it as bait to attract environmental bacteria. Second, they provide enzymes that assist with the breakdown proteins and fats that will help during the ripening process. Next, they contribute their own flavor profile and can help reduce bitterness. And last, they help keep the surface moist through the production of water as a by-product of the metabolization of lactic acid, and this helps keep the rest of the cheese from drying out.

While the enzymes produced by both the yeasts and the surface bacteria do not diffuse to the interior of the cheese (and therefore do not directly soften the paste), their by-products continue to change the pH. As the pH rises from the outside of the cheese toward the center, proteins move away from their isolectric point, where they are *hydrophobic*, or rejecting water, and back toward their water-loving tendency. As this happens the proteins begin fixing water and the paste of the cheese will soften, a process we discussed earlier called resolubilization.

The pronounced odor from washed rind cheeses comes primarily from a variety of sulfur compounds produced by the surface bacteria's metabolization of sulfer amino acids. The ripening process requires oxygen and good air exchange to support the continued respiration and metabolic activities of the surface flora.

Because of their high moisture content and high pH (at the end of ripening, they are often above 6.5), these cheeses have a higher risk of supporting pathogenic bacteria than do drier, more acidic cheeses because of raw or incompletely pasteurized milk or from postcontamination. In addition, the fact that they are often washed by smearing surface flora from older cheeses onto younger ones means that there are increased chances of spreading contamination from one wheel to a large number of cheeses.

ALL ABOUT MAKING WASHED RIND CHEESES

If you read the sidebar earlier about my accidental creation of a really tasty washed rind cheese, you might be wondering if there is any more to the story. Well, there is. We duplicated this cheese for several years, making it only in the winter months and in a very limited production. We called it Wimer Winter (after our little village here and the season, respectively). But then, as my knowledge base improved, I came to better appreciate the food-safety risk of this type of cheese. Since we only produce raw-milk products here on the farm, we decided that the risks probably outweighed the gains and have not made it again, much to the disappointment of Wimer Winter's passionate fans. In this section, however, I'll give you some tips for producing a tasty and safe stinky cheese.

CHOOSING MILK

It is absolutely possible to produce safe washed rind cheese with raw milk in the same way it is possible to make *unsafe* wheels from pasteurized milk. Whichever route you choose, remember to learn what rules apply to your production. By understanding the science of making this type of cheese, you should also be able to ascertain where the risks lie and how to best protect the process and produce a safe cheese. Currently, in the United States, these types can be made from raw milk if aged over 60 days, but the FDA is reviewing these rules, especially where high-moisture cheeses are concerned.

If you decide to work with raw milk, ensuring that it is of the highest quality is extremely important. Part of this is educating yourself about not only bacterial contamination but diseases that are transmittable from animal to human (known as *zoonotic*) that would be prevented through pasteurization.

The components of the milk itself will have an effect on the finished texture of the cheese. High-fat milk, either from sheep or seasonally produced, may somewhat reduce the resolubilization, or softening, of the paste.

CHOOSING STARTER CULTURES

You can choose from a wide variety of mesophilic cultures for this category of cheeses. Even some of the blends that include a dash of thermophilic bacteria, for adding qualities during aging, are a good choice; in fact, that is what I used in Wimer Winter. I have grouped the cultures in the following table by similarity of bacteria strains. Each company's versions will vary in proportion of each, however, so keep that in mind.

TABLE 11-1. STARTER CULTURE CHOICES FOR WASHED RIND CHEESES

Name	Contains	Manufacturer	Small-Batch Dosage:* 3–4 Gal (12–15 L/26–34 lb)	Large-Batch Dosage per 26 Gal (100 L/220 lb)	Characteristics and Application
Flora Danica	LL, LLC, LLD, LMC	Packaged by home cheese supply co.	1 packet/4 gal (15 L/34 lb)	N/a	Starter and aroma cultures packaged for small batches
Flora Danica or CHN 11, 19	LL, LLC, LLD, LMC	Chr. Hansen	¼ tsp (0.4 g)	10U	Starter and aroma cultures
Meso Aromatic B	LL, LC, LD, LM	Abiasa	¼–½ tsp (0.6–1.2 g)	10 g	Starter and aroma cultures
MO 030R	LL, LC, LD, LM	Clerici Sacco	1/16 tsp (0.3 g)	0.5–4U	Starter and aroma cultures
MM 100 Series (Choozit) (includes MM 100, 101)	LL, LLC, LLD	Danisco	¼ tsp (0.4 g)	4U	Starter and milder aroma
Mesophilic	LL, LLC	Packaged by home cheese supply co.	1 packet	N/a	Starter, some aroma, packaged for small batches
MA Series (Choozit) (includes MA 11, 14, 16, 19)	LL, LLC	Danisco	⅛ tsp (0.3 g)	4U	Starter and some aroma production
Meso III	LL, LC	Abiasa	½ tsp (1.2g)	10 g	Starter and some aroma production
MO 030	LL, LC	Clerici Sacco	1/16 tsp (0.3 g)	4U	Starter and some aroma production
MA 4000 Series (Choozit) (includes 4001, 4002)	LL, LLC, LLD, ST	Danisco	⅛ tsp (0.3 m)	4U	Starter and aroma, plus one thermophile for adjunct flavor during aging

*Remember that culture is always best measured by weighing and dividing into unit doses based on the strength of the culture, instead of using volume or set weight.

TABLE 11-2. SURFACE-RIPENING BACTERIAL CULTURES FOR WASHED RIND CHEESES

Name/Dosage per package	Manufacturer	Dosage (when added to milk)	Appearance and Characteristics
LR (FR 11) **2- and 10-dose package**	Danisco Choozit	1/16 tsp (0.3 g)/3–4 gal or 2–5 doses/264 gal (1000 L)	Light orange, aromatic
SR3 (FR 23) **2- and 10-dose package**	Danisco Choozit	1/16 tsp (0.3 g)/3–4 gal or 2–5 doses/264 gal (1000 L)	Very bright orange, fast growth
LB (FR 10) **10-dose package**	Danisco Choozit	1/16 tsp (0.3 g)/3–4 gal or 2–5 doses/264 gal (1000 L)	Ivory, aromatic

TABLE 11-3. YEAST CULTURES FOR WASHED RIND CHEESES

Name/Dosage per package	Manufacturer	Dosage (when added to milk)	Characteristics and Application
Geotrichum Candidum **50 gallons/package**	Packaged by home cheese supply co.	Follow directions on package	Variable depending on company
GEO 15 (Choozit) **2- and 10-dose package**	Danisco	1 skewer tip/2 gal (8 L) or >2 doses/264 gal (1000 L)	Yeastlike growth, neutralizes surface pH for red bacteria growth
KL 71 (Choozit) **2- and 10-dose package**	Danisco	1 skewer tip/2 gal (8 L) or >2 doses/264 gal (1000 L)	*Kluyveromyces lactis*
DH (Choozit) **2- and 10-dose package**	Danisco	1 skewer tip/1 gal (4 L) or >1 dose/264 gal (1000 L)	*Debaromyces hansenii*

CHOOSING BACTERIA RIPENING CULTURES

Often the red bacteria-ripening cultures you choose will be a matter of what is available through your supplier. I have listed some of the more common varieties that come in reasonably sized dosages for the smaller producer. Remember, even if you add these, you might well end up with some interesting opportunistic strains growing on your cheese.

CHOOSING YEAST RIPENING CULTURES AND COAGULANT

You may not need to add any yeast cultures to your cheese recipe, as environmental yeasts might be enough to help prepare the cheese for the growth of the red bacteria. But if you want to include some, the following table lists several yeast cultures, including several *Geotrichum candidum* strains.

Often it is convenient and effective to choose a culture blend that contains both red bacteria and yeasts. Table 11–4 has two blends that include both types of ripening culture; these can make the adding of cultures and mixing of wash solutions a bit easier.

Any type of coagulant works well with washed rind cheeses. The type used can be the personal preference of the cheesemaker. Many of the finest Spanish sheep's milk washed rind cheeses use thistle rennet.

CUTTING AND STIRRING THE CURD

Curd is cut a bit larger than with some cheeses—usually about ¾ inches (2 cm). Remember, the curd will be higher in moisture than a semihard cheese but not so high that the wheel cannot be lightly pressed.

TABLE 11-4. SURFACE-RIPENING BLENDS OF YEASTS AND BACTERIA FOR WASHED RIND CHEESES

Name	Contains	Manufacturer	Dosage (when added to milk)
ARN (Choozit) 2- and 10-dose package	*B. linens* orange, *Arthrobacter nicotianae*, *B. linens* ivory, *Geotrichum candidum*	Danisco	1/16 tsp/2 gal (8 L) or >1 dose/264 gal (1000 L)
PLA (Choozit) 2- and 10-dose package	*B. linens* orange, *Arthrobacter nicotianae*, *Geotrichum candidum*	Danisco	1/16 tsp/2 gal (8 L) or >1 dose/264 gal (1000 L)

Cooking temperatures usually do not exceed 95°F (35°C), which helps ensure a moist curd that can continue to acidify during pressing. Often, the temperature at which stirring occurs is the same as the ripening temperature. As with bloomy rinds, washing the curd will create a less soluble paste and stabilized curd that will retain body and create a springier finished texture.

DRAINING AND PRESSING

A very light weight is used to press this variety of cheese, usually not more than equal to the weight of the cheese. The resulting cheese should have a smooth rind but can retain some mechanical openings in the paste. The light weight will allow for a slower drainage of whey and, consequently, a bit more loss of minerals as the pH drops, and minerals can still leave the curd along with the draining whey.

SALTING

Salting is usually done in fully saturated brine, although some varieties call for dry salting or longer brining in a less saturated solution. Salting time is proportionate to the size of the cheese. If the volume to surface area ratio is similar—no matter the total weight of the cheese—brining time is the same. You might think that these cheeses need more salt initially, but because of the brine washes they will receive during the initial stages of aging, they are brined for a relatively short time. Dry salting can also be done and usually takes place over one to two days. Some recipes call for brine salting in a 16 to 18 percent salt brine instead of a fully saturated solution. In that case the time suggested for the cheese to remain in the brine is about half again as long as for a fully saturated brine.

TABLE 11-5. SUGGESTED BRINING TIMES FOR WASHED RIND CHEESES

Weight (rounded)	Height	Diameter	Brine Time (Saturated brine)
1 lb (0.5 kg)	1¼ inch (3 cm)	6 inches (15 cm)	20 min
4.5 lb (2 kg)	1½ inch (4 cm)	11 inches (28 cm)	60 min

AFFINAGE

Traditionally, stinky cheeses were produced by taking scrapings from an older cheese with established surface microflora and mixing that with water or a liquor and some salt. The resulting mixture, sometimes referred to as a "morge," was smeared and wiped onto the younger cheeses. This process effectively spread the bacteria and yeasts to the new generation. While this practice is still used, it has come under scrutiny for the gap it creates in the food-safety chain—if one cheese is contaminated, you are spreading that contamination to a larger number of wheels.

It is generally believed that once a population of desirable surface-ripening bacteria have become established in an aging room (either through the initial purposeful inoculation of several cheeses or through native, wild bacteria, such as happened with our initial Wimer Winter stinky cheese) washing with plain brine or a yeast solution alone will attract the desired bacteria to the cheese. If you are making these cheese types regularly, you can inoculate the milk for the first batches by adding the recommended dose and developing the rinds; then stop inoculating. Wash new cheeses with a

A Reblochon-style cheese after its first flip during draining

light brine, and see if they develop the desired surface flora without having to add it to the milk or brine.

If you decide to inoculate the cheeses via the wash, you will need to rehydrate the ripening cultures before using. To do this, mix a quart of sterile water with ½ teaspoon salt, then add the culture powders. Mix well and refrigerate, and let the solution rehydrate 24 hours. *Be sure to add the salt to the water first, as the dehydrated bacteria cells will be ruptured if added to plain water.* Then using a clean, smooth cloth, wash the solution onto the surface of the cheeses at least twice a week until you can see the color begin to turn pinkish-orange and the surface of the cheese feels a bit sticky.

SUPPORTING AND WRAPPING

Many types of washed rind cheeses are supported during aging with a wrap around the curved side. From a belt of cheesecloth or lace, to several wraps with raffia, to a girdle of spruce bark, all of these wraps serve to support the cheese to a finished texture that is oozing so much it is eaten with a spoon. Small wooden boxes are even used to contain some of these cheeses that would otherwise disintegrate as they ripen or are transported.

Wraps are usually applied during the first stages of aging. Cloth wraps encircle the cheese once, with the end tucked under itself. Don't worry if the cloth is a little loose in the beginning; as the cheese ages, it will sink a bit and widen. Raffia is looped several times around the cheese at staggered heights and tucked into itself. Spruce bark strips are boiled to sanitize them, then often soaked in an inoculated solution to help develop the desired flora on the surface of the cheese. They are typically pinned a bit loosely to the salted cheeses using a wooden toothpick or other sanitized securing pin that is removed later.

How the cheese is wrapped after affinage is complete will depend on how soft and sticky the cheese is. Some cheesemakers "dry off" the cheeses by washing the surface flora off the cheese and putting it in a drier environment for a few days. Others wrap the cheese in a greaseproof paper before wrapping again. Plastic wrap is not a good choice; it sticks to the cheese, as will plain paper.

When shipping stinky cheeses, it is best to package and box them separately from others, or their smell will transfer to other cheeses. It is even a good idea to label the box as containing a "fragrant, strong-smelling cheese"—I have heard of shipping companies assuming that something has gone bad and pulling the box from their shipments. It sounds slightly amusing, but it won't be if that is your income that is marked as unfit for travel.

JOS VULTO'S *OULEOUT*—A WASHED RIND CHEESE

For the turophile new to the delights of stinky cheeses, it might be surprising to find that this family of cheeses is vast and filled with great variety, as well as similarity. It is quite difficult to find cheeses that you can say, without a doubt, are the exact kin of another. Taleggio is the classic northern Italian washed rind cheese, made from pasteurized (usually) cow's milk. This ancient PDO cheese is made in square forms and is named after the Val Taleggio region from which it originated.

French Munster from the Alsace region of France is a much-copied—in name if not flavor—washed rind AOC cheese with a flavor as strong as its pedigree. The US version, on the other hand, is mild and colored, not by bacterial growth, but by paprika or another colorant. If you are looking for a cheese similar to the French Munster that's made in the United States, my cheesemaker friend and former cheesemonger, Meghan McKenna (see chapter 10 for more) suggests Meadow Creek Farm's Grayson. Meghan notes that "these Virginia cheesemakers use rotational grazing and seasonal milking practices to really elevate their

Jos Vulto's washed rind cheese *Ouleout*

Jersey milk to something spectacular." In this recipe cheesemaker Jos Vulto (see sidebar profile story that follows) shares his recipe for a washed rind cheese.

INGREDIENTS

Milk: 2 gal (8 L/17 lb) whole milk
Culture: 1/16 tsp (0.2 g) MM 100, 1 skewer tip MA 4001, 1/16 tsp (0.2 g) TA 50, 2 skewer tips each LBC 80 and GEO 15 (if using raw milk, you can omit the LBC 80), and 1 skewer tip SR3
Calcium chloride (optional): Maximum ½ tsp (2.5 ml) calcium chloride diluted in ⅛ cup (30 ml) cool, nonchlorinated water
Rennet: Scant ¼ tsp (1 ml) single-strength rennet diluted in ¼ cup (60 ml) cool, nonchlorinated water
Salt: Approximately 2 tbsp (30 g) pure salt (goal is 2% salt by weight of drained cheeses)

STEPS

Prepare Equipment: Make sure all equipment is cleaned and sanitized and that your cheesemaking space is free from possible contaminants. Refer to chapter 6 for tips on proper equipment preparation.

Prepare Milk: Warm milk to 70°F (21°C).

Culture: Sprinkle cultures on top of milk. Let set 3 to 5 minutes, then stir gently for 3 to 5 minutes.

Additions: Stir in calcium chloride.

Ripen: Warm to 90°F (32°C), and ripen for 1.5 to 2 hours. The goal pH drop is about 0.13.

Rennet and Coagulate: Stir in rennet solution using an up-and-down motion, then still the milk. Cover, and hold at 90°F (32°C) until a clean break is achieved. The goal coagulation time is 40 minutes. The flocculation factor is 4 with goal flocculation time of 10 minutes.

Cut: Cut curd into vertical columns of ¾ inch (2 cm). Rest 10 minutes. Make horizontal cuts at ¾ inch (2 cm). Let rest for 5 minutes. Maintain a temperature of 90°F (32°C).

Stir: Stir gently for 30 minutes. Maintain a temperature of 90°F (32°C). Let settle for 10 minutes.

Drain: Drain off whey to the level of the curds. Ladle or scoop curds into basket forms, with an ideal height-to-diameter ratio of 1:4 for finished cheese. Turn after 5 minutes, 30 minutes, and 6 hours. Total drain time is approximately 12 hours. The goal curd pH is 5.33.

Salt: Unmold cheeses onto draining mat, and sprinkle with salt and let sit at room temperature for 2–4 hours.

Affinage: Age at 55°F (13°C) and 90 to 95 percent RH. Beginning on the third day of aging, wash cheeses with a 3 percent brine every other day for three washings. Then wash with *B. linens* solution twice weekly for 3 weeks (see page 200 for more on mixing a washing solution). Continue to age until desired texture is achieved. Turn daily during affinage.

LARGE-BATCH GUIDELINES

Milk: 12 gal (46 L/100 lb) whole milk (for larger batches increase ingredients proportionately)
Culture: ¼ tsp (0.6 g) MM 100, 1/16 tsp (0.1 g) MA 4001, ¼ tsp (0.6 g) TA 50, ⅛ tsp (0.25 g) LBC 80 (when using pasteurized milk), ⅛ tsp (0.6 g) GEO 15, and 1/16 tsp (0.3 g) SR3
Calcium chloride (optional): 10–15 ml (about 1 tbsp)
Rennet: 4.5–9 ml single-strength rennet (2.25–4.5 ml double strength)
Salt: 2% by weight of drained cheeses

JOS VULTO, BROOKLYN AND WALTON, NEW YORK

Jos Vulto, a Holland native who grafted his life to New York City in the early '90s, happened upon cheesemaking in almost the same way that the first cheesemaker in history likely did—luck.

Jos came to the United States as a resident artist at what is now the Museum of Modern Art/PS1. However, having been raised in the country, 15 years in the city of New York left him feeling as though something fundamental was missing. He and his wife bought a dilapidated farm in upstate New York and spent weekends there, dreaming of someday making it their permanent residence.

As Jos's awareness of food quality and its production increased, he began growing vegetables on his land and purchasing local raw milk. One week, upon returning to the farm, he found that a container of forgotten raw milk in the back of the refrigerator had fermented, and lo and behold, it smelled like cheese. It was an epiphany. Jos thought that if cheese, of a sort, can spontaneously "occur," perhaps with a bit of knowledge and the best intentions, he could make his own artisan cheese.

Utilizing the website of Dr. David B. Fankhauser at the University of Cincinnati Clermont College, Jos made his first encouraging batches of cheese. He soon discovered the website of cheese mentor Peter Dixon, and his craft improved quickly. As there were few resources at the time for a budding cheesemaker, Jos started a blog to detail his adventures and experiments (www.heinennellie.blogspot.com). (The name of his blog is a combination of his father's and mother's names with the Dutch conjunction "n.") The blog also attests to Jos's obvious skill and art as a cheesemaker.

On his blog Jos documents his search for the perfect aging space for his cheeses. After frustration with using old refrigerators, he hit upon a wonderful solution—an underground space beneath the sidewalks of Brooklyn. The space in question was part of the metalworking company that Jos was a partner in at the time. After a cleaning, tiling, and installing shelves, it proved to be a fantastic little cheese cave. While humidity was stable at 90 to 95 percent, the temperature control needed assistance. Jos used a small heater in the winter and a CoolBot (see chapter 4 for more on these devices) to effectively regulate the temperature.

His cheeses were received with such great enthusiasm by his friends, as well as by cheese professionals, that it made him wonder if he couldn't develop a small artisan creamery on his property in rural New York. The appeal was trifold—what if the business allowed him and his wife to live outside the city, would provide personal gratification, and might even bring jobs to an otherwise struggling community? He knew that the region has plenty of wonderful, grass-fed milk, and isn't that the basis for a successful cheese business?

The market research and business planning that Jos did indicated that building on their property would be too costly; instead he rented a space and has spent over 18 months converting it to a small creamery. With his license not too far off, we can expect some pretty amazing cheeses in the near future from Jos Vulto and his Walton, New York, cheesemaking enterprise: Vulto Creamery.

Home cheesemaker Jos Vulto at his vat

REBLOCHON-STYLE WASHED RIND CHEESE

Author and well-known cheese expert Max McCalman was an invaluable resource for finding similar cheese types for most of the recipes in this book, and one particularly interesting point he makes is that genuine Reblochon is truly a stand-alone cheese. This AOC cheese takes its name not from its area of origin, as is more common for historical cheeses, but from the verb *reblocher*, which roughly translates to "squeeze the udder a second time."

The story goes that when the landlord or overseer would come to collect the milk, the cows would intentionally not be fully milked out. Once the landlord was gone, the farmer would milk the cows "a second time," and this extra-rich milk was used to make the cheese. (It is true with goats and sheep that milk given toward the end of each milking is higher in butterfat than that given at the beginning of each milking.) Max suggests that the Swiss raw cow's milk cheese Stanser Rotelli and the raw sheep's milk Stanser Reblochon are noteworthy cousins of the French original.

INGREDIENTS

Milk: 2 gal (8 L/17 lb) whole, high-fat milk (40% or higher)
Culture: ⅛ tsp (0.3 g) MM 100, 1⁄16 tsp (0.2 g) TA 50, and 1⁄16 tsp (0.2 g) PLA or similar blend including *Geotrichum* and *B. linens*
Calcium chloride (optional): maximum ½ tsp (2.5 ml) calcium chloride diluted in ⅛ cup (30 ml) cool, nonchlorinated water
Rennet: ¼ tsp plus (1.6 ml) single-strength rennet diluted in ¼ cup (60 ml) cool, nonchlorinated water
Salt: Dry pure salt or heavy brine

STEPS

Prepare Equipment: Make sure all equipment is cleaned and sanitized and that your cheesemaking space is free from possible contaminants. Refer to chapter 6 for tips on proper equipment preparation.

Prepare Milk: Warm milk to 93°F (34°C).

Culture: Sprinkle cultures on top of milk. Let set for 3 to 5 minutes, then stir gently for 2 to 5 minutes.

Ripen: Maintain temperature at 93°F (34°C), and ripen for 1 to 1.5 hours.

Additions: Stir in calcium chloride solution for 1 minute then wait 5 minutes before adding rennet.

Rennet and Coagulate: Add rennet solution and stir with an up-and-down motion for 1 minute. Still milk, and maintain 93°F (34°C) until a clean break is attained. The goal coagulation time is 20 to 25 minutes. If using flocculation factor, multiply floc time by 2.5.

Cut: When curd is ready to cut, cut mass in ¾-inch (2-cm) cubes. Rest for 5 minutes.

Stir: Stir gently for 10 minutes. Let curd settle, and hold at 93°F (34°C) until whey pH reaches 6.4 to 6.3. This could take 45 minutes to an hour. Stir periodically to prevent matting.

Drain: Drain whey to level of curds, and ladle curd into well-perforated forms. Turn after 5 minutes, then again after 30 minutes. After the second turn add a light weight about half the weight of the cheeses. Turn after 2 hours. Press for 4 to 6 hours to a curd pH of 5.4 to 5.3. The room temperature should be about 81°F (27°C) and decreased to 72°F (22°C) by the end of draining.

Salt: Salt with 2 percent of weight of drained cheese. Alternatively, brine in 23 percent solution for 30 minutes per 1.5 pounds (700 g) of the weight of each wheel.

Affinage: Dry for 4 days at 65°F (18°C) and 95 percent RH. Turn daily. After 4 days, if yeast is not felt on surface (sticky, slippery feel), wash with rehydrated *Geotrichum* or PLA solution.

Age for 2 weeks at 55°F (13°C) and 95 percent RH. Wash twice weekly with light brine, with or without annatto. Turn daily. Can add *B. linens* to wash as needed. Washing should continue as needed to keep rind clear of unwanted molds. Age to desired texture.

LARGE-BATCH GUIDELINES

Milk: 12 gal (46 L/100 lb) high-fat milk (for larger batches increase ingredients proportionately)
Culture: 2U MM 100, 1U TA 50, and 0.25U PLA
Calcium chloride (optional): 10–15 ml (about 1 tbsp)
Rennet: 7–9 ml single-strength rennet (3.5–4.5 ml double strength)
Salt: 2% of the weight of drained cheeses

Mini-Grayson, a semi-soft, surface-ripened washed rind cheese by Meadow Creek Dairy, Virginia

PHOLIA FARM'S WASHED RIND CHEESE

Our washed rind cheese Wimer Winter was a raw-milk, aged stinky that came about by accident. (Red Hawk is a wonderful washed rind from Cowgirl Creamery in California that also came about thanks to serendipity and is still made without using any added ripening bacteria during the process.) While we no longer make Wimer Winter, it was a local (and Max McCalman) favorite. Max said it reminded him of Chevrotin des Aravis, a raw-milk goat cheese produced in the Haute-Savoie region of France. Because of the high fat content of our farm's Nigerian Dwarf goat's milk, our washed rind cheese had the texture of a double or triple crème, similar to the cow's milk cheese Red Hawk.

INGREDIENTS

Milk: 2 gal (8 L/17 lb) whole milk
Culture: ¼ tsp (0.4 g) Flora Danica (or equivalent blend) and one skewer tip SR3
Calcium chloride (optional): maximum ½ tsp (2.5 ml) calcium chloride diluted in ⅛ cup (30 ml) cool, nonchlorinated water
Rennet: ¼ tsp plus (1.5 ml) single-strength rennet diluted in ¼ cup (60 ml) cool, nonchlorinated water
Salt: Heavy brine

STEPS

Prepare Equipment: Make sure all equipment is cleaned and sanitized and that your cheesemaking space is free from possible contaminants. Refer to chapter 6 for tips on proper equipment preparation.

Prepare Milk: Warm milk to 88°F (31°C).

Culture: Sprinkle cultures on top of milk. Let set for 3 to 5 minutes, then stir gently for 2 to 5 minutes.

Ripen: Maintain temperature at 88°F (31°C) for 1 hour.

Addition: Stir in calcium chloride solution if using. Wait 5 minutes.

Rennet and Coagulate: Add rennet solution, and stir with an up-and-down motion for 1 minute. Still milk, and maintain 88°F (31°C) until a clean break is attained. The goal coagulation time is 45 minutes. If using flocculation factor, multiply floc time by 3.5.

Cut: Cut curd mass into ⅜-inch (1-cm) cubes. Let rest for 5 minutes.

Stir and Cook: Stir gently and increase temperature in vat slowly to 98°F (37°C) over 30 minutes. Let settle in the vat.

Drain: Drain whey to level of curds. Ladle or scoop curds into unlined, well-perforated forms. Ideal size ratio of drained cheese is 2:5 (height to diameter). Turn cheeses after 5 minutes, then again after 30 minutes. After second turn add light pressing weight of about half that of the cheeses. Turn again after 2 hours. If rind has not closed, add more weight. Ideal room temperature during draining is 72°F (22°C).

Salt: Unmold when curd pH has reached 5.3 to 5.4 (usually after about 4 to 6 hours of pressing). Brine cheeses in heavy brine for 2 hours per 2-pound (1-kg) cheese.

Affinage: Age at 55°F (13°C) and 95 percent RH. Wash twice weekly with light brine made from ale (2 tsp salt per 12 oz beer) for the first 4 to 6 weeks. Turn daily. If red surface is slow to develop, add a bit of SR3 to the ale brine. Washing should continue as needed to keep rind clear of unwanted molds. Age to desired texture.

LARGE-BATCH GUIDELINES

Milk: 12 gal (46 L/100 lb) whole milk (for larger batches increase ingredients proportionately)
Culture: 5U Flora Danica (or equivalent blend) and 1⁄16 tsp (0.2–0.5 dose) SR3
Calcium chloride (optional): 10–15 doses (about 1 tbsp)
Rennet: 9 ml single-strength rennet (4.5 ml double strength)
Salt: Heavy brine

THINKING OUTSIDE THE VAT

SCENARIO 1: All the batches of cheese you have made lately have a starting milk pH of lower than usual by 0.10 (e.g., instead of 6.60, it starts at 6.50). You should rule out which of the following?

A. The milk is contaminated with other bacteria or wasn't cooled properly after milking.
B. Your pH meter isn't calibrated properly or your calibrating (buffer) solutions have become contaminated.
C. This is normal when the moon is full.
D. You used to check the start pH when the milk was cold, and now you are checking it after it reaches the goal ripening temperature.

SCENARIO 2: After you cut the curd, it seems as though it might be too soft and fragile to stir. The recipe, though, says to begin stirring immediately after cutting. You should do which of the following?

A. Wait for a brief time, say, 5 to 10 minutes, to let the curds heal.
B. Follow the instructions; they probably are right.
C. Make a note in your logbook that the curd seemed soft, and go ahead and stir.
D. Make a note in your logbook, and let the curds heal for a bit. Consider adjusting the cook time to let them lose more whey slowly.

1. Answer: A, B, and D.
2. Answer: A and D.

Flora Nelle, one of the blue cheeses from the historic Rogue Creamery, Oregon

12: BLUE CHEESES

FINISHED TEXTURE: Semisoft to semihard
METHOD OF COAGULATION: Rennet
METHOD OF RIPENING: Internal bacteria, internal and external molds

I don't think any other cheese variety has quite the ability to divide people into such distinctive camps of lovers and haters (except maybe the stinky cheeses of the previous chapter)—although I have seen a few miracle conversions from those who hate blue cheese to those who can't get enough. Our youngest daughter, Amelia, an otherwise confirmed turophile, *hated* blue cheese and wouldn't even try it. Until, that is, she got a job working at the world-renowned Rogue Creamery in "The Cheese Shop." Within 3 days she was bringing home, hoarding, and ingesting large wedges of all the different varieties of blue cheese for which the Rogue Creamery is famous. Whether this is proof of her succumbing to peer pressure or just goes to show that many people who hate blue cheese just haven't ever tried really *great* blue cheese, I am not sure!

Blue cheese has been produced for centuries, and undoubtedly the first blues were accidental and dependent on wild, native-mold adventitious organisms. Fortunately for those of us that love blue cheese, those formerly wild *Penicillium roqueforti* molds can now be purchased in little packets and used intentionally to make amazing, moldy cheeses.

I will be honest with you; this category of cheesemaking is my area of least accomplishment. In the beginning of my career, I made some very good blue cheese, and then I, like many other cheesemakers, discovered the amazing ability of *P. roqueforti* to aggressively invade other cheese types through spore inoculation of the milk during cheesemaking and wheels in the aging room. These guys should not be taken lightly! Making blue can be a challenge if you make other cheese types with an open texture, unless you can adequately segregate the process.

In this chapter you will learn to make several variations of blue cheese, including one hybrid between a white mold surface-ripened and blue internal mold–ripened cheese. First, let's go over some science on the topic, tips for making, and tips for aging and storing blue cheeses.

THE FUNDAMENTAL CONCEPTS BEHIND BLUE CHEESE

Many blue cheeses are made in a process quite similar to the feta types you learned about in chapter 9. Some recipes call for short ripening times and some for as long as 5 hours. The curds are usually not heated beyond the initial ripening temperature and are stirred only briefly or intermittently. Most have a long draining period in the forms and are not pressed. As with feta, this allows the high-moisture curd to develop acid slowly during draining, leading to a crumbly open texture. The open texture is important to create what are often called "caves" in which the blue molds can grow.

The blue molds (which actually have quite a range of colors, of which a greenish tint is often more pronounced than blue) are either added to the milk before coagulation or sprinkled onto the curd during draining—sometimes both. These blue molds, all strains of *P. roqueforti*, can grow in quite low oxygen environments, unlike most molds. They are very active during the cheese's aging, breaking down fat and proteins to produce complex aromas and a distinctive texture. There are different strains of *P. roqueforti* that have more or fewer strengths to accomplish these changes. Your culture supplier may only carry one strain, however.

Yeasts also figure prominently in the aging of blue cheeses. Most arrive spontaneously via the milk and the environment, especially in brine-salted blues. But they can also be purposely added during the cheesemaking process.

Fat and protein breakdown is higher in blue cheese varieties than in any other cheese type, thanks to the activity I mentioned a moment ago. High-fat milks, such as that from sheep and Nigerian Dwarf goats, make particularly good blue cheese milks. In fact, the French classic Roquefort is made only from the rich milk of Lacaune sheep. In the production of some other blue cheeses, cream is separated, homogenized, and added back into the milk, resulting in further fat breakdown and therefore more flavor and aroma development.

Most blue cheeses are pierced at specific times during aging to allow the molds access to more oxygen once they have used up the supply inside the cheese. This also allows for carbon dioxide to leave the cheese. Once the blue mold's growth has developed to the desired extent, cheeses are often moved to a colder aging environment. Some blues have natural rinds, while others are wrapped or even waxed to keep the surface white and without a rind or to keep the cheese from drying if the aging environment is not humid enough.

A DEEPER LOOK AT THE SCIENCE OF MAKING BLUE CHEESE

As I mentioned earlier, the process for making many blue cheeses starts out in a very similar fashion to feta—with a moist, uncooked, unpressed curd. The goal is a fairly high pH at the start of draining and a low pH at the end. The average pH for blue cheese at the end of draining is 4.6 to 4.7, although some data reports that Italy's iconic blue cheese, Gorgonzola, is a bit higher at 5.15 to 5.3. As it ages, the pH of blue cheese rises, much as in surface-ripened bloomy rinds—more on that in a moment. This moist, unpressed curd develops enough acid to provide a substrate (as lactate) on which the yeasts and molds will be nourished. It also creates an open texture mottled with "caves" in which *P. roqueforti* can begin to thrive and spread throughout the cheese.

During the first stages of aging, the molds metabolize acid (as lactate and citrate) as well as any milk sugars (e.g., lactose, glucose, and galactose). During this process they produce carbon dioxide (CO_2) as well as speed up the rate of proteolysis. This breakdown of amino acids produces ammonia (NH_3) and begins raising the pH of the cheese from the interior out. (You might remember in chapter 10 that surface mold–ripened cheeses experience a pH rise in the opposite direction, from the outside in.) Yeasts also ferment lactose, producing gas, which assists in creating and maintaining an open texture, necessary for the growth and spread of *P. roqueforti*.

Lipolysis occurs to a greater extent in blue cheese than in any other cheese type. The breakdown of fats to fatty acids produces distinctive flavors and aromas in blue cheese, such as methyl ketones produced from free fatty acids. The particular profile of fatty acid types in the milk source (particularly in goat's milk) will create distinguishing characteristics for that particular cheese. Keep this in mind when using milk types other than what's called for in the original recipes. Fat breakdown is so critical to the development of flavor

Piercing a blue cheese with a 4-millimeter sanitized knitting needle

that in some recipes the milk is homogenized prior to production, or the cream is separated, homogenized, and returned to the vat. The homogenization, as you learned in chapter 2, damages the fat globules, making them more susceptible to lipolysis. But in the case of blue cheese, this can be a good thing!

As with the breakdown of fat, proteolysis is also more profound than in most other cheese types, contributing to blue cheese's distinctive texture as well as its aroma. Breakdown of proteins comes not only from the proteases (if you recall, these are the enzymes that act on proteins) in starter culture and coagulants but also from *P. roqueforti*.

Salt, or sodium chloride (NaCl), plays a prominent role in the creation of blue cheese as well. The salt content of most blues is typically a bit higher than that of other cheeses, at between 2 and 5 percent (even-higher-salt cheeses include feta and Romano types, which can average 4 to 7 percent). The high salt concentration in the outer portion of the cheese moves slowly toward the center during aging (this is not true in the case of milled and salted curd processes, such as in the making of England's king of blues, Stilton).

Blue molds grow well in a salt concentration of from 1 to 3 percent, so initially the growth is at the center of the cheese. As the NaCl concentration moves toward the center of the cheese, it changes the way the mold grows. Instead of spreading and growing fuzzy (that being the mycelia portion of mold), it is instead stimulated to grow more *conidia*, or spores. This is important for two reasons: First, the conidia are the part of *P. roqueforti* that have the blue-green color, and second, the mycelia grow to form a thick mat called a mycelium that feels rubbery and unpalatable when eaten. So for both visual and textural reasons, this shift in growth is important for the development of a superior blue cheese. In the case of milled curd that is salted, the level of salt throughout the cheese is the same from the beginning, meaning a different pattern of mold growth from the beginning of aging.

While blues start out with a low pH in the 4.6 to 5.1 range, it will rise during aging to an average of 6.5 at the center of the cheese and 5.9 at the surface. When we studied bloomy rinds in chapter 10, we learned that their pH rises as well, but from the surface *inward*. You've probably already figured out that where the mold growth is strongest, on either cheese type, is where the pH rises the most. Does this make sense? As with bloomies, the consumption of lactic acid (as lactate) and the breakdown of proteins leading to the production of ammonia (NH_3) has a profound effect on the pH of the cheese. It isn't just fun to know this little fact; it's important when considering the safety of a cheese—as the pH rises, the cheese loses its primary defense against several contaminants. Even if there were very few pathogenic bacteria in the cheese milk, they can be reintroduced during aging. Don't be afraid of your cheese, though; just be knowledgeable about the process, and take steps to avoid the risks.

TIPS FOR MAKING BLUES

When you are learning to make blue cheeses, note how the ripening time and the stirring time will usually be opposite of each other in length. For example, a recipe with a ripening time of 4 to 5 hours may only have a stirring-in-the-vat time of 10 to 20 minutes, while a recipe with a combined ripening and coagulation time of 2 hours may have a stirring time of 1 hour. When you ponder these differences, try to remember that it is all about creating the right texture and developing just the right amount of acid. A typical pH goal at the start of draining is 6.3 to 6.1. Yet the curd must have the right consistency to both knit together and retain an open texture. As you learn to "read the curd," you will be able to modify recipes as the conditions under which you are making them change.

CHOOSING MILK

Blue cheeses were historically made from raw milk, but some that now have PDO status are required to be made from only pasteurized milk. Stilton is probably the best known with this condition. A newer British cheese, Stichelton, is modeled after the traditional method and uses raw milk. As with other aged cheeses, superior-quality raw milk offers native lipase enzymes and nonstarter lactic acid bacteria (NSLAB) that will aid in the creation of a unique and complex end product. If you are making cheese for home use and are comfortable with the quality of the milk, you can use raw milk. For commercial sales, you must follow the rules of your jurisdiction. In the United States the current rule requires 60 days of aging. For blue cheese, this is not a problem, as they often are best when aged for longer periods of time.

When you are working with raw milk, I suggest adding the starter culture at 80°F (27°C), then increasing the heat to the goal ripening temperature. For milk that you have pasteurized, you can take a similar step by adding the culture at just below 102°F (39°C), then cooling the milk to the ripening temperature. (Please note, this little trick applies to the use of freeze-dried, direct vat set [DVS] cultures only!) This will allow the DVS culture time to rehydrate and reactivate (which typically takes 30 minutes) and can cut a little time off the total ripening time.

You can use any type of milk—cow, goat, sheep, or a combination thereof—to make these recipes. You will be departing from the traditional styles, but who cares? As you have probably figured out by now, my goal is to help you make your own styles of cheese, not duplicate another.

CHOOSING STARTER CULTURES

Mesophilic starter cultures (often with added thermophilic culture), along with aroma- and gas-producing cultures, are the best choice for blue cheeses. The gas producers will help create the open texture so necessary for good mold growth. Thermophiles can help slow acid production in the vat but contribute enzymes for protein breakdown during aging. Table 12–2 lists some common blends and their dosage rate. If you find that acid development is too quick, reduce the rate next time.

CHOOSING MOLD CULTURES

P. roqueforti comes in several strains that vary in color and strength (that is, how rapidly they break down proteins and fats leading to a more pronounced blue cheese flavor). If you are purchasing in small doses, there is often only one choice available, but if you want to experiment with other strains, contact other culture suppliers and see if they offer any other choices not listed in their catalogs. If you start feeling a bit overwhelmed by all of these options, talk to the culture supply company; they will most likely be very helpful with making selections. Again, the small producer is often limited to what is packaged in smaller-size doses.

Table 12-2 lists some of the more readily available *P. roqueforti* strains. It is also possible to use a finished blue cheese from which to harvest mold spores of unknown strain and use as an inoculant; however, this would not be considered a "good manufacturing practice" for a commercial producer, nor probably an ethical practice, considering how blue cheese makers zealously guard their mold strains! One beloved Greek blue cheese called Kopanisti is traditionally cultured using bits from a matured version.

For small home-size batches, you can use the tip of a sanitized meat skewer to measure small amounts. Usually two to three tips will give you a fairly accurate measurement for a 3- to 4-gallon (12–15 L) batch. You can add more with no ill effect, but mold cultures are fairly expensive, so it is economical to measure properly. Packets usually contain ten doses—while only one to two doses are enough to inoculate 264 gallons (1,000 liters) of milk. As with other large culture packets, it is a good idea to divide these large doses into smaller packets measured out by grams.

Some versions of blue mold are available in a liquid form. These are more often used by commercial cheesemakers, as they do not have a long shelf life and are meant to inoculate larger single batches. Remember to dose and measure molds in a separate space if possible to make sure that you are not inoculating your cheese area! (See chapter 2 for a refresher on measuring and dividing cultures.)

CHOOSING YEASTS

You might want to consider adding a yeast adjunct culture to your milk, especially if working with pasteurized milk and dry salting the curd (more on that later). If you decide to experiment with adding some yeast, try Danisco's KL 71 (*Kluyveromyces lactis*) or its equivalent from another culture house.

TABLE 12-1. STARTER CULTURE CHOICES FOR BLUE CHEESE

Name	Contains	Manufacturer	Small-Batch Dosage:* 3–4 gal (12–15 L/26–34 lb)	Large-Batch Dosage: 26 gal (100 L/220 lb)	Characteristics and Application
Flora Danica	LL, LLC, LLD, LMC	Packaged by home cheese supply co.	1 packet	N/a	Starter and aroma cultures packaged for small batches
Flora Danica or CHN 11,19	LL, LLC, LLD, LMC	Chr. Hansen	¼ tsp (0.4 g)	10U	Starter and aroma cultures
Meso Aromatic B	LL, LC, LD, LM	Abiasa	¼–½ tsp (0.6–1.2 g)	10 g	Starter and aroma cultures
MM 100 Series (Choozit) (includes 100, 101)	LL, LLC, LLD	Danisco	¼ tsp (0.4 g)	10 g	Starter and aroma culture
MA 4000 Series (Choozit) (includes 4001, 4002)	LL, LLC, LLD, ST	Danisco	⅛ tsp (0.3 g)	5U	Starter and aroma, plus one thermophile for adjunct flavor during aging
LM 57 (Choozit)	LMC	Danisco	Add to culture blend at rate of 25%	Add to culture blend at rate of 25%	Aroma and gas mesophile; can be added to cultures in red
TA 50 Series (Choozit) (includes 51, 52)	ST (slow-acid producer)	Danisco	Replace ¼–⅜ of mesophilic blends	Replace ¼–⅜ of mesophilic blends	Thermophilic; use to replace ¼–⅜ of mesophilic blends in yellow and MM series

*Remember that culture is always best measured by weighing and dividing into unit doses based on the strength of the culture, instead of using volume or set weight.

CHOOSING COAGULANT

You can use any rennet variety to make blue cheese, although many cheesemakers swear by the superiority of traditional veal (calf) rennet for helping create a less bitter flavor. As with so many other things to do with cheesemaking, you can experiment and see what works for your milk and appeals to your taste buds, as well as your ethical standards.

CUTTING AND STIRRING

Curd for blues is cut a bit larger than for other semi-hard cheeses, usually between ½ and 1½ inches (1–4 cm). In some recipes the curd is not cut at all but is ladled directly into draining cloths. The cut curd is not heated beyond the ripening temperature and is stirred either only briefly or intermittently. The goal during this phase is enough even moisture loss from the curd to allow it to hold its shape during draining and create an open-textured wheel that will provide aeration for the mold spores.

DRAINING

There are several different approaches to draining curd for blue cheese: ladling curd directly into the forms; draining first in the vat; draining in the vat and hand-stirring after adding an application of a small dose of salt; and draining in cheesecloth bundles until firm, then breaking or milling the curd before placing it in hoops.

TABLE 12-2. *PENICILLIUM ROQUEFORTI* MOLD CULTURE CHOICES FOR BLUE CHEESES

Name/ Dosage per package	Manufacturer	Dosage* per 26 gallons (100L/220 lb)	Characteristics and Application
P. roqueforti	Packaged by home cheese supply co.	1/16 tsp (.05–0.1 dose) rehydrated in 1/4 cup (60 ml) milk/3–4 gal (12–15L)	Blue-green, slow to medium growth
PRB 6/5D	Danisco Choozit	Liquid form: follow instructions	Bright green, medium to fast growth
PR 3/10D	Chr. Hansen	0.5–1 dose	Bright green, strong blue flavor, creamy texture; high proteolysis, medium lipolysis
PRB 18/5D	Danisco Choozit	Liquid form: follow instructions	Light blue
PJ/10D	Danisco Choozit	0.5–1 dose	Middle green, typical blue flavor, fast growth
PA/10D	Danisco Choozit	0.5–1 dose	Dark green, mild blue flavor, very fast growth, Roquefort styles
PR4/10D	Chr. Hansen	0.5–1 dose	Blue-green, medium blue flavor, creamy texture; high proteolysis, medium lipolysis
PS/10D	Danisco Choozit	0.5–1 dose	Blue-green, mild blue flavor, medium-fast growth
PV/10D	Danisco Choozit	0.5–1 dose	Blue-green, strong blue flavor, very fast growth, Roquefort and Gorgenzola styles

*Small-dosage rates are in the recipes, but generally only 1 to 2 skewer tips are needed per 3 to 4 gallons of milk.

Each method creates a different-textured curd, along with a different rate of acid development (which, of course, has a great influence on texture). When working with the curd during draining, pay particular attention to the room temperature and its effect on drainage time and pH development. Remember the final goal is one of texture and proper acidity, so make that more important than suggested time parameters.

The size and perforations of the forms must be chosen to accommodate the moisture and texture of the curd. For all blue cheeses a straight-sided form, usually taller than its diameter, is preferred. Let's go over the three basic curd types and the best form choices for those examples.

For curd that is partially drained in the vat and when an open texture of the rind is desired, as in our first recipe in this chapter, use a basket-type form, closed at the bottom, with many small perforations on the sides. If the perforations are small and of adequate number, no cheesecloth is needed to line the forms.

For curd that is well drained in the vat, choose a form of equal height (or a bit taller) and diameter with both ends open and without perforations on the sides. The forms are lined with cheesecloth of fine weave and placed on matting that can drain rapidly.

How to tie a "stilton" knot: Gather three of the corners, hold the fourth corner, and wrap it tightly around the neck of the three corners; make each wrap a bit lower on the neck and close to the curd to keep the knot snug.

Curd that has been milled and salted will have a drier, firmer texture and should be hooped in a form with a height-to-diameter ratio of 2:1. Both ends should be open, and the sides can have a few small perforations. The extra height will allow for more curd to fill each form, providing weight to knit it together.

Cheeses in open-ended forms that are lined can be turned by allowing the cheese to slide out of the bottom, removing the cloth and draping it over the top of the empty form, then sliding the cheese back into the hoop. For open-ended, unlined forms, turning is done similarly to that of Brie—with a mat being placed on top of the form and the entire form (along with top and bottom mat) flipped. For unlined, perforated forms with a closed bottom, the cheese can be upended into a new, empty form. The empty form that the cheese was flipped from can even be left sitting on top of the filled form to provide support of extra curd while the cheese drains and settles in size (this is especially helpful if you don't have forms that are tall enough).

SALTING

Blue cheeses can be brined or dry salted. If the mold spores are added directly to the milk, as opposed to sprinkled in the layers of curd, and the curd is fairly well knit, it can be brined. Dry salting, however, is far more common than brining, as it is easier to control the gradient of salt concentration. Do not choose brining for blue cheeses that have a very open structure, as it is likely that the brine will quickly penetrate the wheel and cause the salt content to be too high for the blue molds to grow.

In some cases a small amount of salt is added to the curd at hooping and again to the outside of the wheel after unmolding. Blue curd that has been milled is completely salted prior to placing in the form. Other types are dry salted after unmolding in one or more applications over a period of days.

Typically, blue cheeses are salted at a rate of 3 to 5 percent of the weight of the cheese. If you are dry salting, you can measure out the total amount, then use that over the period of recommended applications. Remember salt's effect on bluing—the more salt, the less blue growth—so you can vary your salt percentage based on your goal for the cheese.

Rubbing up the sides of a Stilton-type blue cheese

AFFINAGE

Earlier, I talked about how blue cheese recipes vary greatly in the length of ripening and coagulation, as well as time in the vat. The styles also differ quite a bit when it comes to aging temperatures, times, and time of piercing. While recipes may give you temperature and humidity goals as well as average times for aging, you must learn to "read the cheese" (as you did the curd earlier) and interpret what its needs are. This ability will not come immediately, so be patient, take good notes, and keep making cheese!

Most blue cheeses spend a week or more aging at 50 to 55°F (10–13°C) with a fairly high relative humidity (RH) of 90 percent. During this time the inoculated blue mold spores will begin to germinate and grow in the open texture of the cheese. The salt gradient will initially protect the outside surface from mold growth of all kinds. As the salt moves to the interior, it will moderate the growth of the blue molds. During this time the cheeses should be turned daily and inspected for mold growth.

> The ability to "read the cheese" during aging will not come immediately, so be patient, take good notes, and keep making cheese!

PIERCING

After an initial period of one to three weeks, the cheeses are usually pierced with a sterilized knitting needle or skewer of 3.5 to 4 millimeters to allow for new air exchange. Sometimes they are pierced twice during this initial aging period. It is a good idea to learn to use a trier to do a core sample to determine the extent of the bluing occurring inside the cheese (see chapter 4 for more on taking core samples). Remember that since the piercing will allow for increased mold activity, you can control the amount of blue growth by the amount of piercing.

Once the cheese has the desired amount of bluing, it is usually moved to a cold aging environment to slow or stop the blue growth and allow for further proteolysis and lipolysis (the breakdown of proteins and fats) leading to texture and flavor development.

> Be very aware that the best blue cheeses in the world have been produced only because of years of effort, research, and possibly multiple wrong turns on the part of the cheesemakers.

KEEPING RINDS PRISTINE

In chapter 4 I introduced you to the natural antifungal mold inhibitor *natamycin*. Derived from the fermentation of a natural (meaning not engineered or modified) bacteria called *Streptomyces natalensis*, this mold inhibitor is widely accepted for food use and is sometimes used in the industrial-scale production of rindless blue cheeses. Natamycin is also the active ingredient in the so-called "cream" wax (such as the brand *Paracoat*) coating that many small-scale cheesemakers use to inhibit mold growth on pressed cheeses. So don't be too surprised if, despite all your attempts, your rindless blue does not look as lily white and pristine as the images you see of some large-production blues.

WRAPPING AND PROTECTING

If the cheese will have a natural rind, it is left unwrapped and kept in a humid, cool to cold environment. Some rind maintenance may be necessary if the cheese becomes too moist and even slimy. In this case you'll need to reduce the humidity and wipe or scrape the surface of the cheese as necessary.

For blues for which a white, clean surface is desirable (a rindless cheese), the wheels can be tightly wrapped in foil or even waxed. It is also possible to vacuum seal the cheese. Sometimes, a totally sealed cheese will not look blue when first opened, and then, once oxygen makes contact with the molds, spores will form and color will appear. A vacuum-sealed or waxed cheese may accumulate some moisture under the wrapping. If the moisture becomes excessive, you should open the wrapping, let the cheese breathe and dry for a brief period, then rewrap if desired.

Several well-known blue cheeses are wrapped in leaves that have been soaked in alcohol. Some examples are the Spanish blues, Cabrales and Valdeón, and the new kid on the block (but instant classic) Rogue River Blue from my friends and neighbors at the Rogue Creamery. If you decide to wrap your blue in leaves, just be sure to select a variety that is edible! First, wash the leaves in plain water, then place them in a sealable bag with enough alcohol (choose a drinking alcohol such as whiskey, bourbon, brandy, or other high-proof choice) to cover the leaves. Soak the leaves in the alcohol in a cool place for at least a couple of weeks. During this time the alcohol will soak into the plant structure and provide antiseptic and preserving properties.

Rindless blue cheese can be packaged in plastic wrap, vacuum sealed, or wrapped in cheese paper. Naturally rinded cheeses do not do as well if sealed for extended periods without some air exchange, so choose double-layered cheesepaper, greaseproof or parchment paper, or freezer-wrap paper for these kinds of blue.

RINDLESS BLUE CHEESE

Roquefort is arguably the most famous blue cheese in the world and I would say easily the most well-known sheep's milk cheese. I can still remember my parents teaching my sister and me that the "Roquefort" salad dressing served to us on our iceberg lettuce salads at the local restaurant should actually be called "blue cheese" dressing, because Roquefort was a specific type of French blue cheese, which our salad dressing was definitely not. Pretty hip for country folks, weren't we? Roquefort, which has been an AOC cheese since 1925, is made exclusively from the raw milk of Lacaune sheep.

Danablu (also called Danish Blue) was purposefully modeled after Roquefort but is milder and made from cow's milk, as is another French cheese called Bleu des Causses. Beenleigh Blue is an English rindless blue made from raw, organic sheep's milk that Max McCalman notes as similar and worth tasting. (My list of "must tries" continues to grow!)

INGREDIENTS

Milk: 4 gal (15 L/34 lb) high-fat milk (about 6% butterfat is ideal)
Culture: ¼ tsp (0.6 g) MM 100 (or equivalent blend) and 3 to 4 skewer tips *P. roqueforti* rehydrated in ¼ cup (60 ml) room-temperature milk
Lipase (optional): ⅛ tsp (0.7 g) lamb lipase (or other lipase). Lipase will help duplicate original Roquefort, which is made with rennet paste (contains lipase).
Calcium chloride (optional): maximum 1 tsp (5 ml) calcium chloride diluted in ¼ cup (60 ml) cool, nonchlorinated water
Rennet: ⅜ tsp (1.6 ml) single-strength rennet diluted just before use in ¼ cup (60 ml) cool, nonchlorinated water
Salt: 3 tbsp (45 g) pure salt

STEPS

Prepare Equipment: Make sure all equipment is cleaned and sanitized and that your cheesemaking space is free from possible contaminants. Refer to chapter 6 for tips on proper equipment preparation.

Prepare Milk: Warm milk to 89 to 90°F (32°C).

Culture: Sprinkle cultures on top of milk. Let set for 2 to 5 minutes, then stir gently for 3 to 5 minutes.

Ripen: Maintain temperature at 89 to 90°F (32°C), and hold it there for 30 to 35 minutes.

Additions: Stir in lipase solution, then stir in calcium chloride solution. Wait 5 minutes.

Coagulate: Stir in rennet solution with an up-and-down motion for 1 minute. Still the milk. Hold temperature at 89 to 90°F (32°C) until clean break is achieved. The goal coagulation time is 2 hours.

Cut: Cut curd mass into ⅜- to 1-inch (1–3-cm) cubes. Rest for 5 minutes.

Stir: Stir curd gently two to five times over the next 40 to 60 minutes. Maintain at 89 to 90°F (32°C). Let curds settle.

> TIPS: During stirring the goal is to keep the curds from matting and allow them to develop acid and shrink a bit. By the end of the stirring phase, curds will be soft and high in moisture content, but they should be even in texture.

Drain: Drain whey to level of curds. Place curds in a cloth-lined colander, and sprinkle with 1 teaspoon (5 ml) salt. Mix gently. Ladle or sprinkle curd into tall, cloth-lined forms. Set follower gently on top of curd with no pressure or weight applied. Drain at room temperature of 68 to 73°F (20–23°C) for 2 to 3 days until curd pH is at 4.8. Turn daily.

> **TIPS:** If you are using a vat from which the whey can be fully drained, the curds can be drained and salted in the vat, then placed in forms. Forms should have at least an equal diameter-to-height ratio but can be a bit taller than the diameter.

Salt: Unmold onto a draining rack. Rub each side of wheel with salt daily for 2 days. Room temperature should start at about 72°F (22°C) and decrease to 68°F (20°C) by the end of this period.

Affinage: Move to aging room at 50 to 55°F (10–13°C) and 90 percent RH. Turn daily. After 8 to 10 days, pierce vertically using a sanitized 3.5 millimeter knitting needle or similar-size meat skewer. Pierce three to four holes per square inch (6.5 sq cm). Pierce again after 2 more weeks of aging. Continue to age at 50 to 55°F and 90 percent RH. When adequate blue growth has occurred (check by doing a core sample), the cheese can be wrapped in foil or waxed and moved to a colder environment at about 37°F (3°C) and aged for several months.

LARGE-BATCH GUIDELINES

Milk: 12 gal (46 L/100 lb) milk
Culture: 2–4U MM 100 (or equivalent blend) and ¼ dose *P. roqueforti*
Calcium chloride (optional): 10–15 ml (about 1 tbsp)
Lipase (optional): ¼–½ tsp (1.4–2.8 g)
Rennet: 5–6 ml single-strength rennet (2.5–3 ml double strength)
Salt: Goal is 2.5–3% of weight of drained cheese

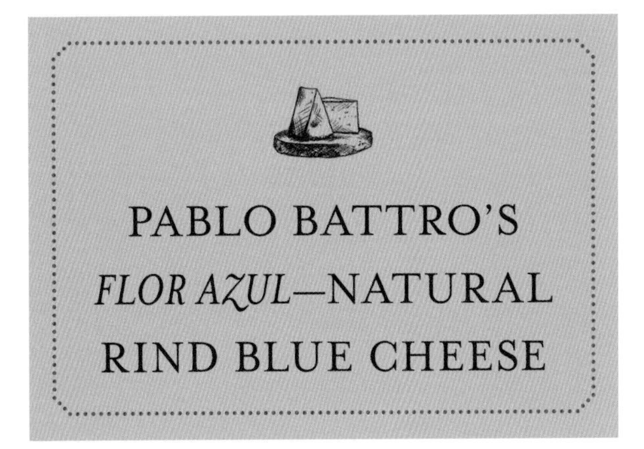

PABLO BATTRO'S *FLOR AZUL*—NATURAL RIND BLUE CHEESE

Gorgonzola, Italian DOC, DOP, can go back to back with Roquefort in a battle for granddaddy of the blues. But let's not pit these two greats against each other! Gorgonzola is made from cow's milk and, unlike its rindless French neighbor, has a natural rind. The original Gorgonzola cheeses were made by draining the curd in two stages before placing it in the form. The evening milk was made into curd and drained on cloth. The morning milk was made into another batch of curd, briefly drained, then placed into the bottom and sides of the forms, creating an open well. The evening curd was placed into the center well, and the process continued. Some modern recipes still call for these steps.

The evening curd, being longer drained, has a more open texture and makes a great canvas for the blue mold growth. This traditional version of Gorgonzola is usually aged longer and called "piccante," although many labeled as piccante are made from a single batch of curds. Another version, younger and less robust, is called "Gorgonzola dolce."

Cashel Blue, from Ireland, is a similar, naturally rinded blue made from pasteurized cow's milk, while Oregonzola is a rindless, raw cow's milk blue that is made in a similar style in the United States by Oregon's Rogue Creamery. For our rinded-blue recipe that follows, Pablo Battro has recreated some of the same qualities as the Italian original, while at the same time allowing his Flor Azul to reflect the qualities the Argentine pampas (grasslands) bring to the milk. In addition to being a cheesemaker, Pablo is the author of several fine Spanish language books on cheesemaking, including *Quesos*

Flor Azul, named after the beautiful flower *Cichorium intybus*, or wild chicory, by Argentine cheesemaker Pablo Battro. Photo courtesy of Pablo Battro

Artesanales. Pablo suggests that the cheese is best made in a large batch, so that each wheel weighs at least 9 pounds (4 kg), but the home-size recipe of 4 gallons (15 L) will yield a wheel closer to 3 or 4 pounds.

INGREDIENTS

Milk: 4 gal (15 L/34 lb) whole milk
Culture: ¼ tsp (0.4 g) Flora Danica, ¹⁄₁₆ tsp (0.1 g) ST 50, and scant ¹⁄₁₆ tsp (0.3 g) *P. roqueforti* (if possible, choose a strain with medium strength and proteolysis)
Calcium chloride: maximum 1 tsp (5 ml) calcium chloride diluted in ¼ cup (60 ml) cool, nonchlorinated water
Rennet: ½ tsp plus (3 ml) single-strength rennet diluted just before use in ¼ cup (60 ml) cool, nonchlorinated water
Salt: 3 tbsp (45 g) pure salt

STEPS

Prepare Equipment: Make sure all equipment is cleaned and sanitized and that your cheesemaking space is free from possible contaminants. Refer to chapter 6 for tips on proper equipment preparation.

Prepare Milk: Warm milk to 90°F (32°C).

Culture: Sprinkle cultures on top of milk. Let set for 2 to 5 minutes, then stir gently for 3 to 5 minutes.

Ripen: Maintain temperature at 90°F (32°C) for 30 minutes.

Additions: Stir in calcium chloride solution for 1 minute, then wait 5 minutes before adding rennet.

Rennet and Coagulate: Stir in rennet solution with an up-and-down motion for 1 minute. Still milk. Hold temperature at 90°F (32°C) until clean break is achieved. The goal coagulation time is 25 to 30 minutes.

Cut: Cut curd mass as follows: Cut across vat in vertical cuts 2 inches (5 cm) apart, then let rest for 10 minutes. Make vertical cuts in opposing direction at same intervals, then let rest 10 minutes. Make horizontal cuts at same intervals, then let rest 10 minutes.

Stir: Hold at 90°F (32°C) for 45 minutes. Stir occasionally to prevent matting.

> TIPS: During stirring the goal is to keep the curds from matting and allow them to develop acid and shrink a bit. By the end of the stirring phase, curds will be soft and high in moisture content, but they should be even in texture.

Drain: Drain whey to the level of curds. Line a large tray with muslin or organdy. Transfer curd to draining table or tray. Set it at an angle so the whey can drain away from the curds. Maintain the room temperature at 72°F (22°C), and let curds drain for 1 hour. The pH will continue to drop during this time. You can cover the tray with a plastic container to help maintain the temperature. The container or cover should not touch the curds.

Drain in Forms: Loosen curds with your fingers, then scoop or pour curds into a tall, cloth-lined form. Drain. Turn after 40 to 60 minutes, then again every 4 hours (if possible) to a goal curd pH of 4.7 (this may take up to 24 hours). Room temperature should be 72°F (22°C). You can use an extension to forms (a smaller form set on top of a larger one) to help fill curd to a height that will settle to the top of the larger form during draining. Traditionally, these cheeses are taller than they are wide.

Salt: When curd pH is 4.7, unmold onto your draining mat. Apply salt to all sides.

> TIPS: If the cheese is tall and seems to need support, use a sushi-type reed mat or even a plastic draining mat and wrap it around the cheese. Secure with sanitized shoestrings or another band. These supports can remain around the cheese for a good part of the aging or until not needed. Goal salt used is 3% of the weight of curd.

Affinage: Move cheeses to an aging space with a temperature of 51°F (11°C) and 90 percent RH. Turn daily. After 7 days use a sanitized 4 millimeter knitting needle or meat skewer of a similar size to pierce the cheeses vertically. Pierce three to four holes per square inch (6.5 sq cm). Age for 7 more days without turning. Pierce again from the other end, and turn. Leave the cheeses unturned for 7 more days, but lift and shift their positions on the rack to prevent sticking. Then turn as needed to prevent sticking to shelves. Age for 60 to 75 days, or until the desired taste and texture is attained.

LARGE-BATCH GUIDELINES

Milk: 12 gal (46 L/100 lb) milk
Culture: 2–4U Flora Danica, 1U TA 50, and ¼ dose *P. roqueforti*
Calcium chloride: 10–15 ml (about 1 tbsp)
Rennet: 7–9 ml single-strength veal rennet (3.5–4.5 ml double strength)
Salt: 3% of weight of curd

PABLO BATTRO, LINCOLN, ARGENTINA

Author's note: We first met Pablo when he and his wife Eugenia came to the United States to speak to the Oregon Cheese Guild. One of Pablo's sons is a cheesemaker in Oregon (Mariano Battro, Mariposa Creamery). Pablo and Eugenia kindly invited Vern and me to visit them in Argentina, which we did in December of 2010, visiting many other creameries in their country and having a memorable visit with this amazing cheesemaker and his lovely, gracious wife. In addition to supplying a Gorgonzola-type recipe, Pablo reviewed the entire chapter on blue cheese, contributing his expertise and advice. Here, in his own words:

"Forty years ago I lived in Patagonia, and like many other families we had a crowd of milk-drinking kids. At that time the answer was to get a cow, and so entered 'La Negra' (Blackie), and it's twenty years ago today that marks the start of my cheesemaking.

"Many years back Patagonia was arid outback country, no libraries, much less books to learn by, no one to ask, 700 kilometers (about 435 miles) of dirt road to a bigger town; it wasn't easy. As there began to be a surplus of daily milk, I got some rennet powder from a local store, put some fresh milk in a pan in the kitchen, added some of the powder to my liking, and voilà, I had made cheese—at least it looked like cheese! It was a thrilling experience.

"After a considerable period of trial and error, things began to look good, and my 'cheese' began to get approval from family and friends. So I bought two more cows: 'La Reina' (Queenie) and 'La Blanca' (Whitie) and, with the help of my sons, built a milking parlour with a homemade re-creation of a milking machine, and so began my serious efforts in this activity.

Argentine cheesemaker and author Pablo Battro at his cheese studio

"I made my cheese in a common kitchen pot, which later became a 100-liter stainless steel vat of my making, which was heated by two burners. Years later I moved into a run-down outhouse, with a 300-liter vat, enough heating to pasteurize, homemade presses, a room to mature the cheese, all the necessary legal permits, the works, and was on my way.

"I produced two types of cheese, *Mariposa* ('Butterfly'), a semihard cheese with short ripening time, and *Cinco Esquinas* ('Five Corners,' the name the district was known by), a raw-milk semihard cheese with a five-month maturing time. The cheese sold very well in local fairs and later on in grocery stores and even in a local supermarket.

"Then came a Gouda-type cheese from sheep's milk, which won a gold medal in 2000, and a feta-type cheese that was 30 percent sheep's and 70 percent cow's milk. I went on adding different varieties of cheese as years piled on, each cheese a new challenge.

"As time went by this hobby of mine became my profession, my mainstay. Articles, talks, conferences, books, consulting—all of it fell in line naturally. It's my profession, and I love it."

MILLED CURD STILTON-STYLE BLUE CHEESE

Stilton, like many English cheeses, has curd that is milled, meaning it is allowed to form large clumps, which are then broken into small pieces just before pressing. This process alone sets it apart. I still remember the first time I was able to taste a great Stilton; it was a revelation. To me naturally rinded blues combine the best of all the mold- and bacteria-ripening possibilities in one powerful cheese. While PDO Stilton is a pasteurized cow's milk cheese, a recent newcomer—but a product meant to honor Stilton's past—is a raw-milk blue called Stichelton. This new but arguably more traditional cheese was created by Randolph Hodgson—a well-known cheese affineur, monger, and head of London's Neal's Yard Dairy—and cheesemaker Joe Schneider. While some US cheeses, such as Jasper Hill's Bayley Hazen Blue—made with raw Ayrshire cow's milk from Vermont—seem to be inspired by the English classic, their actual make process is different.

Stilton-style curd ripening in bundles in a small vat. Whey is slowly squeezed and drained from curd balls by the tightening of the stilton knot.

INGREDIENTS

Milk: 2 gal (8 L/17 lb) whole milk
Culture: 1⁄16 tsp (0.2 g) MA 4000, 2 skewer tips LM (or equivalent blend), and 3 skewer tips *P. roqueforti* mold culture rehydrated first in ¼ cup (60 ml) 89°F (32°C) milk from vat
Lipase (optional): ⅛ tsp (0.7 g) lipase dissolved in ⅛ cup (30 ml) cool, nonchlorinated water
Calcium chloride (optional): maximum ½ tsp (2.5 ml) calcium chloride diluted in ¼ cup (60 ml) cool, nonchlorinated water
Rennet: scant ¼ tsp single-strength rennet diluted just before use in ¼ cup (60 ml) cool, nonchlorinated water
Salt: 2 tbsp (30 g) pure salt (goal is 2.5% of weight of drained curd)

STEPS

Prepare Equipment: Make sure all equipment is cleaned and sanitized and that your cheesemaking space is free from possible contaminants. Refer to chapter 6 for tips on proper equipment preparation.

Prepare Milk: Warm milk to 70°F (21°C).

Culture: Sprinkle cultures on top of milk. Let set for 2 to 5 minutes, then stir gently for 3 to 5 minutes.

Ripen: Increase temperature to 89°F (32°C) over 15 to 30 minutes.

Additions: Stir in lipase solution, then stir in calcium chloride solution. Wait 5 minutes.

Rennet and Coagulate: Stir in rennet solution with an up-and-down motion for 1 minute. Still milk. Hold temperature at 89° (32°C) until clean break is achieved. The goal coagulation time is 60 minutes. If using flocculation as a guide, multiply flocculation time by 2 to get the ideal total length of coagulation.

Cut: Cut curd mass into ¾- to 1¼-inch (2–3 cm) cubes. Rest for 10 minutes.

Ripen in Bundles: Line colander with cheesecloth, and scoop or ladle curds into cloth. Tie cloth closed with a stilton knot. Place bundle back into pot. Cover and let rest with whey draining around bundles. Tighten knot on bundle every hour for 2 hours or until curd is firm. Maintain temperature of curd at 89°F (32°C). Goal curd pH at end of this period is approximately 6.36.

Unwrap curd bundles on drain board or in pot tilted to drain. Keep warm (88 to 89°F [31–32°C]) and turn curd balls every 60 minutes until the curd pH reaches 4.8 to 4.6. It can take up to 7 hours to reach the goal pH.

Mill and Salt: Pour off whey, open bundles, and break curd into 1-inch (2.5-cm) chunks. Add salt, and mix well. The goal salt content is 2.5% of the weight of the milled curd.

Drain in Forms: Place curd in a tall, warmed, unlined mold, and set on a draining mat. Turn at 30 minutes, 3 to 4 hours, then 12 hours. Turn daily for 2 to 4 days; room temperature can decrease to 68°F (20°C) by the second day.

> **TIPS:** Stilton-type wheels have many openings formed from being milled, salted, and hooped without pressure. Using a tall hoop will help utilize the weight of the curd to form the wheel.

Unmold: When wheel has consolidated its shape and will hold its form, remove from mold. "Rub up" the sides, top, and bottom to create a smooth finished surface.

> **TIPS:** Rubbing up is not simple; you can expect lots of curd bits to want to crumble off the sides of the wheel. Make sure the room and wheel are not too cold, and mostly, be patient!

Affinage: Move wheel to aging space at 50 to 55°F (10–13°C) and 90 percent RH. Turn daily for first month.

At 3 to 5 weeks of age, pierce (with a sanitized 4 mm knitting needle or a meat skewer of equal size) from sides using a slight up or down angle. Move to 44°F (7°C) aging environment for 6 to 15 months or until desired flavor and texture has been attained. Turn to prevent sticking to shelves.

LARGE-BATCH GUIDELINES

Milk: 12 gal (46 L/100 lb) milk
Culture: 2–4U MA 4000, 1U LM, and ¼ dose *P. roqueforti* rehydrated in 1 cup (240 ml) milk from vat
Calcium chloride (optional): 10–15 ml (about 1 tbsp)
Lipase (optional): ¾ tsp (4.2 g)
Rennet: 7–9 ml single-strength rennet (3.5–4.5 ml double strength)
Salt: 2.5% of weight of milled curd

ANN HANSEN'S BLUE-BRIE CHEESE

I mentioned earlier my enjoyment of the combination of blue and surface ripening that occurs with a naturally rinded blue cheese. Well, behold the glory that is the blue mold interior with the soft, surface-ripening work of white molds—the German Cambozola (sometimes called Blue Brie). While it might be considered a bit of an industrial, even gimmicky, cheese,

Home cheesemaker Ann Hansen's Cambozola-type cheese. Photo courtesy of William D. Lester

if it's good, who cares? FireFly Farms, in Maryland, produces an award-winning goat's milk cheese they call Mountain Top Bleu (along with several other standout cheeses) that combines surface blues with white molds. In addition, Saga Blue is a large-scale but well-recognized Danish blue-brie cheese that is widely available. These hybrid cheese styles are tricky to make but satisfying to eat for those of us who like to mix and match our molds.

The instructions for this recipe are provided by a passionate home cheesemaker, Ann Hansen, of Ruch, Oregon. Ann, who calls herself a "foodie of the highest order," struggled with other Cambozola recipes until she hit upon this ratio of milk to cream and the steps she included. She says that other recipes had too high a fat content and would not drain properly or develop bluing in the center, as the texture would close and leave no open piercings. In recipes that called for the white molds to be added to the milk, blue was inhibited by the earlier growth of the whites (this can actually be a problem for all blue cheeses when *Geotrichum* and *P. candidum* are present in the milk or aging environment). By piercing and allowing the blue to begin growing, then inoculating with the white molds after that, she was able to create a lovely blue-veined bloomy rind cheese. And she has only been making cheese for 2 years!

I don't have large batch guidelines for this recipe and was reluctant to attempt to provide them without having first made the recipe and become comfortable with its steps. Due to the fact that it calls for added cream, it could be a tricky one to produce commercially unless you have access to cream that is both of high quality and affordable (in other words it translates into a good choice for the commercial producer).

INGREDIENTS

Milk: 1 gal (4 L/8.6 lb) whole milk plus 1 pint heavy cream (not ultrapasteurized)
Culture: ¼ tsp (0.6 g) MM 100, ⅛ tsp (0.6 g) *P. roqueforti*, 5 skewer tips *P. candidum*, and 1 skewer tip *Geotricum*
Calcium chloride (optional): maximum ¼ tsp (1.25 ml) calcium chloride diluted in ⅛ cup (30 ml) cool, nonchlorinated water
Rennet: ¼ tsp (1.25 ml) single-strength rennet diluted just before use in ¼ cup (60 ml) cool, nonchlorinated water
Salt: ½ tsp (2.5 ml) pure salt

STEPS

Prepare Equipment: Make sure all equipment is cleaned and sanitized and that your cheesemaking space is free from possible contaminants. Refer to chapter 6 for tips on proper equipment preparation.

Prepare Milk: Warm milk and cream mixture to 75°F (24°C).

Culture: Sprinkle MM 100 culture on top of milk. Let set for 2 to 5 minutes, then stir gently for 3 to 5 minutes.

Ripen: Increase temperature to 86°F (30°C), and maintain for 20 minutes.

Additions: Stir in calcium chloride solution. Wait 5 minutes.

Rennet and Coagulate: Stir in rennet solution with an up-and-down motion for 1 minute. Still milk. Hold temperature at 86°F (30°C) until clean break is achieved. The goal coagulation time is 1.5 hours.

Cut: Cut curd mass into ½-inch (1.3-cm) cubes. Let rest for 5 minutes.

Stir: Stir gently for 5 to 10 minutes. Maintain temperature at 86°F (30°C); settle for 5 minutes. Curds will be very soft and fragile.

Drain: Drain whey to the level of the curds, then transfer curds to a cloth-lined colander. Cover and let drain for 20 minutes.

Ladle: Prepare the draining setup by covering a draining rack with a cheese-draining mat, then covering that with a fine-weave cloth such as organdy. Set two Camembert/Brie–type forms (open at both top and bottom) on top of the layered draining setup. Fill the forms halfway with curd, and sprinkle the surface of the curds with half of the *P. roqueforti* blue mold culture powder. Let drain 15 minutes. Ladle the remaining curd into the forms.

Drain: Drain for 6 hours. Room temperature should be about 72°F (22°C). Flip by placing another layered draining setup on top of forms, then turning entire stack as a whole. Drain 8–12 hours, and allow the room to cool to 68°F (20°C) by the end of draining. The goal pH at the end of draining is 4.7 to 4.8.

Salt: Unmold onto a draining mat, and sprinkle tops with ¼ teaspoon (1.25 g) salt. Wait 5 minutes, turn, and sprinkle other side with another ¼ teaspoon salt. You don't have to salt the sides if less than 1.5 inches (3.8 cm).

Affinage: Drying Phase: Move to an aging space at 50 to 55°F (10–13°C) and 90 percent RH. Turn daily. If drying is too slow, utilize a small low-velocity fan to help.

Piercing: On day four (when surface has dried some), pierce horizontally eight to ten times. When piercing, penetrate the cheese completely; use a small, sterilized needle that makes a hole large enough to stay open (3.5 mm).

Inoculate: Prepare white mold cultures following the rehydration instructions on page 180. After 12 hours of rehydration, mist onto the surface of the pierced cheeses. An alternative to misting is to sprinkle the mold powders on the cheese, then, using a sanitized and rinsed cloth, dab across the surface of the cheese.

Age: Increase RH to 95 percent. Turn daily and watch for blue mold growth around the pierced holes. White molds should appear on the surface of the cheese by day 8 or 9.

Wrap: When white mold growth is even and adequate, wrap in cheese paper, and continue to age at 55°F (13°C) or move to a refrigerator for longer aging.

THINKING OUTSIDE THE VAT

SCENARIO 1: You notice that during draining and pressing it is taking longer than it should to reach the goal pH, even though the pH in the vat and at draining were right on target. You should consider trying which of the following?

A. Increase the room temperature.
B. Use more starter culture the next time.
C. Make sure the curds are a bit higher in moisture at the beginning of draining and pressing.
D. Increase the pressure.

SCENARIO 2: Four weeks into aging your blue cheese, and a week after piercing, you take a core sample of the cheese and find some blue veins, but not as many as you would like. You can do which of the following?

A. Get over it.
B. Pierce the cheese again.
C. Rub the cheese with more salt.
D. Sprinkle the surface with more *P. roqueforti* mold spores.

1. Answer: A and C.
2. Answer: B.

Aged, smoked provolone-type cheeses

13: STRETCHED AND KNEADED PASTA FILATA CHEESES

FINISHED TEXTURE: Semisoft, semihard, and hard
METHOD OF COAGULATION: Rennet
METHOD OF RIPENING: Internal bacteria

These types of cheeses will really get you into hot water, literally! Stretched and kneaded cheeses, also known by their Italian name, pasta filata, go through the unique process of being dunked and soaked in hot water or whey, kneaded or stretched (a texturizing process), then shaped.

The cheeses in this chapter are all fantastic melting and stretching cheeses when young and become great melting-only cheeses when aged. Some are produced to be consumed when very young (such as mozzarella types) while others (such as provolone and kashkaval) can be aged for a year or more. Because industrial provolone has been widely available in the United States, it is, unfortunately, not very much appreciated. Once you make your own, however, you will have a new appreciation for these amazing cheeses.

THE BASICS OF STRETCHED CURD CHEESES

In chapters 1 and 2 you learned about the network of proteins that forms during cheesemaking, specifically during coagulation. The extra step of stretching and kneading uses heat and movement to elongate and create parallel alignments of these networks. Think about string cheese—the protein networks have been stretched and lined up so that when the cheese is cool you can peel off strings of cheese.

Before curd can stretch, however, there must be specific changes in the protein structure. For those changes to occur, the curd must reach the magic pH level of about 5.2. Through the development of acid, calcium is removed from the protein structures, allowing for the formation of the right kind of protein network for stretching. (You can read more about how calcium and other minerals interact with acid in chapters 1 and 3.) To successfully make these cheeses, you need to be able to monitor the development of acid. A pH meter is the easiest method, but I'll be telling you how to perform a stretch test on your curd that will tell you the same thing (this is the way they did it in "the old days").

The high-heat treatment in whey or hot water essentially (but not by legal definition) pasteurizes these cheeses. Any culture remaining will be killed as well—one more reason it is important to be sure to have the proper acid development before you try to stretch the curd. Some of the coagulant used will be deactivated, too, causing changes in the breakdown of protein during aging. But the enzymes remaining from the starter culture should provide plenty of protein breakdown power.

A DEEPER LOOK AT THE SCIENCE OF STRETCHING AND MELTING

When we visited Argentina we were on a quest for the perfect Provoleta, the nation's iconic grilled (as in barbecue) cheese dish. The amazing thing about Provoleta is that it's not supposed to melt. Supposedly, it is a version of provolone, which *does* melt.

Cheese generally both stretches and melts—but keep in mind these are actually different properties that happen at about the same time but are regulated by slightly different interactions (that I'll explain in a bit). The primary factor influencing stretch and melt has to do with the development of a suitable protein network. Other factors, such as fat composition, moisture content, protein breakdown (proteolysis),

WHAT ABOUT QUICK (A.K.A. 30-MINUTE) MOZZARELLA?

Many cheesemakers' first homemade cheese is quick mozzarella. And it is a great cheese! Essentially, acid, usually in the form of citric acid, is used to quickly acidify the milk to a pH of about 5.2. Then a small amount of rennet is used to coagulate the curd, which coagulates very quickly at such a low pH. The thickened mass is cut into cubes and heated slightly to shrink the curd a bit. The curd is drained, then heated in the reserved whey (or sometimes in a microwave—not my favorite method) with some added salt and stretched and formed.

The resulting cheese is quite satisfying and so fast to make that it is often worth the loss of the flavor that comes from developing acid through starter culture. Expect a lot of white whey (colored by some lost butterfat, protein, and other components) leaking from the curd during stretching because the protein network is not as strong as that formed during a long, slow rennet coagulation. It's also easier to end up with a tough texture using the quick method, but with some practice you can make some darned good, quickly ready-to-use cheese.

Fortunately for you, a talented home cheesemaker named Christy Harris has provided a recipe for a somewhat hybrid version of, shall we say, "quicker" mozzarella. Her recipe includes both added and developed acid. You will find the directions at the beginning of the recipe section later in this chapter.

calcium content, and temperature also influence melt and stretch characteristics.

To stretch the casein network that was formed during coagulation, cheese must restructure itself into a network that has the ability to maintain its cohesion when pressure (stretching) is applied. Stretch not only allows your pizza cheese to form long, delicious elastic strings, but it also allows propionic acid bacteria to form large eyes in the paste of an Emmental (Swiss) cheese. For stretch to occur the cheese must reach a warm to hot temperature. For cheese to develop large eyes, it must spend some of its aging time in a warm room—the same reason mozzarella must be dunked in hot water before it can be stretched. Stretch requires an intact network of proteins in the cheese. Cheese proteins in properly made mozzarella have the ability to slide past each other, then reform, keeping the strand from breaking and allowing you to stretch it into new shapes. This characteristic in science is called "plastic." Sometimes you will hear mozzarella curd referred to as "plasticized," which does not, of course, means you'll see plastic on the ingredients list.

HIGH-TEMPERATURE ADDED-ACID CHEESES

If the casein network will stretch and melt when it gets to the right pH, how do we explain nonmelting cheeses such as paneer and ricotta? These cheeses have added acid, which can easily drop the pH to the desired range for creating a melting and stretching cheese, right? You may remember from chapters 2 and 8 that high heat treatment of milk causes the denaturing of the whey proteins. The altered whey proteins are able to bond with the surface of the caseins and form what you might think of as a "coating" on the cheese proteins. This allows whey proteins—that would otherwise be lost in the whey during draining, to become a part of the curd—increasing both yield and total protein content of the cheese. But it also has the effect of blocking interactions between the casein molecules and limiting their ability to form a network that will allow for the properties of stretching and melting.

Let's review protein interactions for a moment. Remember that the development of acid (reduction in pH) occurs thanks to bacteria producing lactic acid (and remember these guys consume lactose). As the acid (hydrogen ions, +H) rises in cheese milk, two important things happen: The calcium phosphate within the protein micelles attempts to buffer the hydrogen ions and in the process is lost out of the micelle structure. As the pH drops, the caseins grow closer to their isoelectric point (where they can no longer maintain their negative charge). As they lose their charge, they no longer repel each other.

So in addition to the network already formed during rennet coagulation, you have additional casein molecules available to form further interactions with each other. Because the isoelectric point of casein is about 5.2, that is also the pH at which curd can begin to stretch—as long as it's at the right temperature, of course! If the pH drops too far (near 4.6), the casein molecules lose their networking ability and also the ability to stretch.

Melting occurs due to a somewhat paradoxical reason: the number of interactions between the protein molecules decreases. The quality of melting begins at roughly the same time as stretching, but it will continue (although in decreasing quality) if the pH of the cheese drops to 4.6 and below, when stretching will cease. Fat content has a greater influence on melting than on stretching. Fat globules are trapped within the protein network, and when present in a large percentage, they get in the way of protein interactions and therefore increase melting properties.

If you are having a little trouble, as I did, understanding the difference between stretch and melt, just remember that something that melts well can be mixed in with other substances, as when you are making cheese fondue. Cheeses that don't melt well but do stretch will just form a stretchy mass that won't mix in well with the melted cheeses.

Moisture is another factor influencing both properties of melting and stretching. Higher moisture content of curd can improve stretch and melt, but not without the other factors—such as calcium content, pH, and temperature—being correct.

You might have noticed that some cheeses will stretch when they are young but not later. The same goes for melting: Some cheeses melt well at 2 to 4 months of age but will later only soften and separate into a glob surrounded by oil. Think about that protein network that was required to both stretch and melt. As cheese ages, the proteins are being broken down during proteolysis. In the beginning of the life cycle of a cheese such as mozzarella, a short aging of a few days or weeks will improve the texture when it's melted, as more of the interactions between proteins are broken. Over longer periods, however, stretch is lost completely, and after very long aging melt quality is greatly diminished.

THAT ELUSIVE PROVOLETA

Eureka! In writing this chapter I learned that my quest for the perfect Provoleta could not be fulfilled by simply buying provolone, prepping it, and cooking it in the various ways suggested by different instructions. It is *impossible* for real provolone to not stretch and melt! The only one we had in Argentina that performed as a true Provoleta was a goat's milk cheese produced by one of the farms we visited. I remember the cheesemaker's saying that they had not stretched it. So I bet they formed it before the pH was low enough to stretch. I tried this with some of the curd from my provolone-type recipe later in this chapter, and low and behold—Provoleta that cooked on the barbecue without melting through the grill! To make it authentic, don't forget to marinate it first in olive oil and oregano or brush it on during grilling.

MAKING STRETCHED CURD CHEESES

I recently attended a class in which we were attempting to make mozzarella from goat's milk. The instructor had not tried to use goat's milk before, and this milk was from a great source but of unknown components and normal starting pH. When it appeared that the curd was ready to stretch, the best efforts of each student produced only a short-stretching, grainy curd. After much speculation, many concluded that you cannot make mozzarella with goat's milk. I had been making quick mozzarella from our Nigerian goat's milk for years but had not ever tried it with "regular" goat's milk (from any goat not a dwarf), so I immediately

TABLE 13-1. CULTURE CHOICES FOR STRETCHED CURD CHEESES

Name	Contains	Manufacturer	Small-Batch Dosage:* 2–4 gal (8–15 L/17–34 lb)	Large-Batch Dosage: 26 gal (100 L/220 lb)	Characteristics and Application
Thermophilic	ST, LB, LH	Packaged by home cheese supply co.	1 packet	N/a	All-inclusive blend for several Italian and stretched curd–type cheeses
Thermo B	ST, LB	Abiasa	¼–½ tsp (1.5–3 g)	5–10 g	A blend usually used for stretched curd, especially mozzarella but also Parmesan
TM 81 (Choozit)	ST, LB	Danisco	⅛ tsp (0.6 g)	2.5–5U	A blend usually used for stretched curd, especially mozzarella, but also Parmesean
Thermo C	ST, LH	Abiasa	¼–½ tsp (1.5–3 g)	5–10 g	A blend for hard and extra-hard cheeses
TA 60 Series (Choozit) (includes TA 61, 62)	ST (fast-acid producer)	Danisco	⅛–¼ tsp (0.3–0.6 g)	5–10U	Use to make your own blends by combining with LH 100 and/or LB 340
LH 100 (Choozit)	LBL, LH	Danisco	Dosage rate ranges from 10–50% of that of TA culture		Use with TA 60
LB 340 (Choozit)	LB	Danisco	Dosage rate ranges from 10–50% of that of TA culture		Use with TA 60
MM 100 Series (Choozit) (includes MM 100, 101)	LL, LLC, LLD	Danisco	⅛–¼ tsp (0.3–0.6 g)	4U	Added to help develop acid in stretched curd varieties
MA Series (Choozit) (includes MA 11, 14, 16, 19)	LL, LLC	Danisco	⅛–¼ tsp (0.3–0.6 g)	4U	Added to help develop acid in stretched curd varieties

*Remember that culture is always best measured by weighing and dividing into unit doses based on the strength of the culture, instead of using volume or set weight.

went home and tried it—not the quick, but a classic mozzarella recipe—and it worked fine. I hope, by the time you finish this chapter, you will have a good understanding of the science behind making stretched curd cheeses. If you do, you will be able to troubleshoot problems such as occurred in the class I mentioned.

CHOOSING MILK

Stretched curd cheeses can be made from cow, goat, and sheep milk. Even good-quality, homogenized, pasteurized (but not ultrapasteurized) store-purchased milk will make some pretty respectable fresh mozzarella. Texture and flavor, however, will be greatly influenced by the fat and protein content of the milk type. For home use fresh varieties can be made from high-quality raw milk. When producing for commercial sale, you must follow regulations regarding the sale of raw-milk cheeses. Since the curd goes through a high-heat treatment during stretching, the cheeses, even when made from raw milk, are at the low end of the spectrum where risk is concerned. When making an aged variety, you will have the best results using farm-fresh, high-quality raw milk—even though native, as well as added, starter bacteria are killed during the high-heat stretching phase, they will still provide enzymes that will create complexity of flavor during aging.

CHOOSING CULTURE

Cultures used in stretched curd varieties are almost always from the heat-loving thermophile family. While both thermophiles and warm-loving mesophiles are destroyed during the high-heat treatment of curd stretching, they still provide valuable enzymes that come into play during aging. The first three cultures in Table 13–1 are common blends of two or more thermophilic cultures. The next three are simply those bacteria but packaged separately, so you can make your own blends.

CHOOSING COAGULANT

When you are making most aged cheeses, coagulant plays a prime role in providing proteolytic enzymes that assist in the breakdown of proteins during aging. For a stretched curd cheese, however, the high-heat treatment destroys chymosin, so for these cheese types made with traditional or fermented rennet, no enzymes remain. This does not mean they won't be amazing when aged, but they will have a different texture and flavor.

STRETCHING

If you are using a pH meter, you can start checking the curd for its readiness to stretch at about a pH of 5.3. If you don't have a meter (or don't trust the one you have), take a 1-inch (3-cm) cube of curd, immerse it in 170°F (77°C) water, knead it briefly, reheat, knead again, reheat, and gently pull the piece in opposite directions. If you can get it to stretch about one yard (one meter), it is ready.

Curd being stretched

When the correct texture has been reached, the curd can be heated in the hot whey, or water, and stretched. If you have more curd than can be stretched within a short time, you can cool it and save it for later. Cool it rapidly by placing it in sealed plastic bags and immersing them in an ice water bath. The curd will hold its pH and can be milled and stretched when you are ready.

If you find that the pH development down to the stretching range is taking too long (meaning so long that you can't go to bed that night), you can extend it to overnight so it is ready in the morning. Depending on where the pH is at, you can slow or stall the development accordingly. For example, I got tired of waiting at about 4 p.m., and the pH was 5.75. So I bagged up the curd and put it in the fridge. The next morning I took it out and warmed it back up, and soon the pH was close to the target. Just keep this in mind as you make these cheeses—you can control the rate of acid development, after the curd is formed, by using temperature and without greatly altering the resulting cheese.

Unless you are a veteran curd stretcher, you will want to protect your hands during stretching. I like to wear heavy dishwashing-type gloves. If you find this is not enough protection, you can either wear a pair of thin cotton gloves underneath or have a bowl of ice water on hand into which you can dunk your hands. The initial stretching can often be done with two wooden spoons, but forming and shaping must be done by hand (unless you are a large-scale cheese company).

You'll want to handle the curd very gently during its first stretch, as it is quite prone to losing butterfat during this time. Be sure to heat it thoroughly before attempting to stretch—this is why the recipes call for cutting the curd—and milling into strip before stretching, as these will heat more evenly than a larger mass of curd. To tell if you have stretched the curd enough, look for a sheen to develop on the surface of the mass and long easy strands to form during stretching; soon you will know that the protein networks have been lined up sufficiently to give you the desired texture.

MAKING SHAPES

Here comes the fun part of making stretched curd cheeses! The elasticity and structure of these cheeses makes them prime candidates for sculpting into beautiful shapes and forms. As you experiment with these shapes, be sure to give yourself plenty of slack as far as mastering them goes. Your first shapes may look a bit like a kindergartner's clay pinch pot, but before long they will seem a bit closer to museum-grade sculptures.

Large Balls and Loaves

These are probably the easiest shapes to form by hand. After the curd reaches the desired sheen and texture, form large balls by taking the hot curd in two hands with your thumbs up. Sweep your thumbs from the center of the mass outward while at the same time using your fingers to tuck and push the curd into itself from underneath. This will form a smooth loaf on the outside, with a tucked, pinched area on the bottom. If the tucked area does not close well, dunk that portion back into the hot liquid and seal it by pinching the outside layer of curd together and over the seam.

Using Molds

Using hot curd that has been stretched to the desired texture, form it into a large ball shape (see photo), then place it into the form, without cheesecloth, with the seam side (where you pinched the ends together) at the bottom. Briefly submerge the cheese-filled form into the hot liquid to help the cheese assume its new shape. Once it looks as though the curd has assumed the shape of the form and the paste has closed tightly, you can dunk the entire thing into cool or salted water, then remove it from the form and float it in fully saturated brine. You can even set the entire form into the brine. Don't forget to sprinkle some salt on top.

Small Balls

Bocconcini (the Italian name for small, egg-shaped balls) are pinched off from larger formed balls. After forming a larger shape, reheat it briefly if it has cooled too much, and then, using your dominant hand, squeeze off an egg-size portion of curd, let that fall into the cool water, and repeat on another section of the larger shape. The small shapes will have a smooth side and a side showing where it was pinched off. Small balls are generally formed using fresh versions of these cheeses. They can be stored in very light whey brine to help maintain their soft texture until use.

When first working the heated curd, gently and repeatedly fold the edges into the center forming a ball. This same technique is used to form a finished loaf shape.

Steps in forming a cheese braid

Pear, Teardrop, and Gourds

These shapes are traditionally hung in pairs over rafters or pegs as they age. *Caciocavallo*, similar in make to provolone and kashkaval, is made in these shapes. The Italian name, cheese on horseback, originates from the custom of tying two cheeses in a fashion so that they could be slung over a horse's back, like a pair of saddlebags, and transported to market. This shape is very endearing and quite easy to make. Start as with large balls, and when the ball is smooth, continue by forming a thick neck. Make sure the neck is sturdy enough to support the round end, or when you hang them, they will "lose their heads" (I made this mistake on one of my first attempts). Brine these shapes in cool brine to help firm them up before hanging. They do not, contrary to the etymology of the name, have to be done in pairs. Hang with sanitized twine.

Braids

Treccia (again, an Italian name for braids) are fun and fairly easy to form. After the curd texture is shiny and smooth, form a long, even rope by rolling it between your hands. A thickness of 1.5 to 2 inches (3.8–5.1 cm) is fine for fresh versions; if aging, make it a bit thicker. Lay the rope out into the shape of a lower-case "e" on its side (this effectively divides the single rope into three even sections). Cut the rope to separate the short end (the middle of the "e"). This will give you a long section in the shape of a horseshoe and a short section in the middle. Overlap the short section over the top center of the horseshoe shape and tuck it under. Then begin braiding by crossing the outside sections over the middle (see photos). When you come to the bottom, tuck and pinch the ends under so they are hidden. Gently place the braid into brine or cool water (depending on whether you have already salted the curd). Braids can be smoked and aged briefly, but don't age for long periods (as you do for the shapes with fewer nooks and overlaps).

Skeins

Skeins (or hanks) are easier to form than they look. Take a hot ball of curd and gently form a ribbon 1 to 1.5 inches (2.5–3.8 cm) from one side. Flatten and continue

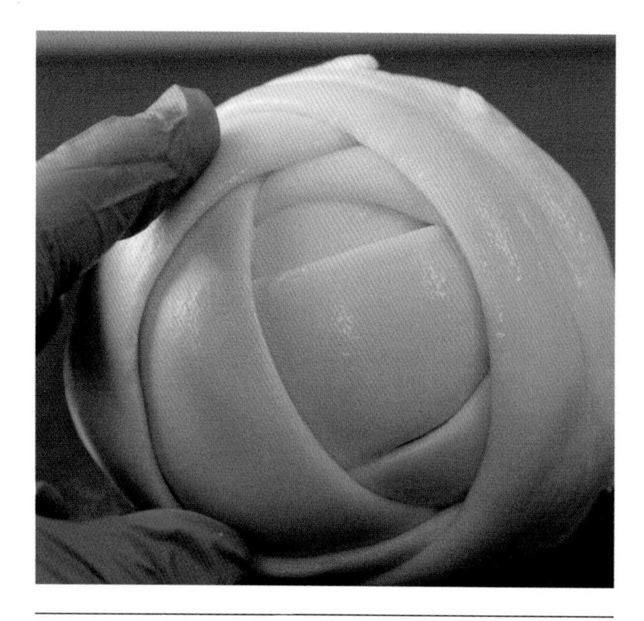

A skein of asadero-type cheese

to work this ribbon off the hot ball of curd. Lay the ribbon out on the counter or cutting board, and continue to work it longer until you have used up the ball of curd. The ribbon can overlap and bend as you lay it out. When you have used up the ball of curd, the ribbon will have cooled enough to roll into a ball and keep its shape (if it is too hot, it will reform with the rest of the skein). Skeins and other shapes such as braids do not age as well, as there are too many entry points for molds, and they cannot be properly waxed or sealed, so use this shape for fresh versions.

SALTING

Stretched curd cheeses can be salted either by adding salt to the hot liquid in which you are heating the curd or by using a saturated brine. If you salt during stretching, you can cool the cheeses in a very lightly salted water; if not, put them in a cool, fully saturated brine.

Aged pasta filata cheeses are typically a bit saltier than other semihard or hard cheeses, which are usually about 2.5 to 3 percent salt, whereas cheddar and its kin are about 2 percent. Brining times are very dependent upon the size of the shapes. Cheesemaker guru Peter Dixon recommends the following:

- Shapes under 5 lb (2.25 kg):
  2 hours per lb (about ½ hr/kg)
- Shapes from 5 to 10 lb (2.25–4.5 kg):
  3 hours per lb (about 40 min/kg)
- Shapes over 10 lb (over 4.5 kg):
  4 hours per lb (about 1 hr/kg)

When brining, remember that the hot shapes will warm the brine beyond its ideal of 50 to 55°F (10–13°C). So if you are doing many shapes, it is a good idea to first cool them in plain water, then submerge them in the brine tank.

STORING FRESH VARIETIES

Mozzarella types are meant to be eaten fresh or after aging for just a few weeks. They are higher in moisture than some of the other stretched curd types and are therefore not great candidates for long aging. To store fresh versions for a limited time, make a very light brine of 1 quart water, ¼ ounce kosher salt (or other cheese-quality salt), 1 milliliter calcium chloride, and ¼ milliliter white vinegar. (See chapter 2 for more on making brine solutions.)

AFFINAGE

After the shapes are salted, they can be aged at 85 percent RH and 50 to 55°F (10–13°C). If you are going to smoke them, they can be cool-smoked prior to aging (see chapter 5 for more on smoking cheeses). If you aren't aging with a natural rind, you can wax or vacuum-seal the cheeses and age at 55°F (13°C) or lower.

The cheeses can be eaten after a couple of months or aged a year if large enough. Use your trier (see chapter 4 for technique on core sampling) before you cut into the cheese to make sure it is ready for harvest!

PACKAGING

Fresh varieties can be packaged in plastic tubs or resealable plastic bags with storage brine. However, these are difficult to ship. Heat-sealed pouches are a good choice for shipping. Aged varieties can be wrapped in waxed or freezer paper.

CHRISTY HARRIS, PROVIDENCE HILL FARM, KANSAS

When I was interviewing Christy for this segment, she mentioned an early influence that we both share—the beloved children's book *Heidi*. Even though I grew up with cows as my first "dairy" love, reading the story of Heidi made me long for high Alpine meadows, a simple cottage, a table set with hearty bread and slabs of goat cheese, and (of course) a herd of personable goats—each with its own tinkling bell. Christy's "Heidi" persuasions finally led her to her own herd of goats and to cheesemaking in 2005 with the acquisition of two Nubian does. She also formed a partnership with another farm in which she made cheese from their milk, and both shared in the bountiful results.

Christy, who also now teaches cheesemaking classes and blogs on the subject, has approached her cheesemaking craft, and especially her mozzarella making, with an eager thirst for perfection. In the beginning she read and tried every recipe she could find, learned how to monitor pH, and faithfully documented each batch. Her pursuit of the perfect string cheese led her to the recipe she has so graciously shared for this book.

Between her blog (http://artisanfarmsteadliving.blogspot.com) and classes, Christy offers some great advice to cheesemakers of all levels. She says that learning to make cheese "is a bit like learning a new language; it is awkward in the beginning, but once you incorporate it into everyday life, it can become second nature." She also suggests that every new cheesemaker "take their time, be persistent, and be forgiving" of both themselves and the innate nature and changeability of milk. Great advice from a great cheesemaker.

Home cheesemaker Christy Harris with one of her Nubian goat kids. Photo courtesy of Whitney Link

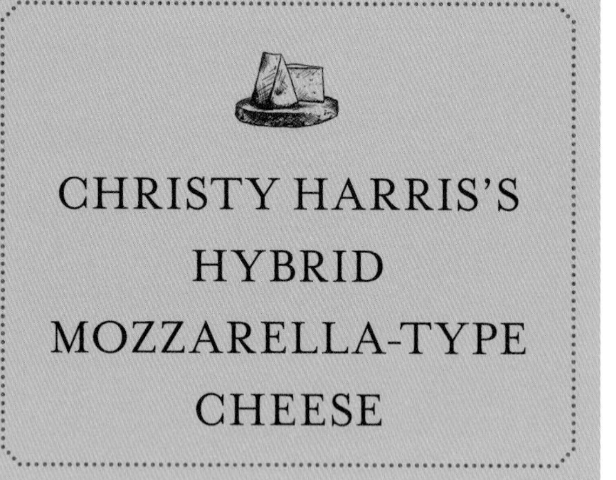

CHRISTY HARRIS'S HYBRID MOZZARELLA-TYPE CHEESE

While most quick versions of mozzarella rely completely on the addition of citric acid to bring the milk quickly to the pH at which it will stretch when heated, this recipe is a hybrid. It uses some starter culture as well as added acid to develop the pH. This takes a bit longer, but the flavor will be more complex, and according to Christy, the results will be more consistent—meaning the stretchability and texture will be more consistent than when a more abbreviated version is made. As with another recipe that was provided by other cheesemakers, and not tested by me, I have not included large batch guidelines for this particular cheese.

INGREDIENTS

Milk: 4 gal (15 L/34 lb) whole milk
Culture: ¼ tsp (1.5 g) TA 60, and ¼ tsp (0.4 g) MM or MA
Calcium chloride (optional): maximum 1 tsp (5 ml) dissolved in ¼ cup (60 ml) cool, nonchlorinated water
Acid: 2–3 tsp (10–15 g) citric acid diluted in ¼ cup (60 ml) cool, nonchlorinated water
Rennet: 1 tsp (5 ml) single-strength rennet mixed just before use in ¼ cup (60 ml) cool, nonchlorinated water
Salt: Heavy brine

STEPS

Prepare Equipment: Make sure all equipment is cleaned and sanitized and that your cheesemaking space is free from possible contaminants. Refer to chapter 6 for tips on proper equipment preparation.

Prepare Milk: Warm 2 gallons of the milk to 96°F (36°C).

Culture: Sprinkle cultures on top of milk. Let set for 2 to 5 minutes, then stir gently for 3 to 5 minutes.

Ripen: Maintain temperature at 96°F (36°C), and ripen for 45 minutes.

Additions: Combine citric acid solution with 2 gallons cold milk. Combine cold milk with warm, cultured milk, and bring temperature to 92°F (33°C). Stir gently. The goal milk pH before rennet addition is 6.1.

> **TIPS:** You can use more or less of the citric acid solution to adjust the pH at this time. This will help standardize the coagulation time.

Additions: Add calcium chloride solution and stir for 1 minute, then wait for 5 minutes before adding rennet.

Rennet and Coagulate: Stir in rennet solution with an up-and-down motion for 1 minute. Still milk. Hold temperature at 92°F (33°C) until firm curd forms; the goal coagulation time is 30 minutes.

> **TIPS:** Depending on the pH of the milk when the rennet is added, the coagulation time will be shorter or longer.

Cut: Cut curd mass into ⅜-inch (1-cm) cubes. Let rest for 5 minutes.

Stir and Heat: Gently stir and heat curd slowly over 30 minutes to 115°F (46°C). Turn off heat, and hold temperature for 15 more minutes. Stir gently to prevent curd from matting. Let set for 5 minutes.

TIPS: While the mesophilic cultures will have ceased to produce acid at this temperature, the thermophilic bacteria will continue to grow and acidify the milk.

Drain: Drain the whey to the level of the curds. Pour curds into a warm, cloth-lined colander. Cover, and keep curd mass at 102°F (39°C) until curd pH reaches 5.1 to 5.3 or curd passes the stretch test (see the sections "Making Shapes" and "Stretching" earlier in this chapter). Turn curd mass every 30 minutes.

TIPS: Remember to squeeze excess whey from curd sample before checking pH.

Mill and Stretch: Cut curd mass into 1-inch (2.5-cm) cubes. Place in a stainless steel bowl, and cover with 180°F (82°C) water. Using heat-resistant gloves, gently form curd into balls (see the section "Making Shapes" earlier in this chapter) and stretch. Once the texture is shiny and smooth, form shapes. To make string cheese, stretch and roll curd into a long, thin roll, then cut into strips. Brine salt in a heavy brine for 2 to 3 hours, depending on size of shapes.

Finishing: Wrap in plastic wrap, or store in very light brine. Keep in refrigerator, and use within 3 weeks.

Fresh mozzarella with tomatoes and basil in a Caprese summer salad

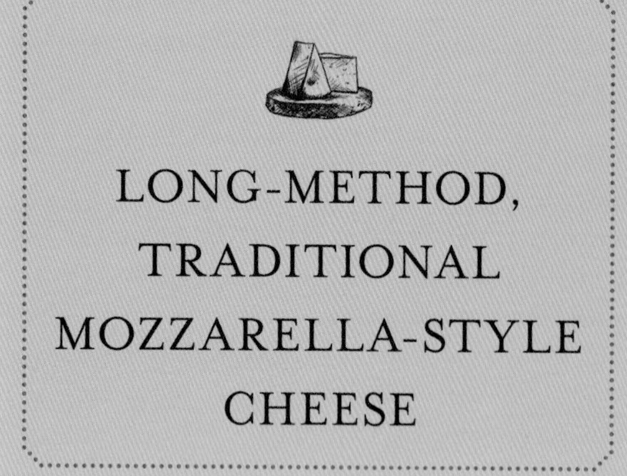

LONG-METHOD, TRADITIONAL MOZZARELLA-STYLE CHEESE

Let's pause for a moment to contemplate the respect that mozzarella—perhaps the most well-known added- and developed-acid stretched curd cheese—deserves. It is tough for a cheese whose current, and highly popular, derivation—it is known as low-moisture mozzarella cheese, or pizza cheese—actually has a long and esteemed history. Mozzarella cheesemaking techniques can in fact be considered the reason that the second most popular Western cheese (second only to cheddar, perhaps) exists. But more on that in chapter 14.

The original mozzarella was produced from the rich milk of water buffalo that had been reared in the south of Italy since the second century AD. Until the tragic, purposeful destruction of these gentle herds during World War II, *mozzarella di bufala*, especially in its fresh, tender state (also called *il fior di latte* or "the flower of the milk") reigned supreme. Even though the water buffalo herds have been repopulated, for most people mozzarella, now mostly made from cow's milk, simply means a handy snack for your kid's lunch bag or a gloppy, greasy, yet delicious mess for your pizza.

In Latin America mozzarella descendants such as *oaxaca* and *asadero* (from Mexico) are also mainstays for cooking. Though it is usually made from cow's milk, goat's milk works well, too. These members of the pasta filata, or stretched curd, family appear in many shapes, from braids and balls to skeins and loaves. Cheeses made in this style are usually meant to be served fresh or with only brief aging. While perhaps not as complex in flavor or challenging to make as some cheese types, they are fun and provide a staple product for many households. Watching the magic of chemistry that occurs when curd hits the right acid level and temperature—being transformed from a semisolid to a plastic state—is one of those thrilling, pivotal moments in the development of a cheesemaker's passion. If you are looking for a way to teach a bit of science to kids (and adults) while at the same time teaching them about our food supply, homemade mozzarella might just be the ticket!

INGREDIENTS

Milk: 2 gal (8 L/17 lb) whole or partly skimmed milk
Culture: ¼ tsp (1.5 g) Thermo B or TM 81 (or equivalent from chart)
Lipase (optional): 1/16 tsp (0.3 g) dissolved in ⅛ cup (30 ml) cool, nonchlorinated water
Calcium chloride (optional): maximum ½ (2.5 ml) diluted in ⅛ cup (30 ml) cool, nonchlorinated water
Rennet: 1/4 tsp plus (1.5 ml) single-strength rennet diluted just before use in ⅛ cup (30 ml) cool, nonchlorinated water
Salt: 2 tsp (10 g) pure salt

STEPS

Prepare Equipment: Make sure all equipment is cleaned and sanitized and that your cheesemaking space is free from possible contaminants. Refer to chapter 6 for tips on proper equipment preparation.

Prepare Milk: Warm milk to 80°F (27°C).

Culture: Sprinkle cultures on top of milk. Let set for 2 to 5 minutes, then stir gently for 3 to 5 minutes.

Ripen: Increase temperature to 90°F (32°C), then maintain for 60 to 90 minutes. The goal milk pH drop is 0.2, or to about 6.4. If milk is raw or mesophilic bacteria are added, the pH will drop more quickly.

Additions: Stir in lipase solution, then stir in calcium chloride solution. Wait 5 minutes.

Rennet and Coagulate: Stir in rennet solution with an up-and-down motion for 1 minute. Still milk. Hold temperature at 90°F (32°C) until clean break is achieved. The goal coagulation time is 20 to 30 minutes. Because the milk pH will have dropped a bit, the coagulation time will be shorter.

Cut: Cut curd mass into ⅜- to ½-inch (1–1.3-cm) cubes. Let rest for 10 minutes.

Stir and Heat: Stir gently and increase temperature to 95 to 98°F (20–21°C) over 15 minutes. Curds should shrink some and be easy to stir.

Ripen: Let the curds settle under the whey and ripen 30 to 60 minutes to a curd pH of 6.0 to 6.1.

TIPS: If pH is dropping too slowly, increase the temperature by about 2°F (1°C) more. Stir until heat is stable, then hold and ripen for longer. Remember to squeeze excess whey from curd sample before checking pH.

Drain: Drain the whey (which can be used later, ideally within a couple of hours, for whey cheese such as ricotta). Save about 2 quarts for stretching later. Place curds in a warm colander set over warm whey or water, and let drain and ripen. Turn slab every hour or so, and ripen to a curd pH of 5.1 to 5.3 or when a test piece passes the stretch test (see the section "Making Shapes" on page 234).

TIPS: No matter what your pH meter says, if the curd is not ready to stretch, don't attempt it. You will end up with hard or grainy cheese and no stretch. If you continue to have trouble obtaining a stretch, begin checking curd with the stretch test before you think it is ready and confirm the temperature of curd during stretching.

Mill and Stretch: Cut or crumble the slab into 1- to 1.5-inch (2–3.8-cm) cubes. Add 2 tsp of salt to reserved whey, and heat to 170 to 180°F (77–82°C). Lower some of the milled curd into the hot whey (you can use a sieve or small colander), and allow to heat thoroughly. While wearing heavy gloves, begin to manipulate the heated curd as shown in images on pages 233 and 234. Reheat, stretch, and shape as needed. Once shapes are formed, they should be immersed in cold water (or whey) at the same proportions as the heated whey, with a small amount of salt.

Finishing: Wrap in plastic wrap, or store in very light brine. Keep in refrigerator, and use within 3 weeks.

LARGE-BATCH GUIDELINES

Milk: 12 gal (46 L/100 lb) milk
Culture: 2–4U Thermo B or TM 81 (or equivalent from chart)
Calcium chloride (optional): 10–15 ml (about 1 tbsp)
Lipase (optional): ⅜ tsp (2.1 g)
Rennet: 7–9 ml single-strength rennet (3.5–4.5 ml double strength)
Salt: 3–4 tbsp (45–60 g)

SEMIHARD TO HARD AGED PASTA FILATA CHEESE

Once you have mastered stretched curd fresh cheeses, you may want to move up to varieties meant to age for extended periods of time. Just a few steps in the making of these varieties will allow these types to become quite different from their fresh cousins. Besides the ability of these cheeses to age, the traditional varieties are made from what the historical agriculture provides with regard to the species of milking animals being used. In addition, local traditions and tastes dictate the length of aging, flavorings (including smoking), and the shapes into which the cheeses are formed. Provolone historically is made from cow's milk and takes a rounded or elongated shape, caciocavallo (also from Italy) a jug shape, Ragusano a rectangular shape, and so on. With the Roman Empire's expansion came a spread of food culture as well. Consequently, cheeses in this family are widespread and diverse. When you make this recipe, remember that you can adapt it and style it to your own milk, tastes, and environment.

INGREDIENTS

Milk: 2 gal (8 L/17 lb) whole milk
Culture: ¼ tsp (1.5 g) Thermo C **or** ⅛ tsp plus (0.4 g) TA 60 and 1/16 tsp (0.2 g) LH 100 (or the equivalent from chart).
Lipase (optional): 1/16 tsp (0.4 g) dissolved in ⅛ cup (30 ml) cool, nonchlorinated water
Calcium chloride (optional): maximum ½ tsp (2.5 ml) diluted in ⅛ cup (30 ml) cool, nonchlorinated water
Rennet: ¼ tsp plus (1.5 ml) single-strength rennet diluted just before use in ⅛ cup (30 ml) cool, nonchlorinated water
Salt: 2 tsp (10 g) pure salt

STEPS

Prepare Equipment: Make sure all equipment is cleaned and sanitized and that your cheesemaking space is free from possible contaminants. Refer to chapter 6 for tips on proper equipment preparation.

Prepare Milk: Warm milk to 80°F (27°C).

Culture: Sprinkle cultures on top of milk. Let set for 2 to 5 minutes, then stir gently for 3 to 5 minutes.

Ripen: Increase temperature to 88 to 90°F (31–32°C), then maintain for 1 hour.

Additions: Stir in lipase solution, then stir in calcium chloride solution. Wait 5 minutes.

Rennet and Coagulate: Stir in rennet solution with an up-and-down motion for 1 minute. Still the milk. Hold temperature at 88 to 90°F (31–32°C) until clean break is achieved. The goal coagulation time is 45 minutes.

Cut: Cut curd mass into ⅜- to ½-inch (1–1.3-cm) cubes. Let rest for 5 minutes.

Stir and Heat: Stir gently, and begin raising temperature slowly to 95 to 98°F (35–37°C) over 10 to 15 minutes. Continue heating and stirring to 118°F (48°C) over 30 more minutes. Hold to curd pH 6.0, stirring occasionally as needed to prevent matting. Let settle for 5 minutes.

> **TIPS:** Remember to squeeze excess whey from curd sample before checking pH.

Drain and Ripen: Drain whey (can be used for whey cheese such as ricotta). Save about 2 quarts for stretching later. Push curds to one side of the bottom of the pot, and tip it so whey will drain away from the curd mass. Maintain a curd temperature of 104°F (40°C). Turn slab every hour or so, and ripen to a curd pH of 5.1 to 5.3 or when a test piece passes the stretch test (see the section "Making Shapes" on page 234).

TIPS: No matter what your pH meter says, if the curd is not ready to stretch, don't attempt it. You will end up with hard or grainy cheese and no stretch. If you continue to have trouble obtaining a stretch, begin checking curd with the stretch test before you think it is ready, and confirm the temperature of curd during stretching.

Mill and Stretch: Cut or crumble the slab into ½-inch (1.3-cm) pieces. Add 2 tsp of salt to reserved whey, and heat to 165 to 170°F (74–77°C). Lower some of the milled curd into the hot whey (you can use a sieve or small colander), and allow to heat thoroughly. While wearing heavy gloves, begin to manipulate the heated curd as shown on pages 233 and 234. Reheat, stretch, and shape as needed. Once shapes are formed, they should be immersed in cold water (or cold whey) with a small amount of salt.

Finishing: Wrap in plastic wrap, or store in very light brine. Keep in refrigerator, and use within 3 weeks.

LARGE-BATCH GUIDELINES

Milk: 12 gal (46 L/100 lb) milk
Culture: 2–4 g Thermo C **or** 2–3U TA 60 plus 0.5–1U LH 100
Calcium chloride (optional): 10–15 ml (about 1 tbsp)
Lipase (optional): ¼–½ tsp (1.4–2.8 g)
Rennet: 7–9 ml single-strength traditional rennet suggested (3.5–4.5 ml double strength)

BURRATA, BURRINO, AND MANTECA

These stuffed, pasta filata cheeses are quite popular right now. *Burrata* is mozzarella curd stuffed with a mixture of curd and ricotta, curd and cream, or cream ricotta. *Burrino* and *manteca* are made using provolone curd and are filled with butter—sometimes butter made from the creamy whey produced during the making of the provolone curd.

Should you decide to make these delicacies yourself (and talk about a way to knock the socks off your dinner party guests!), save some curd, prior to stretching, from the making of any of the cheeses in this chapter. Prepare a mixture of chopped-up bits of more curd, ricotta, cream, and butter. Mix it to a fine consistency, and chill. I made a cream ricotta using the high-fat whey from making the provolone-style recipe, as follows:

1. Bring 8 cups of whey to 190 to 195°F (88–91°C).
2. Add ½ to 1 cup of heavy cream. Turn off heat.
3. Add 1 tablespoon citric acid dissolved in a bit of water.
4. Let set.
5. Filter through cheesecloth (don't expect the same type of curd formation from this high-fat ricotta as you would get with a milk or whey ricotta that has far more protein than fat). It will be very fine.
6. Drain for several hours.
7. Taste; if too tangy, add a bit of baking soda and taste again (the baking soda will neutralize the acid for a milder taste).
8. Salt to taste and chill thoroughly.

After your mixture tastes just right and is well chilled, prepare your curd for stretching. While the hot water heats up, form ping pong to golf-ball-size shapes from the ricotta mixture. Stretch the curd, and form a flattened, rounded shape at least ½ inch (1.25 cm) thick. Quickly place the ball of ricotta in the center, and gather the hot curd around it and together to form a neck (for manteca and burrino) or to close the packet into a ball (for burrata). Chill quickly in cold, lightly salted water. You can serve them individually with a salad or drizzled with seasonings of your choice.

Filling a manteca-type cheese with whey ricotta, forming it, and later as an aged cheese

GREEK-STYLE AGED STRETCHED CURD CHEESE

It is a bit tough, and perhaps I am prejudiced in favor of my own Greek ancestry, to separate out this Greek derivation of the pasta filata family members we have already covered. But I couldn't help myself. Traditionally, *kasseri* is made from sheep's milk, in a fashion very similar to that of its other Mediterranean cousins. Some differences may occur, as in the recipe here, during draining and in the final salt content. But remember, these are things that you can dictate as the cheesemaker.

INGREDIENTS

Milk: 2 gal (8 L/17 lb) whole high-fat milk
Culture: ¼ tsp (1.5 g) Thermo C and 2 skewer tips MA or MM **or** ⅛ tsp plus (0.4 g) TA 60 and ⅛ tsp (0.3 g) LH 100 and 2 skewer tips MA or MM
Lipase (optional): 1⁄16 tsp (0.35 g) dissolved in ⅛ cup (30 ml) cool, nonchlorinated water
Calcium chloride (optional): maximum ½ tsp (2.5 ml) diluted in ⅛ cup (30 ml) cool, nonchlorinated water
Rennet: ¼ tsp plus (1.5 ml) single-strength rennet diluted just before use in ⅛ cup (30 ml) cool, nonchlorinated water
Salt: 1 tbsp (15 g) or more pure salt

STEPS

Prepare Equipment: Make sure all equipment is cleaned and sanitized and that your cheesemaking space is free from possible contaminants. Refer to chapter 6 for tips on proper equipment preparation.

Prepare Milk: Warm milk to 85°F (29°C).

Culture: Sprinkle cultures on top of milk. Let set for 2 to 5 minutes, then stir gently for 3 to 5 minutes.

Ripen: Increase temperature to 95°F (35°C), then maintain for 1 hour.

Additions: Stir in lipase solution, then stir in calcium chloride solution. Wait 5 minutes.

Rennet and Coagulate: Stir in rennet solution with an up-and-down motion for 1 minute. Still the milk. Hold temperature at 95°F (35°C) until clean break is achieved. The goal coagulation time is 35 minutes.

Cut: Cut curd mass into ¼-inch (6-mm) cubes. Let rest for 5 minutes.

Stir and Heat: Stir gently, and begin raising temperature slowly to 99 to 104°F (37–40°C) over 30 minutes. Hold temperature, and continue to stir until the pH of the whey is pH 6.1 to 6.0, stirring occasionally as needed to prevent matting. Settle 5 minutes.

Drain and Ripen: Drain whey (can be used for whey cheese, such as ricotta). Save about 2 quarts for stretching later. Gather curd in a whey-moistened cheesecloth, and transfer to a sanitized working surface.

Knead the curd gently until the texture is smooth. Tie up in the cloth, and place a sanitized cutting board on top of the cloth-wrapped ball. After 15 minutes add half the weight of the curd mass on top of the board. After 1 hour add more weight to total the same weight as the curd. Let press until the pH of the curd is 5.1 to 5.3 and the curd will pass the stretch test. During pressing, the room temperature should be no cooler than 72 to 75°F (22–24°C). Total pressing time is usually 4 to 6 hours.

Mill and Stretch: Unwrap, and cut curd mass into ½-inch (1.3-cm) strips. Heat reserved whey to 165 to

170°F (74–77°C). Lower some of the milled curd into the hot whey (you can use a sieve or small colander), and allow to heat thoroughly. While wearing heavy gloves, begin to manipulate the heated curd as shown previously. Reheat, stretch, and shape into a ball. While still warm, press the cheese into unlined forms.

Salt: Allow to cool in the form—you can rinse the cheese while it's in the form with cold water to assist the cooling process. Once the cheese is cool, begin rubbing with dry salt, and place back into the form. Continue to salt and turn the cheese over a 3- to 4-day period.

Finishing: Rinse the cheeses with cool water, dry with a clean cloth, and place in the aging room at 55°F (13°C) and 85 percent RH. Age a minimum of 4 months for the best flavor.

LARGE-BATCH GUIDELINES

Milk: 12 gal (46 L/100 lb) milk
Culture: 2–4 g Thermo C and 0.5U MM or MA **and** 2–3U TA 60 0.5–1U LH 340, and 0.5U MM or MA
Lipase (optional): ¼–½ tsp (1.4–2.8 g)
Calcium chloride (optional): 10–15 ml (about 1 tbsp)
Rennet: 7–9 ml single-strength; traditional rennet suggested (3.5–4.5 ml double strength)
Salt: 6 tbsp (90 g)

THINKING OUTSIDE THE VAT

SCENARIO 1: You are making Christy Harris's mozzarella recipe and find that when you add 2 teaspoons of citric acid, it takes the milk 1 hour to coagulate instead of the goal of 30 minutes. You should try which of the following?

A. Use more rennet the next time.
B. Use a bit more citric acid the next time.
C. Don't worry about it.
D. Keep the milk warmer than recommended during the coagulation phase.

SCENARIO 2: You core sample a 3-month-old provolone-style cheese and find the curd to be bitter. You should consider which of the following?

A. You shouldn't have used microbial rennet; it causes bitterness.
B. Use less culture next time.
C. The cheese may just be going through a phase in aging where bitterness can occur; you should let it age for longer, then resample.
D. The milk was contaminated; bitterness is a common sign of this problem.

1. Answer: B.
2. Answer: C.

Semihard and hard cheeses make up a huge category of wheels from all over the world. Pictured here is Pholia Farm's Takelma, a St. Paulin–style cheese.

14: SEMIHARD TO HARD CHEESES

FINISHED TEXTURE: Semihard to hard
METHOD OF COAGULATION: Rennet
METHOD OF RIPENING: Internal bacteria

Prepare yourself for a big chapter! I thought about breaking this family of cheeses into separate chapters for the many branches that represent the different styles, but they are so similar in so many ways that I hope it will make sense to have them share this space. This is the group of cheeses that I am the most comfortable working with and teaching; they are my specialty, you might say.

The finished texture of these cheeses is greatly dependent on their age. Even some that may be best as a semihard, such as Gouda types, can be consumed in their younger, near-semisoft stage. I have grouped them based more on their ideal aging time and texture (with flavor being the priority) than on the complete range of possible consumption dates.

GENERAL TIPS FOR MAKING SEMIHARD AND HARD CHEESES

We'll start with some general tips for making that will apply to all the varieties in this chapter. Then I'll break it down into separate cheese types. This branch of the family tree includes simple, basic cheeses, tomme styles, and manchego; washed curd cheeses such as Colby, St. Paulin, and Gouda; textured types such as cheddar and Cheshire; and cheese with eyes—for instance, Emmental and the Gruyère group. You are going to see quite a bit of crossover between these groups, another reason I put them in the same chapter—for example, Gouda is a washed curd cheese, but it also often has eyes, and Emmental is a cheese with eyes that sometimes is also a washed curd. So keep in mind that many of these cheeses have techniques that can be combined with others to produce a different result. Once again, it will be my goal to teach you the variety of techniques available, not necessarily to duplicate a masterpiece but eventually to create your own.

CHOOSING MILK

When milk of superior quality can be used and/or regulations allow, semihard and hard cheeses can be made using raw milk, often with wonderful results. Some semihard varieties, however, are made only from pasteurized milk, either because of legal standards of identity or their short aging time. Whether you choose to work with raw or pasteurized milk, try to keep some of the tips (such as amounts and types of culture) in mind to help either approach reach its best conclusion.

CHOOSING STARTER CULTURE

Table 14–1 includes most of the mesophilic culture types and blends that are readily available for home and small artisan cheesemakers. These types can be used for all the recipes in this chapter except the final "cheese with eyes" types; they have their own chart. The final blend on this first table, MA 4000, includes some thermophiles (ST) that will add complexity during aging and a flavor and aroma producer (LLD).

CHOOSING RIPENING CULTURES

Table 14–2 contains several ripening and mold cultures that might come in handy for adding depth of flavor and creating interesting natural rinds. If you have an established aging room, it may have plenty of natural microflora that will provide interesting surface ripening. If not, consider adding some of these cultures to help establish a population and help ripen the cheeses.

TABLE 14-1. STARTER CULTURE CHOICES FOR SEMIHARD AND HARD CHEESES

Name	Contains	Manufacturer	Small-Batch Dosage:* 3–4 gal (12–15 L/26–34 lb)	Large-Batch Dosage: 26 gal (100 L/220 lb)
Mesophilic Blend	LLC, LL	Packaged by home cheese supply co.	1 packet	N/a
MA Series (Choozit) (includes MA 11, 14, 16, 19)	LLC, LL	Danisco	⅛ tsp (0.3 g)	10U
MM 100 Series (Choozit) (includes MM 100, 101)	LL, LLC, LLD	Danisco	⅛ tsp (0.3 g)	10U
MA 4000 Series (Choozit) (includes MA 4001, 4002)	LL, LLC, LLD, ST	Danisco	1/16 tsp (0.1–0.2 g)	4U
TA 60 Series (Choozit) (includes TA 61, 62)	ST	Danisco (fast-acid producer)	⅛ tsp (0.3 g)	Add 1 part ST to every 4 parts mesophilic

*Remember that culture is always best measured by weighing and dividing into unit doses based on the strength of the culture, instead of using volume or set weight.

DRAINING AND PRESSING

You will find that the cheeses in this chapter go from needing very little weight to be properly formed to needing about the most any press or setup might have to offer. It is close to impossible to provide exact and correct pressure numbers because the size of the wheel and the design of the system will vary. These cheeses will provide you with the opportunity to implement the observation recommendations of chapter 2 regarding how a rind should look during pressing. That may sound like a tall order, but it isn't as complicated as it might at first seem. I will provide as many tips as possible, but remember that every new cheese you make has the possibility of a new challenge—so stay alert!

THE FUNDAMENTALS OF SIMPLE-PROCESS PRESSED CHEESES

This category of cheese offers a great way to get started making some amazing hard cheeses without a lot of time and equipment. Both of these recipes can age from 6 to 12 months, with the cheese improving with age, or they can be consumed younger if you prefer a softer, moister texture.

The first recipe is a fairly common approach to what many of us simply call a *tomme*. While there is no one cheese type that is a "tomme," the name usually covers a group of cheeses made in this simple style, from various types of milk, and in many sizes throughout the mountainous regions of Europe, especially the Pyrenees.

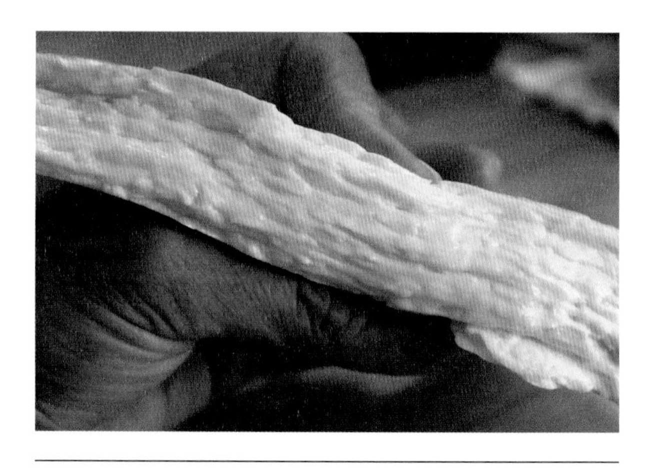

Typical "chicken breast meat" texture of cheddared curd

TABLE 14-2. RIPENING AND MOLD CULTURES FOR SEMIHARD AND HARD CHEESES

Name	Contains	Manufacturer	Small-Batch Dosage:* 3–4 gal (12–15 L/26–34 lb)	Large-Batch Dosage: 26 gal (100 L/220 lb)	Use
LM 57 (Choozit)	LM (LMC)	Danisco	Add at ¼ the rate of main culture	Add at ¼ the rate of main culture	Aroma culture
LBC 80 (Choozit)	LBC	Danisco	1/16 tsp (0.1–0.2 g)	0.2 dose	Bacteria used to assist flavor and aroma development, especially in pasteurized milk varieties; can help with eye formation for Swiss varieties
Mycodore	*Tricothecium domesticum* ssp. *cylindrocarpon*	Danisco Choozit	1/16 tsp (0.1–0.2 g)	0.2 dose	White, grey, brown molds especially for mixed mold rinds, tomme, Caerphilly
Mycoderm	*Verticillium lecanii*	Danisco Choozit	1/16 tsp (0.1–0.2 g)	0.5 dose	White mold, especially for tomme styles, Caerphilly

*Remember that culture is always best measured by weighing and dividing into unit doses based on the strength of the culture, instead of using volume or set weight.

Tomme types are made using mesophilic culture, often with a bit of thermophilic added. Tommes are often pressed with no more than a weight equal to that of the curd, making them easy for cheesemakers without a mechanical press to produce. Our cheese Elk Mountain, which made *Wine Spectator*'s top 100 list a few years back and has been featured in many cheese books (even being called a "rock star"), is made in this fashion.

The following recipe is quite similar to the first in cooking and pressing but is cut and stirred for half an hour without any temperature increase. This creates a different curd texture in the end product. The recipe is modeled after manchego, which is made from sheep's milk using traditional rennet or rennet paste, thus making it a piquant, sharp-flavored cheese.

During cooking and stirring, both of these recipes have curd that can be checked for readiness using the standard texture test (see "Checking Curd for Readiness" in chapter 2, page 52). The curd should be even in texture and will form a consolidated mat when pressed in your hand. This mat should then also be easy to break back apart. You can also monitor whey pH. For both cheeses it should be 6.42 to 6.45 at draining (remember that curd pH will be a bit lower, if you are able to check curd).

Neither of these recipes will require a mechanical press to form a closed rind wheel. Adjust pressure and weights as needed to follow the standard pressing guidelines of chapter 2. Both are pressed until a goal pH of 5.3 to 5.4 is reached. As with other cheese types, the smaller the wheel, the closer to the lower pH it should be, since salt uptake in the brine will reach the center sooner than with a larger wheel.

Both of these styles offer options for aging—you can wax them, or you can develop a natural rind by brushing, oiling, or washing with ale (which is what we do for Elk Mountain). Some common flaws with this type of cheese include too much acid development in the vat, leading to a sour, crumbly paste; too much initial pressure during pressing, leading to internal cracks and whey retention; and too low humidity during aging, causing more drying than this style of cheese can survive.

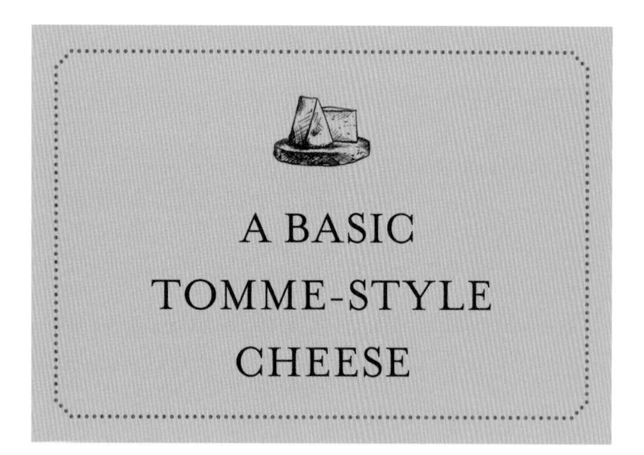

A BASIC TOMME-STYLE CHEESE

Now that I have been making cheese for some time, I am often confounded by the overarching belief that to make an aged, pressed cheese one actually needs a rather expensive and complicated cheese press. For that reason I wanted to begin this section with one of the most common, and loved, cheese styles that does not require the use of a fancy press. The cheeses in this undefined and broad category were made in the most basic of circumstances. In fact, I think of the term "tomme" as a catchall for simple, honest cheeses made from whatever milk the cheesemaker has at hand; without fancy antics or orchestration; and with pure, simple, yet paradoxically complex tasting results. From the classic French Tomme de Savoie to my own Elk Mountain to Marcia Barinaga's sheep's milk Baserri, this family of cheese is open minded and all inclusive.

INGREDIENTS

Milk: 2 gal (8 L/17 lb) whole milk
Culture: ⅛ tsp (0.1–0.5 g) MA 4000 (or equivalent blend)
Calcium chloride (optional): maximum ½ tsp (2.5 ml) calcium chloride diluted in ¼ cup (60 ml) cool, nonchlorinated water
Rennet: ¼ tsp plus (1.5 ml) single-strength rennet diluted just before use in ¼ cup (60 ml) cool, nonchlorinated water
Salt: Heavy brine

STEPS

Prepare Equipment: Make sure all equipment is cleaned and sanitized and that your cheesemaking space is free from possible contaminants. Refer to chapter 6 for tips on proper equipment preparation.

Prepare Milk: Warm milk to 80°F (27°C).

Culture: Sprinkle cultures on top of milk. Let set for 2 to 5 minutes, then stir gently for 3 to 5 minutes.

Ripen: Increase temperature to 88 to 90°F (31–32°C), then maintain for 15 minutes.

Additions: Stir in calcium chloride solution. Wait 5 minutes.

Rennet and Coagulate: Stir in rennet solution with an up-and-down motion for 1 minute. Still milk. Hold temperature at 88 to 90°F (31–32°C) until clean break is achieved. The goal coagulation time is 45 minutes. If using flocculation as a guide, multiply time to flocculation by 3.5 for total time from rennetting to cutting.

Cut: Cut curd mass into ⅜-inch (1-cm) cubes. Rest 5 minutes.

Stir and Heat: Stir gently, and begin to increase temperature to 100°F (38°C) over 30 minutes. Hold temperature, and stir slowly until curd passes texture test (see "Checking Curd for Readiness" in chapter 2, page 52). This will take 15 to 30 minutes. Goal whey pH at end is about 6.4. Curds will be springy and even in texture.

Drain and Press: Drain whey to the level of the curds, then gently hand-press the curds into a solid mass as close to the size of the mold as possible while they are still covered with whey. Drain off the rest of the whey. Place curd mass inside cloth-lined mold, and gently press to spread the curd into the mold. Fold cloth over curd, and set follower on top without additional weight. Room temperature should be 72 to 80°F (22–27°C).

After 15 minutes, turn and add a light weight, about half that of the curd. After 30 minutes turn. After 1 hour turn and add weight to equal that of the curd. Goal curd pH at the end of pressing is 5.2 to 5.4. This could take 4 to 8 hours.

TIPS: Observe the rind, which should gradually close (see chapter 2 for more on this process). The larger the cheese the closer to 5.4 it should be, as it will take longer for the brine to penetrate the curd and halt acid production.

Salt: When goal pH is reached, unmold the cheese and place in a heavy brine for 3 to 4 hours per pound of weight (or 6 to 8 hours per kg) of the wheel. Turn during brining or dry salt the top.

Affinage: Dry the surface of the wheel with a clean, sanitized cloth or with paper towels. Move cheese to an aging environment at 55°F (13°C) and 85 percent RH. Age for 4 to 6 months. Rind can be brushed, oiled, or brine-washed depending on the desired look.

LARGE-BATCH GUIDELINES

Milk: 12 gal (46 L/100 lb) whole milk (for larger batches increase ingredients proportionately)
Culture: 2U MA 4000 (or equivalent blend)
Calcium chloride (optional): 10–15 ml (about 1 tbsp)
Rennet: 7–9 ml single-strength rennet (3.5–4.5 ml double strength)
Salt: Heavy brine

Elk Mountain, a simple, tomme-style cheese

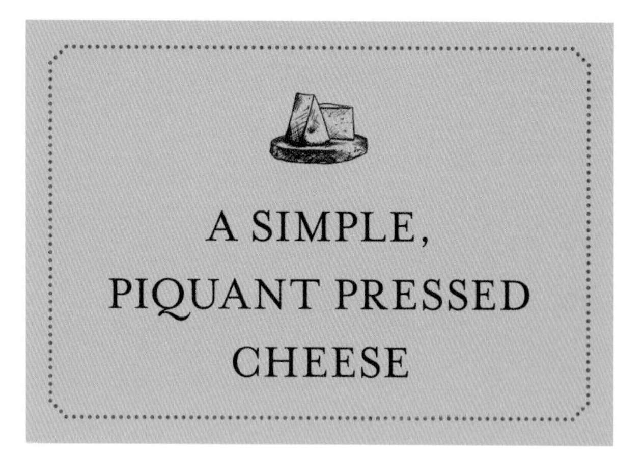

A SIMPLE, PIQUANT PRESSED CHEESE

I would like you to think of this category as a nation's possession of the simple tomme cheese style. The cheeses in this group are also fairly simple, yet their use of the vastly complex milk of ewes (female sheep) provides a canvas of beautiful possibility. While manchego is arguably the most well-known Spanish cheese, keep in mind that very few small countries with a strong food culture will export their national food treasures. Steve Jenkins, the amazingly frank and honest cheese author/monger/anticelebrity, has been a hero of mine for some time. In his seminal work, *Cheese Primer* (which was my first-ever cheese book), he lets readers know that if they look to Roncal (made in the Pyrenees from raw sheep's milk) and Zamorano they will find cheeses of similar style but of superior taste to the mass-produced-for-export wheels of manchego.

INGREDIENTS

Milk: 2 gal (8 L/17 lb) whole, high-fat milk
Culture: ¼ tsp (0.6 g) MA 4000 (or equivalent blend)
Lipase: 1 skewer tip lipase (necessary for traditional piquant flavor)
Calcium chloride (optional): maximum ½ tsp (2.5 ml) calcium chloride diluted in ⅛ cup (30 ml) cool, nonchlorinated water
Rennet: ¼ tsp plus (1.5 ml) single-strength rennet diluted just before use in ¼ cup (60 ml) cool, nonchlorinated water, use calf rennet for the most traditional flavor
Salt: Heavy brine

STEPS

Prepare Equipment: Make sure all equipment is cleaned and sanitized and that your cheesemaking space is free from possible contaminants. Refer to chapter 6 for tips on proper equipment preparation.

Prepare Milk: Warm milk to 80°F (27°C).

Culture: Sprinkle cultures on top of milk. Let set for 2 to 5 minutes, then stir gently for 3 to 5 minutes.

Ripen: Increase temperature to 90°F (32°C), then maintain for 30 minutes.

Additions: Stir in lipase solution, then stir in calcium chloride solution. Wait 5 minutes.

Rennet and Coagulate: Stir in rennet solution with an up-and-down motion for 1 minute. Still milk. Hold temperature at 90°F (32°C) until clean break is achieved. The goal coagulation time is 30 minutes. If using flocculation as a guide, multiply the time to flocculation by 2.5 for the total time from rennetting to cutting.

Cut and Stir: Cut curd mass into ⅜- to ¾-inch (1–2 cm) cubes over 5 minutes. Continue to cut over next 20 minutes to rice-grain size, 1/16 to ⅛ inch (1–2 mm). Stir for 10 minutes.

> TIPS: Cutting and stirring at 90°F (32°C) takes 35 minutes total.

Stir and Heat: Continue to stir gently, and increase heat to 100°F (38°C) over 30 minutes. Settle 5 minutes.

Pitch: Press under the whey for 5 to 10 minutes to form a single mass.

Drain and Press: Drain whey to curds. Press mass into cloth-lined mold cover with follower, and let sit without weight for 20 to 30 minutes. Room temperature should be about 75°F (24°C); if not warm enough, you can set forms in the pot on a rack. Flip, and add a light weight for 30 minutes. Flip, and add same weight for 30

Removing a braided plait of raffia from the top of a manchego-style cheese

minutes. Flip again, adding more weight to close rind. By the end of pressing, weight is approximately four times the weight of the cheese. The goal curd pH at the end of pressing is 5.3 to 5.4.

Salt: Immerse in a heavy brine for 4 hours per pound (8 hours per kg). The goal salt content is 2%.

Affinage: Dry rinds with a clean cloth or paper towels. Place in an aging environment at 55°F (13°C) and 85 to 90 percent RH for 6 to 12 months. Turn daily for the first 2 months, then weekly. Rind can be brushed, oiled, or waxed.

LARGE-BATCH GUIDELINES

Milk: 12 gal (46 L/100 lb) whole, high-fat milk
Culture: 2U MA 4000 (or equivalent blend)
Lipase: ¼ tsp (1.4 g)
Calcium chloride (optional): 10–15 ml (about 1 tbsp)
Rennet: 9 ml single-strength rennet (or 4.5 ml double strength), use calf rennet for the most traditional flavor
Salt: Heavy brine

THE FUNDAMENTALS OF MAKING WASHED CURD CHEESES

Washed curd cheeses are some of the most popular varieties in the world, especially for the "entry level" cheese connoisseur. Mild in flavor, and creamy of texture when young, they do not challenge the palate. Yet when aged, they can develop quite complex flavors and a hard texture. Consequently, for the cheesemaker they afford some flexibility, whether you are consuming them yourself, sharing, or selling.

For all washed curd cheeses, it is extremely important that you know that the water you're using is safe for consumption. It is usually recommended that the water be boiled, pasteurized, or UV filtered before being used as curd wash water. Some books recommend a 5 parts per million chlorine dilution as a safety measure.

Whey is removed in varying amounts (which will affect the level of acid in the final curd) and water is added. The temperature of the water varies, and this will affect both the curd texture and the final moisture content of the cheese. Washing is done in one or more stages. Usually, no more than 25 to 30 percent of the whey is removed. Keep in mind that it is not simply a matter of how much whey is removed, but of how much water is added. While most recipes call for the same amount of water to be added back, you can remove more whey and add back less water.

For example, one day Vern mistakenly removed 50 percent of the whey instead of 25 percent. Instead of adding 50 percent water back into the vat, he only added what would have been the right amount, 25 percent of the original volume of the milk. This brought the cheese close to the same goal, and the batch was not lost. When making any kind of washed curd cheese, don't forget that the pH of the water will affect any pH readings taken after it has been added to the curd. Still, developing benchmark pH goals is possible and important.

When a cheese is made using a washed curd technique, you are limiting the level of acid that can be developed in the final cheese. So during pressing there will be a point at which the curd pH will stop. This is called the "terminal pH." In most cheeses the pH will continue to drop to as low as 4.6, depending upon the amount of lactose remaining in the curd, but when you remove a portion of the lactose, you limit this potential. Usually, the goal with a washed curd cheese is a pH of not lower than 5.2.

We'll cover three core approaches to washed curd cheeses—washing with cool, hot, and same-temperature water. In case I haven't said it before, don't forget that there are almost innumerable approaches to these core techniques. Once you grasp the goals, don't be afraid to experiment with some hybrid methodologies.

When making any kind of washed curd cheese, don't forget that the pH of the water will affect any pH readings taken after it has been added to the curd.

The first recipe, Cool-Water Washed Curd Cheese, produces a semisoft to semihard mild cheese similar to the American cheeses Colby and Monterey Jack (known also as either Jack or Monterey). Both of these cheeses, while washed curd, are actually close relatives of cheddar cheese (which you will learn about soon). The curd is cooked and stirred to its finished texture before whey is removed, and cool to cold water is added to bring the curd to between 80 and 93°F (27–34°C). The warm, dehydrated curd absorbs moisture (after it has already lost whey and therefore lactose), creating a lower-acid, higher-moisture curd. You can control the moisture content by varying this final temperature. The warmer the temperature (closer to 93°F [34°C]), the lower the moisture content. In general, the higher the moisture content, the shorter the aging time.

Salt is added to the curd in this recipe, but the cheese can be made by pressing first, then brine salting (I do this for one of our cheeses, Covered Bridge). The addition of salt keeps the pH at a higher level than when brined after pressing. For that reason, as well as the fact that these cheeses are usually consumed young (we age Covered Bridge), they are usually best made of pasteurized milk.

If the curd is to be salted, you will need either a mechanical press to knit it together, or you can try the "belly pressing technique" used by Vella Cheese of Sonoma, California (and others), to make their outstanding Vella Dry Jack. Belly pressing entails forming the loose curds into a mass inside of a sturdy, cheesecloth square. The cloth is tied with a stilton knot (see chapter 12), then squeezed and massaged using your hands and your torso pressing against the edge of the vat. At home you may want to try using a sturdy colander to hold the curd ball while you press and tighten the knot. After a firm ball has been formed, the cloth bundle is pressed under boards (use food-grade plastic cutting boards or similar material). On a larger scale many bundles are pressed by stacking in layers, then weighting the top with heavy weights or cans of water. At home you will want to weight a single bundle in such a fashion that it will not topple over.

The next recipe, Same-Temperature Light-Brine Washed Curd Cheese, uses wash water at the same temperature as the whey. This, as you might imagine, produces a curd that is in between that created using cold water and that using hot water. The lactose is reduced, and some moisture is introduced into the curd, but the curd is not cooled or heated by the water.

The recipe I have included is in the style of St. Paulin but is based on our modifications for a cheese we make called Takelma. St. Paulin is an interesting cheese, in that it was originally designed to be made using pasteurized milk. The wash water also includes a small amount of salt, which further slows acid development. The recipe produces a surprisingly flavorful cheese in a relatively short period of aging.

This cheese does not need a mechanical press to obtain the right texture. As with the hot-water washed curd, use only as much pressure as it takes to obtain your desired goals. We perform a little trick with ours by pressing them first in tall, straight-sided forms, then unmolding them and placing a cutting board on top of the wheels. After a few minutes we flip the wheels and repeat this process to give them nice, curved sides.

The last recipe, Hot-Water Washed Curd Cheese, uses hot water to wash the curd. Similar cheeses include Gouda and Edam. We also make one of our cheeses, Hillis Peak, using a similar method. For most of these styles, the curd is cut, then stirred for a period of time without raising the heat. This allows the curd to lose whey without developing a "skin" and develop a bit of acid (but not much at this lower temperature). After a certain percentage of whey is removed, hot wash water is slowly and carefully added to the vat, often in stages. The hot water cooks the curds (a process often called scalding) and dehydrates them. Since the curds still contain quite a bit of whey, the dilution of the whey in the vat by the water causes the lactose and other constituents in the curd to want to leave the curd and attempt to equalize the solution. So curds washed with hot water do not absorb quite the same amount of moisture as do those from our first recipe. They will lose lactose, however, and end up with a bit higher possible final pH than if not washed.

Hot-water washed curd cheeses do not usually require the high pressure of a mechanical press to properly knit the curd back together, unless it is a salted curd variety. Choose the amount of pressure you apply based on your goal for the finished texture—closed or open paste—while making sure that the rind is closed a few hours into pressing.

While cheeses made using this method are often eaten young, they can also be aged. Small wheels of about 2 pounds (1 kg) can age a year if properly cared for. By that time they are more of an "extra-hard" texture and will often develop amino-acid crystals.

COOL-WATER WASHED CURD CHEESE

Once you start messing around with washing curd, it should become fairly evident how tiny differences can create vastly different cheeses. I have grouped these first few cheeses based on the temperature of the water the curd is washed with. But they can vary in the percentage; maybe throw in a bit of salt, or wash the curd at a different stage during the making, and you will end up with an entirely dissimilar cheese. Monterey Jack is one of only a handful of cheeses that the United States can rightfully claim as an original from our part of North America.

The story of how Monterey Jack (also called either just "Jack" or simply "Monterey") was developed changes a bit depending on whom you ask. The most common legend attributes the name to David Jacks, a shrewd, perhaps even ruthless businessman who established a successful enterprise marketing and selling cheese in the mid-1800s. There is no question that he branded the cheese with his name, but stories attribute the original name to both the Franciscan friars and an enterprising homemaker by the name of Juana Cota de Boronda. Both parties were making a semisoft cheese based on a Spanish *queso del pais* (cheese of the country) that they pressed with wooden house jacks (tools meant to lift heavy structures still used today—think of a car jack used for changing a flat tire) and may have called "jack cheese."

If you happen to visit the amazingly fertile Monterey peninsula, you will find only one indigenous cheesemaker crafting anything similar to Monterey Jack: the Schoch Family Farmstead. But thanks to the US Federal Standards of Identity, which stipulates that to be called Monterey Jack the milk must be pasteurized (you think they did that when the cheese was first made in colonial times?), the Schoch family call their aged raw-milk complex cheese Monterey County Jack. Oh, well, the results are worth the difference in name. Colby, another US original, is also made with cool wash water, and the cheese we make here at our farm called Covered Bridge is washed first with cool water, then, after the curd is drained, I rewarm the curds and wash again with a cool local ale. Colby- and Jack-type cheeses usually depend on a little more starter culture to develop more acid in the vat, but are also salted at a bit higher final pH. You can vary both the starter amount and the level of salt to control the final sweetness of the cheese. In the United States they are usually made from pasteurized milk and sold quite young.

INGREDIENTS

Milk: 2 gal (8 L/17 lb) whole milk
Culture: ¼ tsp (0.6 g) MA or MM
Calcium chloride (optional): maximum ½ tsp (2.5 ml) calcium chloride diluted in ¼ cup (60 ml) cool, nonchlorinated water
Rennet: ¼ tsp plus (1.5 ml) single-strength rennet diluted just before use in ¼ cup (60 ml) cool, nonchlorinated water
Salt: 2–3 tsp (10–15 g) pure salt divided into two doses (goal is 2.2–2.5% of curd weight)

STEPS

Prepare Equipment: Make sure all equipment is cleaned and sanitized and that your cheesemaking space is free from possible contaminants. Refer to chapter 6 for tips on proper equipment preparation.

Prepare Milk: Warm milk to 80°F (27°C).

Culture: Sprinkle culture on top of milk. Let set for 2 to 5 minutes, then stir gently for 3 to 5 minutes.

Ripen: Increase temperature to 88 to 90°F (31–32°C), then maintain for 40 minutes.

Additions: Stir in calcium chloride solution. Wait 5 minutes.

Rennet and Coagulate: Stir in rennet solution with an up-and-down motion for 1 minute. Still milk. Hold temperature at 88 to 90°F (31–32°C) until clean break is achieved. The goal coagulation time is 35 minutes. If using flocculation as a guide, multiply the time to flocculation by 3 for the total time from adding rennet to cutting.

Cut: Cut curd mass into ⅜-inch (1-cm) cubes. Let rest for 5 minutes.

Stir and Heat: Cook and stir slowly and evenly to 102°F (39°C) over 45 minutes. Hold temperature and stir for 60–90 minutes. Whey pH at the end of cooking should be about 6.2. Settle 15 to 30 minutes. Goal curd texture is springy and curds will have shrunk to about half their original size.

Drain: Drain whey to 1 inch (2.5 cm) above curds.

Wash: Stir curds, then add cold water (about 60°F [16°C]) slowly while stirring until temperature stabilizes at 80 to 86°F (27–30°C). Stir for 15 minutes.

> TIPS: Water will raise pH slightly, so readings will be different from those for nonwashed curd cheese. From the time the wash is added, pH readings can be done, and a standard for your water type can be developed.

Drain: Drain whey to level of curds, and stir for 10 minutes. Then drain completely, and stir for 20 minutes. Curd pH should be 5.7–5.8.

Salt: Add salt to curd in two applications. Stir curd well, and let mellow for 5 to 10 minutes between each salting. Final salt content of the cheese will be about 1.7–1.9% due to some loss during pressing.

Mold and Press: Fill cheesecloth with curd, form into a ball, and tie with a stilton knot (see chapter 12 for stilton knot instructions). Knead and press the ball, and continue to tighten the knot. Then press between boards on draining table with knot up. Retighten knot at 15, 30, and 60 minutes. Continue to add weight to achieve closed rind. Maintain room temperature of 72 to 80°F (22–27°C) during pressing. Press for 12 to 24 hours. Final cheese pH goal is 5.4–5.5.

> TIPS: Alternatively, you can press in a screw or other mechanical press. Remember that any cheese that has a salted curd will require a great deal more pressure than one that is not salted. Also the curd is cooler for this cheese, which also means it does not tend to want to knit back together easily.

Affinage: Unmold when rind is closed and cheese is tightly knit. If pressed in the cloth bag, the cheese will have what some like to call a "belly button" from the knot. Move to an aging environment at 55°F (13°C) and 85 percent RH for 2 to 3 months. It can be aged longer for "dry" versions.

> TIPS: If rind is still open by the end of pressing, you can briefly submerge the cheese in hot water and repress. Alternatively, you may want to wax the cheese.

LARGE-BATCH GUIDELINES

Milk: 12 gal (46 L/100 lb) whole milk
Culture: 3U MA or MM
Calcium chloride (optional): 10–15 ml (about 1 tbsp)
Rennet: 7–9 ml single-strength rennet (3.5–4.5 ml double strength)
Whey removal: To 1 inch (2.5 cm) above curds
Salt: 2.2–2.5% of drained curd weight

SAME-TEMPERATURE LIGHT-BRINE WASHED CURD CHEESE

In my cheesemaking career washing with a very light brine is a technique I learned very recently. When experimenting with the recipe for this book, we liked the resulting cheese so much we added it to our main product line. While this was originally modeled on a St. Paulin recipe, we made a few adjustments that resulted in a semihard cheese we call Takelma (following in our tradition of naming cheeses after local landmarks, Takelma is the name of the Native Americans that originally settled this area). The French cheese St. Paulin, while in most cases a somewhat bland, rather industrial cheese, is interesting in that it was one of the first cheeses designed to be made from pasteurized milk. In a reversal of trends, you can now find raw-milk versions. The Danish cheese Havarti is also somewhat bland, mild, and ubiquitous. For a superior US example, check out my cheesemaker friend Rod Volbeda's Willamette Valley Cheese Jersey cow's milk Havarti.

INGREDIENTS

Milk: 2 gal (8 L/17 lb) whole milk
Culture: ⅛ tsp (0.3 g) MM 100 and 1/16 tsp (0.1 g) LM
Calcium chloride (optional): maximum ½ tsp (2.5 ml) calcium chloride diluted in ¼ cup (60 ml) cool, nonchlorinated water
Rennet: ¼ tsp (1.5 ml) single-strength rennet diluted just before use in ¼ cup (60 ml) cool, nonchlorinated water
Salt: Heavy brine

STEPS

Prepare Equipment: Make sure all equipment is cleaned and sanitized and that your cheesemaking space is free from possible contaminants. Refer to chapter 6 for tips on proper equipment preparation.

Prepare Milk: Warm milk to 79°F (26°C).

Culture: Sprinkle cultures on top of milk. Let set for 2 to 5 minutes, then stir gently for 3 to 5 minutes.

Ripen: Increase temperature to 89°F (32°C), then maintain for 1 hour or to 0.05 drop in milk pH.

Additions: Stir in calcium chloride solution for 1 minute then wait 5 minutes before adding rennet.

Rennet and Coagulate: Stir in rennet solution with an up-and-down motion for 1 minute. Still milk. Hold temperature until clean break is achieved. The goal coagulation time is 45 minutes. If using flocculation as a guide, multiply the time to flocculation by 3.5 for the total time from adding rennet to cutting.

Cut: Cut curd mass into ¼-inch (6-mm) cubes slowly over 20 minutes. Maintain at 89°F (32°C). Let rest for 10 to 15 minutes.

Partial Drain and Wash: Remove 1 gallon (4 L) of the whey (50% of total volume). Add 1 gallon (4 L) of 89°F (32°C) water with 1 teaspoon salt (6 g) (0.2% salt by weight of volume in pot).

Stir and Heat: Stir and cook slowly to 100°F (38°C) over 30 minutes. Settle 5 minutes.

Pitch: Press under whey for 10 to 15 minutes to form a consolidated mass.

Drain: Drain off the whey, and press curd mass into a cloth-lined mold. Knead gently by hand to form an even shape in the mold. Add weight equal to that of the curd. Press for 15 minutes. Flip, and press for 30 minutes. Flip, add more weight (about double the cheese weight), and press for 1 hour. Turn, and continue to press until the cheese pH is 5.2 to 5.3. This may take 4 to 8 hours. Room temperature should be 72°F (22°C). By the third flip the rind should be closed. Add more weight as needed.

Salt: Unmold and immerse in a heavy brine for 2 to 3 hours per pound (4 to 6 hours per kg) of the wheel.

Affinage: Pat dry, and move to an aging environment at 55°F (13°C) and 85 to 90 percent RH for 2 to 3 months or more. Can be washed with brine and annatto or another wash or waxed.

LARGE-BATCH GUIDELINES

Milk: 12 gal (46 L/100 lb) whole milk
Culture: 2U MM 100 and 0.5U LM (or approx.)
Calcium chloride (optional): 10–15 ml (about 1 tbsp)
Rennet: 9 ml single-strength rennet (4.5 ml double strength)
Whey removal: 50%
Salt: 0.2% of weight of first wash water plus heavy brine

HOT-WATER WASHED CURD CHEESE

An amazing amount of wonderful cheese comes from the Netherlands. The Dutch are also leading producers of cheesemaking equipment and cultures. Two closely related Dutch cheeses, Gouda and Edam, both named for their original towns of origin and sale, are much copied throughout the world (including our creation, Hillis Peak). While Gouda meant for export has traditionally been waxed, which was originally intended to compensate for unfavorable storage conditions during long transport, fortunately you can now buy many superior, aged Goudas from most fine cheese counters.

The French cheese Mimolette has a very interesting legend attached to its creation. By all accounts it was meant to be a copy of Dutch Edam—and it does indeed take the same oval shape. But it differs greatly in that its rind is purposely inoculated with cheese mites to assist with developing a unique rind and interior flavor. Legend has it that the bright orange color was meant to be a purposeful comment by the French regarding the Dutch ruling body, the House of Orange (*Huis van Oranje-Nassau*). In an interesting turn of cultural exchange, I saw a stack of obscenely orange cheeses for sale in a cheese shop in San Francisco that were labeled "Dutch Mimolette."

Pholia Farm's hot-water washed curd cheese, Hillis Peak, about 5 months into its affinage

INGREDIENTS

Milk: 2 gal (8 L/17 lb) whole milk
Culture: ⅛ tsp (0.3 g) MA 4000 and pinch (0.1 g) LM (or equivalent blend)
Calcium chloride (optional): maximum ½ tsp (2.5 ml) calcium chloride diluted in ¼ cup (60 ml) cool, nonchlorinated water
Rennet: ¼ tsp (1.25 ml) single-strength rennet diluted just before use in ¼ cup (60 ml) cool nonchlorinated water
Salt: Heavy brine

STEPS

Prepare Equipment: Make sure all equipment is cleaned and sanitized and that your cheesemaking space is free from possible contaminants. Refer to chapter 6 for tips on proper equipment preparation.

Prepare Milk: Warm milk to 80°F (27°C).

Culture: Sprinkle cultures on top of milk. Let set for 2 to 5 minutes, then stir gently for 3 to 5 minutes.

Ripen: Increase temperature to 88 to 90°F (31–32°C), then maintain for 45 minutes. Normal milk pH drop is 0.1 to 0.5.

Additions: Stir in calcium chloride solution. Wait 5 minutes.

Rennet and Coagulate: Stir in rennet solution with an up-and-down motion for 1 minute. Still milk. Hold temperature until clean break is achieved. The goal coagulation time is 45 minutes. If using flocculation as a guide, multiply the time to flocculation by 3.5 for the total time from adding rennet to cutting.

Cut: Cut curd mass into ⅜-inch (1-cm) cubes. Let rest for 5 minutes.

Stir: Stir gently for 15 to 20 minutes, maintaining 88 to 90°F (31–32°C). Settle 5 minutes. Curd should shrink some and be of even texture but still quite soft.

Partial Drain and Wash: Remove 2 to 2.5 quarts whey (25 to 30% of total volume). Add hot water at about 160°F (71°C) slowly to increase temperature over 10 minutes to 101°F (38°C). Hold and stir to final texture, 15 to 20 minutes. Let settle for 5 minutes. Stir constantly while hot water is being added to not overheat any of the curds.

Drain and Press: Drain whey to level of curds. Press by hand to consolidate curd into single mass. Drain off remaining whey. Fill cloth-lined molds, and hand-press curd into forms. Press with half the weight of cheese for 15 minutes. Flip, then increase weight to equal cheese for 30 minutes. Flip again, then increase weight to double that of the cheese for 1 hour. Flip again, and increase weight if needed to close rind. The goal room temperature is 72°F (22°C). Press to goal curd pH of 5.2 to 5.3, which usually takes 4 to 8 hours.

Salt: Unmold, and immerse in a heavy brine for 3 to 4 hours per pound (6–8 hours per kg) of cheese in a wheel.

Affinage: Pat dry with paper towels, and move to an aging environment at 55°F (13°C) and 85 percent RH for 3 to 12 months. Cheese can be waxed, oiled, spice rubbed, and so on.

LARGE-BATCH GUIDELINES

Milk: 12 gal (46 L/100 lb) whole milk
Culture: 1.5U MA 4000 series and 0.5U LM (or equivalent blend)
Calcium chloride (optional): 10–15 ml (about 1 tbsp)
Rennet: 7–9 ml single-strength rennet (3.5–4.5 ml double strength)
Whey removal: 25–30%
Salt: Heavy brine

THE FUNDAMENTALS OF MAKING TEXTURED AND MILLED CURD CHEESE

Now we start entering a category of cheeses that are favorites in Britain and the United States: stirred and milled curd cheeses and the particular favorite, the all-famous cheddar. Almost in contrast to their popularity is the complexity of their makes. There are seemingly more steps in making a stirred or milled curd cheese than in most others. But keep in mind that once they are in the forms and pressing, they are almost finished, without the need for brine or dry salting and the days of monitoring that those techniques bring. In addition, while the process is long, other chores can be accomplished between the make steps. So for a kitchen-bound farm wife (yes, historically, it was usually women making the cheese), producing a cheddar could be integrated into the daily routine without too much trouble.

Cheddar and other cheeses, where the process includes dry salting of the curds, are thought to have evolved in England and specifically the county of Cheshire, where rock salt is readily available. The British cheese, Cheshire, has been made for over a thousand years, making it even older than cheddar. The influence of these milled and dry-salted cheeses is seen in British blue-veined cheeses as well (see the Stilton-type recipe earlier in chapter 12).

Traditionally made cheddar will have a bit of a different texture and flavor than its industrial cousin, being more open, drier, and—for the fan of artisan cheeses—have more depth and complexity. In fact, it can be quite a transition for those making and tasting artisan, naturally aged (not vacuum sealed) cheddar, when their only frame of reference is for smooth, dense, usually highly colored cheeses of the same name.

Cheeses in this category spend an extended period in the vat after the curd is drained. During that time they are held at a warm temperature and the curd is piled, cut in slabs and piled, or stirred while it continues to acidify and dry. This process is called "cheddaring." The weight of the other curds also begins to change the texture of the protein network, similar to what happens to our stretched curd cheeses in chapter 13.

When cheddaring, or stirring small batches of curd while it develops acid, you can use a colander set over a pot of warm water to keep the slabs or curds warm. To keep the top of the curds warm as well as weighted (as they would be if the batch size were larger), use a ziplock bag with warm water to cover the curds.

When the goal acid level is reached, the curds or slabs are cut or broken into smaller pieces. This step is referred to as milling. If you read the blue cheese and stretched curd chapters, you might know that some other cheeses also go through a milling process. After milling, the curd is salted, bringing a halt to acid development and removing more whey. Salt is added in two or more additions so it can be slowly absorbed without excess whey washing it from the curd. These stages of absorption are called the "mellowing" of the curd. When this salted curd is put in the forms, a great deal of pressure is needed to knit it back together.

Some varieties of these cheeses are aged in a bandage wrap, and others have a natural rind. Of course, the majority of cheddar types are simply vacuum sealed and aged at refrigeration temperatures. The world's great cheddars can age for years and grow quite complex—a far cry from the orange grocery store cheese most of us grew up with.

Our first recipe is for what is often called a stirred curd, or farmhouse cheddar. Even many commercial plants now produce cheddar cheese using a similar technique. Some studies have shown that after a certain period of aging, stirred curd varieties reach a texture similar to that of traditionally produced cheddar types. Indeed, many superior cheddars are produced using the stirred curd technique.

Stirred curd varieties (sometimes called "granular cheddar") spend less time developing acid after being drained; instead, they must develop more while still in the vat. For this reason you may need to add a bit more starter culture. You should also be sure to use blends with good acidification power.

Some techniques (especially where mechanization is involved) call for the curds to be stirred continuously, while other methods include periodic stirring and breaking of the curd. The recipe I have included calls for intermittent stirring, but remember, these are all things you can adjust, as long as you reach the goal texture and acid level.

The second recipe is another variation of a stirred-curd cheese that is also milled. is for traditional cheddar. This type of cheese may have one of the most well-documented processes available to cheesemakers. It is easy to find acid development goals and other information about this popular cheese.

If you are a fan of fresh (squeaky) curds, traditional cheddar is the recipe to use—simply, don't press the curds. Or save part of the curds for fresh consumption and press the remaining portion. If you plan to eat fresh curds, I would recommend using pasteurized milk.

Traditional cheddar is often bandage wrapped for long aging. See chapter 4 for more on this technique, as well as other affinage tips for all semihard and hard cheeses.

Fresh, squeaky cheddar curds with pesto

STIRRED CURD CHEESE

If you have read the previous chapter on mozzarella types, as well as the recipe for making a Stilton-type blue cheese, then you are familiar with a technique that involves holding drained curd in the vat while the acid level increases, then tearing or cutting the acidified curd into smaller pieces—also known as milling (a term you're probably familiar with by now)—before proceeding with forming the cheese into shapes. Now that we are working with cheddar types—where this same process is actually called "cheddaring"—you might wonder about the true origins of this technique. As with the cultural give and take that has created other great cheeses, it is believed that during the Roman occupation of Great Britain, the technique for making stretched curd cheeses was adapted to create an entirely new family of cheeses—including primitive cheddar types. Steve Jenkins says that Cheshire (a stirred curd variety) is probably England's oldest cheese.

Another stirred curd from Wales, Caerphilly, is in a group often referred to as the "British Crumblies" because its texture is more friable than that of Cheshire or bandaged cheddar. Caerphilly usually has a mold-ripened surface that often produces some softening of the paste, making this Welsh cheese almost a hybrid surface-ripened and semihard cheese. In the United States, Landaff Cheese by Landaff Creamery in New Hampshire (aged by my friends at The Cellars at Jasper Hill in Vermont) is Max McCalman's suggestion for an authentic, traditional Caerphilly-type cheese. Luckily for all of us, there are increasingly good and true Caerphilly cheeses being made in Wales, and some even make their way to cheese counters in other countries.

While the name "cheddar" for much of the world still brings to mind lightly orange-colored, plastic-sealed, semihard cheese best found in the local grocery store, increasing numbers of cheese lovers have had the privilege of trying a true British cheddar such as Keen's or Montgomery's Cheddar. US versions of superior quality include Cabot Clothbound from Vermont and Beecher's Flagship Reserve from Seattle. Laguiole is a French AOC cheese, made since the nineteenth century, that goes through a process not identical but similar to the texturing that takes place during the making of cheddars and stirred curd cheeses.

INGREDIENTS

Milk: 2 gal (8 L/17 lb) whole milk
Culture: ¼ tsp (0.6 g) MA 4000 (or equivalent blend)
Calcium chloride (optional): maximum ½ tsp (2.5 ml) calcium chloride diluted in ¼ cup (60 ml) cool, nonchlorinated water
Colorant (for Cheshire style): 4 drops annatto coloring diluted in ⅛ cup (30ml) cool, nonchlorinated water
Rennet: ¼ tsp (1.25 ml) single-strength rennet diluted just before use in ¼ cup (60 ml) cool nonchlorinated water
Salt: 2–3 tsp (10–20 g) pure salt (1.85–2.1% of drained curd weight)

STEPS

Prepare Equipment: Make sure all equipment is cleaned and sanitized and that your cheesemaking space is free from possible contaminants. Refer to chapter 6 for tips on proper equipment preparation.

Prepare Milk: Warm milk to 80°F (27°C).

Culture: Sprinkle cultures on top of milk. Let set for 2 to 5 minutes, then stir gently for 3 to 5 minutes.

Ripen: Increase temperature to 86 to 87°F (30–31°C), and maintain for 40 to 60 minutes.

Additions: Stir in calcium chloride solution. Wait 5 minutes. Stir in annatto if using.

Rennet and Coagulate: Stir in rennet solution with an up-and-down motion for 1 minute. Still milk. Hold temperature at 86 to 87°F (30–31°C) until clean break is achieved. The goal coagulation time is 40 to 50 minutes. If using flocculation as a guide, multiply the time to flocculation by 3.5 for the total time from adding rennet to cutting.

Cut: Cut curd mass into ⅜-inch (1-cm) cubes. Let rest 5 minutes.

Stir and Heat: Stir gently for 15 minutes at 86 to 87°F (30–31°C). Continue to stir and increase temperature evenly to 95°F (35°C) over 40 to 50 minutes. Hold temperature, and stir until curd passes the texture test (see "Checking Curd for Readiness" in chapter 2, page 52). The pH of the whey at the end of cooking should be 6.1 to 6.2. Let curds settle in pot for 5 minutes.

Drain and Texture: Drain whey, and push curd to side of pot. Tilt pot so that whey drains away from curd or place curds in a colander over a pot of hot water or whey. Keep curd at 95°F (35°C). Every 15 to 20 minutes, break curd mass into large pieces and turn pile. Goal curd pH at the end of texturing is 5.4.

TIPS: To check curd pH, place a small amount of curd around the tip of the pH meter and squeeze it against the probe.

Mill: Break or cut curd into fine pieces ¼ to ⅜ inch (1 cm) in size. Place in warmed colander.

Salt: Thoroughly stir in salt in two even applications 5 to 10 minutes apart "mellowing."

Mold and Press: Put curds in a cloth-lined mold, and place in a mechanical press. Use enough weight to feel gentle resistance from the curd mass and see translucent to faintly white whey draining. After 30 minutes turn and increase pressure. After 1 hour turn and increase pressure to close the rind. Press for 12 hours or overnight.

Unmold and check rind. If closed, rewrap and press for 12 hours more without cloth to smooth out rind. If rind is not closed, immerse the cheese in 160°F (71°C) water for 1 to 3 minutes, replace in press, and press for 12 hours.

Affinage: Unmold. Cheese can be bandaged (as for cheddar), air-dried, then waxed, or aged with a natural rind. Age at 55°F (13°C) and 85 percent RH for 6 to 12 months.

LARGE-BATCH GUIDELINES

Milk: 12 gal (46 L/100 lb) whole milk
Culture: 3U MA 4000 (or equivalent blend)
Calcium chloride (optional): 10–15 ml (about 1 tbsp)
Rennet: 7–9 ml single-strength rennet (3.5–4.5 ml double strength)
Salt: 1.85–2.1% of curd weight

SAM KOSTER, FARMER AND HOME CHEESEMAKER, NEW ZEALAND

Author's note: Sam and his wife visited our farm in 2011 on one of their many trips to the United States and visits to cheesemakers across the globe. In his own words:

"I am a rural doctor living and working in Opotiki, New Zealand, a remote town of about five thousand people. For 20 years I have owned a seasonal two-hundred-cow dairy. About 5 years ago I began making aged, raw-milk cheeses from the milk of my grassfed cows.

"I give all the cheeses I make to my family or to workers on the farm. To sell cheese in New Zealand, one has to comply with food-safety rules that are designed for larger processors. This means that there is little in the way of artisan cheese for sale in my country. I would love to go commercial, but the cost of compliance is prohibitive.

"I like to produce raw-milk cheese in a fashion that is close to that of traditional cheeses produced by farmers' wives—before mass markets, commodity cheese, and presliced cheese.

"The recipe I make is for a traditional Welsh cheese called Caerphilly. This cheese was first made by busy farmers' wives who were already very overworked, so they kept the process as simple as they could to produce a semihard cheese that could be eaten or sold within a couple of weeks. The local market town was called Caerphilly; hence, the name was given to the cheese, too.

"My wife's uncle lives in Aberaeron, Mid Wales, and on a recent trip there, I managed to visit Caws Cenarth, where Carwyn Adams and his mother Thelma have been making award-winning raw-milk Caerffilli (note Welsh spelling) for 50 years. Here I picked up a few tips and a lot of encouragement. Carwyn was in the middle of a major expansion. He was spreading out from the converted stone barn they had been using and converting old cowsheds to create a much larger facility, so in the near future their wonderfully crafted cheeses will be more readily available.

"The cheese is usually clean white with creamy texture and a noticeably acidic bite but overall quite mild, depending on age. The mildness is said to be why the coal miners of South Wales enjoyed it so much, as it was unlikely to cause indigestion. Its relative saltiness helped replace that lost from sweating down in the hot and dangerous mines. Those of us who don't swing a pick will enjoy Caerphilly as a table cheese, but it is exceptional when toasted, as is the famous Welsh Rarebit."

New Zealand cheesemaker Sam Koster—dairy farm owner, physician, and avid home cheesemaker

MILLED CURD CHEESE WITH CRUMBLY TEXTURE AND MOLD-RIPENED RIND

Traditional Welsh Caerphilly is experiencing a comeback, through increased and improved production both within its native land and elsewhere, as consumers grow to appreciate this unique cheese. Part cheddar, part bloomy rind, and part tomme, Caerphilly combines processes and affinage to create an interesting combination of textures and flavors. The long production process is thought to have evolved around the natural rhythms of farm life and allowed the dairy women (historically, by the way, most cheesemakers were women) to continue with their impressive chore list while at the same time still making cheese. As you make this recipe, think about the flexibility that can be introduced by simply changing a few parameters, such as ripening time (before coagulation) and time developing acid during the stirring of the curds.

The rind of this style of cheese counts on the appearance of several types of mucor molds. These are encouraged to appear spontaneously on the rind through a higher humidity than normal for semihard cheeses (usually 90 to 95 percent instead of the typical 85 percent RH). Mateo Kehler, part of the brother team that owns and runs The Cellars at Jasper Hills (where the cheese pictured, Landaff, is aged), says that they have tried adding the mold culture mycodore to encourage the right kind of mold development on rinds of cheeses such as this, but "it was completely outcompeted by our local flora." A beautiful example of *terroir* in action! In addition, he points out the expensive nature of these molds when purchased for large batches of cheese. When making the following recipe, if you are unsatisfied with the results, experiment with adding some cultures, but also give time and locality a chance to augment the flora in your aging room. You might even want to place a purchased cheese whose rind is desirable into the space as a way to introduce some friendly flora.

A beautiful Caerphilly-style cheese from Landaff Creamery, New Hampshire. Affinage is done by The Cellars at Jasper Hill, Vermont. Photo by Lark Smotherton and courtesy of the Cellars at Jasper Hill

INGREDIENTS

Milk: 2 gal (8 L/17 lb) whole milk
Culture: ¼ tsp (0.6 g) MA 4000 (or equivalent blend) and 2 skewer tips mold blend such as mycoderm or mycodore (optional)
Calcium chloride (optional): maximum ½ tsp (2.5 ml) calcium chloride diluted in ¼ cup (60 ml) cool, nonchlorinated water
Rennet: Scant ⅜ tsp (1.53 ml) single-strength rennet (traditionally, calf rennet is used) diluted just before use in ¼ cup (60 ml) cool, nonchlorinated water
Salt: 2 tsp (10 g) pure salt (salt will be added in two additions, each at 1% of the total weight of the drained curd)

STEPS

Prepare Equipment: Make sure all equipment is cleaned and sanitized and that your cheesemaking space is free from possible contaminants. Refer to chapter 6 for tips on proper equipment preparation.

Prepare Milk: Warm milk to 80°F (27°C).

Culture: Sprinkle cultures on top of milk. Let set for 2 to 5 minutes, then stir gently for 3 to 5 minutes.

Ripen: Increase temperature to 86°F (30°C), and maintain for 45 to 90 minutes.

Additions: Stir in calcium chloride solution. Wait 5 minutes.

Rennet and Coagulate: Stir in rennet solution with an up-and-down motion for 1 minute. Still milk. Hold temperature at 86°F (30°C) until clean break is achieved. The goal coagulation time is 40 to 45 minutes. If using flocculation as a guide, multiply the time to flocculation by 2 for the total time from adding rennet to cutting.

Cut: Cut curd mass into ½-inch (1.3-cm) cubes. Let rest 15 minutes.

Stir: Stir gently for 15 minutes at 86°F (30°C). Continue to stir, and increase heat to 92 to 94°F (33–34°C) over 15 minutes. Hold temperature, and stir as needed to prevent matting until curd texture is even and curd passes the texture test (see "Checking Curd for Readiness" in chapter 2, page 52). This may take up to an hour. The goal whey pH at the end of the cook phase is 6.52 to 6.45.

Drain and Texture: Drain off whey, and push curds to one side of the pot. Tilt the pot so the whey will drain away from the curds, or place curds in a colander over a pot of hot water or whey. Pile curds up against the side of the pot in a semicone shape. Keep the pot covered and curd temperature at 92 to 94°F (33–34°C). Every 15 to 20 minutes, remove excess whey and cut bottom edge of curd pile; place trimmings on top of stack. Continue to remove whey and restack curd until curd pH is about 5.75.

> TIPS: To check curd pH place a small amount of curd around the tip of the pH meter and squeeze it against the probe.

Mill and Salt: Break into 1-inch (2.5-cm) pieces, and sprinkle with 1 teaspoon salt (or 1% of the weight of the curd). Mix until salt is completely dissolved.

> TIPS: Be sure to weigh the curd and use the right amount of salt. Too much salt at this point will prevent the right acid development during the first pressing.

Press: Place curd in cloth-lined forms, and press for 1.5 to 2 hours, using enough weight to consolidate mass but not completely close rind.

> TIPS: Press lightly—curd is still acidifying, and too much whey removal will prevent goal pH from being achieved.

Salt and Press: When pH is about 5.3, unmold and rub with 1 teaspoon salt (goal is 1% of the weight of the curd). Return to the lined mold, and repress for 12 to 24 hours. Use increasing weight to close rind. Rind will be closed but will not be perfectly smooth.

Unmold and Salt: Unmold cheese, and immerse in a heavy brine for about 45 minutes per 2 pounds (1 kg).

Caerphilly-type textured curds ready to press

Affinage: Pat dry, and move to an aging environment at 55°F (13°C) and 90 to 95 percent RH. Turn daily, and pat down molds, but don't brush. This cheese usually ages for only 2 months. The desired rind is thick and feltlike, with softening under the rind but not in the center of the cheese.

LARGE-BATCH GUIDELINES

Milk: 12 gal (46 L/100 lb) whole milk
Culture: 3U MA 4000 (or equivalent blend) and 0.2–0.5U mycoderm or mycodore (optional)
Calcium chloride (optional): 10–15 ml (about 1 tbsp)
Rennet: 9 ml single-strength calf rennet (4.5 ml double strength)

CLASSIC MILLED CURD, ENGLISH CHEDDAR-STYLE CHEESE

While cheddar may historically and in reality be an English cheese, many US citizens think of cheddar as their national cheese. Of course, the industrial cheddar they are referring to bears little similarity to what the following recipe will create, as well as what many US producers are also making: traditional, clothbound, aged cheddar. There are several theories as to why 40-pound blocks of vacuum-sealed, orange-colored cheddar became the king of US cheeses, from the fact that it doesn't need brine tanks (therefore taking up less production space in a cheesemaking facility) to the cheese being ready to go into aging on the same day as its "make" to the fact that it is less labor intensive over the long run (by the next day the vat and forms are ready to make another batch). (Trust the United States to develop a love for a product based on practicality.)

Gordon Edgar, a cheese buddy and author of *Cheesemonger* (Chelsea Green, 2010), is researching the various nuances of the multihued history of cheddar in the United States. Gordon points out another fact that has probably led to cheddar's reign: "Another aspect of Cheddar that appeals to cheesemakers is that even though it needs to be aged before one can sell the finished block, curds can be sold the same day in the Cheddar-makers immediate vicinity. Selling something the day you make it definitely helps with the cash flow issues common to most cheesemakers." Whichever way you like your cheddar—hard and crumbly or sliceable and orange—you have to appreciate this mainstay cheese and its historical impact.

INGREDIENTS

Milk: 2 gal (8 L/17 lb) whole milk
Culture: ¼ tsp (0.6 g) MA 4000 (or equivalent blend)
Calcium chloride (optional): maximum ½ tsp (2.5 ml) calcium chloride diluted in ¼ cup (60 ml) cool, nonchlorinated water
Rennet: ¼ tsp plus (1.5 ml) single-strength rennet diluted just before use in ¼ cup (60 ml) cool, nonchlorinated water
Salt: 2–3 tsp (10–15 g) pure salt (goal is 2% salt by weight of milled curd)

STEPS

Prepare Equipment: Make sure all equipment is cleaned and sanitized and that your cheesemaking space is free from possible contaminants. Refer to chapter 6 for tips on proper equipment preparation.

Prepare Milk: Warm milk to 80°F (27°C).

Culture: Sprinkle cultures on top of milk. Let set for 2 to 5 minutes, then stir gently for 3 to 5 minutes.

Ripen: Increase temperature to 90°F (32°C), and maintain for 30 minutes.

Additions: Stir in calcium chloride solution. Wait 5 minutes.

Rennet and Coagulate: Stir in rennet solution with an up-and-down motion for 1 minute. Still milk. Hold temperature until clean break is achieved. The goal coagulation time is 45 minutes. If using flocculation as a guide, multiply the time to flocculation by 3.5 for the total time from adding rennet to cutting.

Cut: Cut curd mass into ⅜-inch (1-cm) cubes. Let rest for 5 minutes.

Stir: Stir gently, and begin to increase temperature to 102°F (39°C) slowly over 30 minutes. Hold, and stir for

A large batch of cheddar curd cheddaring in the vat

45 to 60 minutes. The goal whey pH at the end of stirring is 6.15 to 6.2. Let curds settle in vat for 5 minutes.

Drain and Texture: Drain the whey completely, and push the curds to one side of the pot. Tilt the pot so the whey drains away from the curds, or place curds in a colander over a pot of hot water or whey. Cover, and keep warm at 95 to 98°F (35–37°C). Let set for 15 minutes. Turn, and let set for 15 minutes more.

Cut the curd mass into two slabs, and stack on top of each other. Turn every 15 minutes by turning each slab over and reversing its order in the stack. Continue until curd pH is 5.3 to 5.35. This will take about 2 hours. The end texture of the curd when torn is often described as "cooked chicken breast" texture.

TIPS: If you have trouble keeping the curd mass warm or the texture seems to not be developing, try partially filling a zipper-top plastic bag with 98°F (37°C) water and place on top of the curd mass.

Use a plastic bag filled with warm water to help keep small batches of cheddaring curds warm and properly weighted.

Mill and Salt: When the texture and pH are correct, cut the slabs into 1-×-2-inch (2.5-×-5-cm) strips. Place in colander, and sprinkle with half the salt. Stir thoroughly, and let set for 5 to 10 minutes "mellowing," then add rest of the salt, stirring well. The goal final salt content is 2%.

Press: Put curd in cloth-lined mold, and place in a mechanical press. Use just enough pressure to feel the resistance of the curd and see some whitish, translucent whey draining from the form. After 15 minutes unmold, turn, and place back in lined mold. Press with more weight, watching that the draining whey is not too opaque. After 30 minutes unmold, turn, and place back in lined mold. Press with more weight. At this point very little whey will be draining. Increase weight as needed to continue to close the rind. Press for 12 to 24 hours.

TIPS: By the end of pressing, the rind should be tightly closed, but you will see faint outlines of the milled curd.

Affinage: Unmold, and move to an aging environment at 55°F (13°C) and 85 percent RH for 12 months or more. This cheese can be used younger, but it develops its best flavor after longer aging. Bandage as instructed in chapter 4 or finish in other manner.

LARGE-BATCH GUIDELINES

Milk: 12 gal (46 L/100 lb) whole milk
Culture: 2U MA 4000 (or equivalent blend)
Calcium chloride (optional): 10–15 ml (about 1 tbsp)
Rennet: 7–9 ml single-strength rennet (3.5–4.5 ml double strength)
Salt: 2% of weight of milled curd

THE FUNDAMENTALS OF MAKING CHEESE WITH EYES

Our last category of cheeses is the big boys from Switzerland, still sometimes called "Swiss cheese" by us North Americans. Besides its being delicious, without cheese with eyes, we would have no universally readable icon for cheese! While there are really two types of cheese with eyes, the Dutch styles with small eyes and the Swiss style with larger eyes, I am covering only the Swiss type here. As the Dutch types are also typically washed curd cheeses, they were included earlier in this chapter.

The Swiss varieties with eyes include distinctive types most commonly represented by Emmental, with very large eyes and huge wheels weighing about 220 pounds (100 kg); and the Gruyère group, which includes Beaufort, Comté, and Gruyère, with smaller wheels, weighing closer to a mere 88 pounds (40 kg), and smaller eyes. These cheeses are not only distinctive because of their appealing eye formations but also their springy texture and sweet, nutty flavor.

Both groups are traditionally made using raw milk produced using very strict hygiene. This helps prevent or limit naturally present propionic acid bacteria (PAB) and other undesirable organisms that can wreak havoc during aging. While Emmental relies upon added propionic acid bacteria, Gruyère styles usually have smaller eyes from native strains of PAB and other eye-forming bacteria.

The first key to good eye formation in the cheese is the ability to retain gas. The characteristics that lead to good gas retention are a firm, elastic, textured paste and a firm rind. Another factor affecting eye formation is mass—large wheels are more likely to develop even, large eyes. (When eyes do not form, the cheese is sometimes called "blind.") While they are not even close to the size of the traditional cheeses, I recommend no less than 4-gallon batches for these cheese types.

The rubbery texture of Swiss-type cheeses is attained through both the high heat at which the curd is cooked and the relatively high pH at which the curd is drained. Way back in chapters 1 and 3, you learned how minerals retained in the cheese will influence the texture: the more minerals, the more supple the curd. During the making of a Swiss-type cheese with eyes, the curd develops very little acid during its cooking, thanks to both the high heat that all but stops acid development and the draining of the whey before too much further acid development can occur. If you remember from earlier in the book, minerals can be washed out

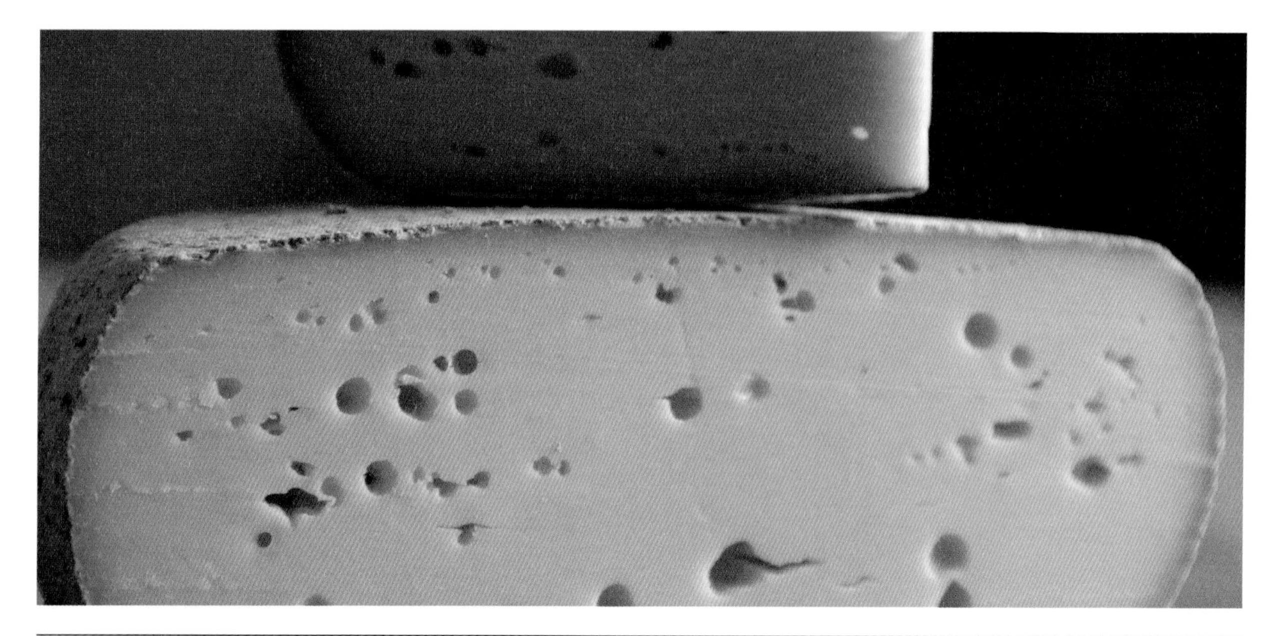

Pholia Farm Pleasant Creek, a Gruyère-style cheese

TABLE 14-3. CULTURE CHOICES FOR CHEESES WITH EYES

Name	Contains	Manufacturer	Small-Batch Dosage:* 2–4 gal (8–15 L/17–34 lb)	Large-Batch Dosage: 26 gal (100 L/220 lb)	Use
Thermophilic	ST, LB, LH	Packaged by home cheese supply co.	1 packet	N/a	All inclusive blend for several Italian-style and other high-temp cheeses
Thermo B	ST, LB	Abiasa	¼– ½ tsp (1.5–3 g)	5–10 g	For Emmental type, can include 1 part LH 100
Thermo C	ST, LH	Abiasa	¼– ½ tsp (1.5–3 g)	5–10 g	A blend for hard aged and extra-hard cheeses
TA 60 Series (Choozit) (includes TA 61, 62)	ST (fast-acid producer)	Danisco	⅛– ¼ tsp (0.3–0.6 g)	5–10U	Use to make your own blends by combining with LH 100 and/or LB 340
LH 100 (Choozit)	LBL, LH	Danisco	Dosage rate ranges from 10–50% of that of TA culture		Use with TA 60
LB 340 (Choozit)	LB	Danisco	Dosage rate ranges from 10–50% of that of TA culture		Use with TA 60

*Remember that culture is always best measured by weighing and dividing into unit doses based on the strength of the culture, instead of using volume or set weight.

with acidic whey as it leaves the cheese. A firm rind is obtained through brining and an initial drying phase at a lower temperature than the propionic acid bacteria can grow at. During this time the paste can stabilize, salt content can equalize, and the rind can dry a bit.

Speaking of brining, propionic acid bacteria are quite sensitive to salt, so it is important to not salt these cheese types beyond their goal level of only 1.1 to 1.2 percent salt (as compared to 2 percent or more for other semihard and hard cheeses).

Let's take a look at the thermophilic bacteria that can be used in making both Swiss styles of cheese with eyes and the propionic acid bacteria strains that can be added. In the ripening culture chart earlier in this chapter, I included LBC 80, a ripening bacteria that can be helpful to add to several types of semihard to hard cheeses. LBC (found naturally in some raw milk) can form some small eyes during aging, so it might be helpful in Gruyère-style cheeses where smaller eyes are desired. The first four cultures listed in the chart are commonly used blends, the next is plain *Streptococcus thermophilus* (ST), and the last two are strains that can be purchased separately so you can make your own blends.

Propionic acid bacteria, for creating eyes, can be purchased in a couple of strains, each with its own rate of gas production. Depending on your aging environment and how well you can regulate the temperature, you may want to experiment with different strains.

TABLE 14-4. PROPIONIC ACID BACTERIA CULTURE CHOICES FOR CHEESE WITH EYES

Name	Gas Production	Producer	Dosage and Packet Size
Propionic	Medium	Abiasa	1–2 g/26 gal (100 L/220 lb) (25 g packet)

HARD CHEESE WITH SMALL EYES

This tasty group of cheeses is quite large and includes some of the world's great AOC cheeses, including French Comté (a.k.a. Gruyère de Comté) and Beaufort and Swiss Gruyère and Appenzeller. Being made from raw milk and aged in just the right conditions leads to spontaneous eye formation from natural propionic acid bacteria usually present in raw milk. In addition, other gas-forming bacteria are often a part of the culture, leading to small eyes in the cheese (these eyes are quite rare in Beaufort). While these cheeses now are each associated with a nation, it is important to remember that they evolved together as cheeses of a region, not a nation. This makes it a bit difficult at times to try to label them as either French or Swiss (at least for those of us who aren't making cheese there). It is perhaps better to simply describe them all as Alpine cheeses.

The cheeses in this group are all made in a large format: 75 to 80 pounds (34–36 kg). Many varieties are free from eyes, but in others they are present. We make our smaller wheels (about 8 pounds) of Pleasant Creek using a Gruyère-style make process.

INGREDIENTS

Milk: 4 gal (15 L/34 lb) whole to partly skimmed milk
Culture: ½ tsp (3 g) Thermo B and 2 skewer tips LH 100 **or** ½ tsp (3 g) Thermo C (or equivalent blend)
Calcium chloride (optional): maximum 1 tsp (5 ml) calcium chloride diluted in ½ cup (120 ml) cool, nonchlorinated water
Rennet: Scant ¾ tsp (3 ml) single-strength rennet diluted just before use in ½ cup (120 ml) cool, nonchlorinated water
Salt: Heavy brine

STEPS

Prepare Equipment: Make sure all equipment is cleaned and sanitized and that your cheesemaking space is free from possible contaminants. Refer to chapter 6 for tips on proper equipment preparation.

Prepare Milk: Warm milk to 90°F (32°C).

Culture: Sprinkle cultures on top of milk. Let set for 2 to 5 minutes, then stir gently for 3 to 5 minutes.

Additions: Stir in calcium chloride solution. Wait 5 minutes.

Rennet and Coagulate: Stir in rennet solution with an up-and-down motion for 1 minute. Still milk. Hold temperature until clean break is achieved. The goal coagulation time is 25 to 35 minutes. If using flocculation as a guide, multiply the time to flocculation by 2.5 for the total time from adding rennet to cutting.

Cut: Cut curd mass into ¼-inch (6-mm) cubes. Let rest in pot for 5 minutes. Cut to ⅛-inch (3-mm) pieces over 5 minutes.

Stir and Heat: Stir gently, and increase temperature slowly to 118 to 120°F (48–49°C) over 1 hour. The goal whey pH at the end of cooking is 6.35. Settle for 5 minutes.

Drain and Texture: Drain the whey to the level of the curds. Gently hand-press curd mass while still in the whey to consolidate it into a single mass. Drain off the rest of the whey.

Press: Place curd mass into cloth-lined mold. Gently press into the mold. Press with weight equal to that of the curd for 15 minutes. Turn, and press again with the same amount of weight for 30 minutes. Turn, and add twice the weight of the cheese; press for 1 hour. Turn again, and add more weight if the rind is not closed. Continue to press to the goal curd pH of 5.3 to 5.4. Ideal room temperature during pressing is 72°F (22°C). Pressing will usually take 4 to 6 hours.

Salt: Unmold and immerse wheels in a heavy brine. Sprinkle tops with several tablespoons dry salt, and brine for 3 to 4 hours per pound of weight per wheel.

Affinage: Pat dry with paper towels, and move to an aging environment at 50°F (10°C) and 85 percent RH for 3 weeks; turn daily. Then age at 59 to 64°F (15–18°C) and 90 percent RH for 2 to 3 months, turning daily for the first month, then every few days, for optimal eye formation. Then continue to age at 55°F (13°C) and 85 to 95 percent RH for 2 to 3 months or more. Turn every week. Rind care can include washing with a *B. linens* wash for the first month (see chapter 11), light brine, or oiling.

TIPS: The eye formation can be skipped, and the cheese can simply be long-aged at 55°F (13°C) and 85 to 90 percent RH for 6 to 12 months. Some eyes will still form.

LARGE-BATCH GUIDELINES

Milk: 12 gal (46 L/100 lb) whole to partly skimmed milk
Culture: 4 g Thermo B and 1g LH 100 **or** 5 g Thermo C (or equivalent blend)
Calcium chloride (optional): 10–15 ml (about 1 tbsp)
Rennet: 10 ml single-strength rennet (5 ml double strength)
Salt: Heavy brine

SEMIHARD TO HARD CHEESE WITH LARGE EYES

Here is the granddaddy of all Alpine cheeses. Emmental (also called Emmentaler) wheels can weigh up to 220 pounds (100 kg) and be as large as the tire on a large truck. There is good reason that you won't be attempting to duplicate this monster exactly at home! Propionic acid bacteria, originally naturally occurring in the raw milk but now added, are responsible for the large eye formation in Emmental. Less salt is used in this variety of cheese to help encourage eye production.

Additionally, these cheeses are made from partly skimmed milk, which helps create a paste that has the texture to encapsulate the gas pockets formed by the bacteria. As with most traditional Alpine cheeses, the curd is made in large copper vats. The copper has a positive effect on the flavor of the cheese. Several US cheesemakers make Alpine-style cheeses using traditional copper vats, including Thistle Hill in Vermont, where the Putnams make their raw Jersey milk Tarentaise.

INGREDIENTS

Milk: 4 gal (15 L/34 lb) whole to partly skimmed milk
Culture: ½ tsp (3 g) Thermo B, 1 skewer tip LH 100 (or equivalent blend, some cheesemakers use Thermo C), and 1/16 tsp (0.3 g) propionic acid bacteria
Calcium chloride (optional): maximum 1 tsp (5 ml) calcium chloride diluted in ¼ cup (60 ml) cool, nonchlorinated water
Rennet: scant ¾ tsp (3 ml) single-strength rennet diluted just before use in ¼ cup (60 ml) cool, nonchlorinated water
Salt: Heavy brine

STEPS

Prepare Equipment: Make sure all equipment is cleaned and sanitized and that your cheesemaking space is free from possible contaminants. Refer to chapter 6 for tips on proper equipment preparation.

Prepare Milk: Warm milk to 90 to 91°F (32–33°C).

Culture: Sprinkle cultures on top of milk. Let set for 2 to 5 minutes, then stir gently for 3 to 5 minutes.

Additions: Stir in calcium chloride solution for 1 minute then wait 5 minutes before adding rennet.

Rennet and Coagulate: Stir in rennet solution with an up-and-down motion for 1 minute. Still milk. Hold temperature until clean break is achieved. The goal coagulation time is 25 to 35 minutes. If using flocculation as a guide, multiply the time to flocculation by 2.5 for the total time from adding rennet to cutting.

Cut: Cut curd mass into ⅛-inch (3-mm) cubes over 5 minutes. Rest 5 minutes.

Stir and Heat: Stir gently, and increase temperature to 122°F (50°C) over 30 to 40 minutes. Hold, and stir until the whey pH is 6.3 to 6.4. Curd texture will be firm and springy.

Drain: Drain whey to the level of the curds, then pour into cloth-lined colander.

Press: Place cloth with curd bundle into mold. Knead, and hand-press curd into mold. Add weight to equal that of the curd, and press for 15 minutes. Unmold, turn, and repress with weight equal to twice that of the curd for 30 minutes. Unmold, turn, and continue to press with as much weight as needed to close the rind after 1 more hour of pressing. Unmold, turn, and continue to press for 12 to 24 hours to a curd pH of 5.3 to 5.4.

Salt: When goal pH has been reached, unmold, and immerse in a heavy brine for 2.5 hours per pound of cheese in the wheel. The goal salt content is 1.1 to 1.2 percent so as not to halt the growth of the propionic acid bacteria, which are salt sensitive.

Affinage: Remove from the brine, pat dry, and move to an aging environment at 50 to 55°F (10–13°C) and 90 percent RH for 10 to 14 days. Turn daily, and wash with a light brine as needed to prevent mold growth. Then move to a warmer environment of 68 to 75°F (20–24°C) and 85 percent RH for 3 to 4 weeks. Turn daily, and continue to wash with a light brine as needed. After eye formation is adequate, move the cheese back to the cooler aging environment, and age for 4 to 8 months. It can even finish aging at refrigeration temperatures of 38 to 40°F (3–4°C).

LARGE-BATCH GUIDELINES

Milk: 12 gal (46 L/100 lb) whole to partly skimmed milk
Culture: 4 g Thermo B plus 1 g LH 100 or 5 g Thermo C, ¼ tsp (0.9 g) propionic acid bacteria
Calcium chloride (optional): 10–15 ml (about 1 tbsp)
Rennet: 10 ml single-strength rennet (5 ml double strength)
Salt: Heavy brine

THINKING OUTSIDE THE VAT

SCENARIO 1: The recipe says to stir and cook the curd to 100°F (38°C) with a goal pH of 5.54. The vat temperature is 99°F (37°C), but the texture is not firm enough. The pH is already at 5.56. You should do which of the following?

A. Raise the temperature to 103°F (39°C), as this will slow the starter culture activity while also helping dehydrate the curds.
B. Stir a bit more rapidly.
C. Lower the temperature to 97°F (36°C), and keep stirring.
D. Accept that the pH will probably be too low, and try to make up for that during draining and pressing by keeping the room much warmer than usual.

SCENARIO 2: The cheese you are making has a goal temperature of 102°F (39°C), and the goal pH is 5.52 at the end of cooking. The temperature in the vat is 95°F (35°C), but the pH is already at 5.54. You should do which of the following?

A. Remove a portion of the whey and replace it with the same temperature water, as this will remove some lactose and slow pH development but not greatly alter texture.
B. Raise the temperature to the goal as quickly as possible.
C. Ignore the pH; it isn't as important as they want you to believe.

SCENARIO 3: You are making a washed curd cheese and realize that you have removed 50 percent of the whey, when the recipe says to remove only 25 to 30 percent. You do which of the following?

A. Add enough water back into the vat to equal the amount that was removed.
B. Add the amount of water that would have been the correct amount.
C. Make a note on the make sheet of what happened.

SCENARIO 4: When you remove your cheese from the press, the pH is at the goal, but the rind has not fully closed—it has a lot of small openings covering the surface. You can do which of the following?

A. Make a note in your logbook, and use more pressure the next time.
B. Ignore them, and know that you may have some mold intrusions in the final cheese.
C. Soak the wheel briefly in really hot water, then press again for a short time.

1. Answer: A and B.
2. Answer: A.
3. Answer: B and C.
4. Answer: A and C.

An 18-month-old Parmesan type by home cheesemaker Cheryl "Carly" Payne, Oregon

15: EXTRA-HARD GRATING CHEESES

FINISHED TEXTURE: Extra-hard
METHOD OF COAGULATION: Rennet
METHOD OF RIPENING: Internal bacteria

One of the most common things I hear new cheesemakers express is, "If I can just make cheddar, mozzarella, and Parmesan, my family (and therefore I) will be happy." These seem to be the domestic trinity of cheese for many people. And truly, these cheeses pretty much cover the major menu needs—a "slicer," a "melter," and a "grater." While it is reasonable to expect some pretty great cheddar and mozzarella from the home creamery, Parmesan is another story—if you are expecting anything that comes close to resembling the world-class, Italian PDO cheese, patience and strict attention to the processes are in order. And oh, yeah, did I mention patience?

Almost all the cheeses that can be classified as extra-hard (meaning a moisture content of about 34 percent or less) originated in Italy. They are all cheeses made using heat-loving bacteria, high cook temperatures, usually partially skimmed milk, and are normally aged for over 2 years. The large scale of the traditional wheels plays a key role in helping them develop their appealing granular (*grana* in Italian) texture and sweet, nutty, complex flavor.

Similar types made at the industrial scale of the United States and other countries are produced using varying standards. Usually, pasteurized milk from multiple sources without animal feed restrictions, such as time spent on pasture and the feeding of silage, is used; the cheeses are normally smaller and are aged for a shorter period of time; and often lipases (enzymes that break down fats and add a strong flavor) are added to simulate the effects of longer aging.

While many of the cheeses from the previous chapter will become extra-hard, with a very low moisture content and crumbly, dry texture, in this final recipe chapter, we'll focus on those that are usually made with the intention of their only being consumed once these traits have been attained.

THE FUNDAMENTAL SCIENCE OF MAKING EXTRA-HARD CHEESE

The grana cheese texture begins by creating a small, low-moisture curd in the vat. The milk is usually not ripened but is instead coagulated just after adding culture, then cut while the curd is still relatively soft. The curd size is quite small and shrinks to an even smaller size during the high-temperature cooking phase. You can expect whiter and more opaque whey during the cooking of these types of cheeses because of the liquefying of fat molecules and their loss into the whey (especially if you start with milk that is much higher in fat than it is in protein).

During the high-heat cooking, most, or all, of the coagulant enzyme will be deactivated. Any mesophilic bacteria in the milk will also be deactivated but will provide enzymes that assist with breakdown of protein during aging. The high heat also slows acidification, but during draining the large wheels will stay warm, and acid development will occur during this time. Both added culture and thermophilic bacteria present in raw milk will assist with acidification during draining and pressing.

The dry nature of the curd and its composition means that aging will occur at a slower rate than it does with other cheeses with more moisture. A long, slow aging at a slightly higher temperature and lower humidity will encourage the development of the desired granular texture and complex, round flavor.

Unlike some of the cheeses we have studied so far, there are not a lot of visible processes that we can discuss, observe, then attempt to control. Most of the magic that happens during the aging of an extra-hard grana-type cheese goes on unseen and at the molecular level. It is only through long aging that these processes can occur (remember that patience I mentioned earlier?).

Most grana-type cheeses rely upon proteolysis, the breakdown of proteins, rather than fats (lipolysis) for the complexity of flavor. The end product of proteolysis in Parmigiano Reggiano, in fact, makes it a very-high-protein cheese rich in free (meaning not joined with others) amino acids. Amino acids are also converted to a large number (at least 56!) of volatile compounds, which you can smell. The largest portion of these compounds comprises esters, ketones, and alcohols. Higher-fat varieties, such as Romano types, develop additional distinctive, piquant flavors thanks to the lipolysis of the fats.

Let's cover some of the choices and options you will be able to address when making these types of cheese. Each decision will have its effect on the outcome of the cheese and will allow you to develop some unique differences in your own version of the same cheese.

CHOOSING MILK

The Italian "Big Three," Parmigiano Reggiano, Grana Padano, and Asiago are all made from raw cow's milk partially skimmed to an average butterfat of 2.8 to 3.2 percent. The Romano family includes cheeses made from whole cow, goat, and, most famously, sheep milk, known as Pecorino Romano. Switzerland has several extra-hard varieties, including Sbrinz, a whole cow's milk cheese. Remember, you can make any of these recipes from other types of milk, but the results will be quite different.

The grana types can achieve the most desirable flavor and texture when made from high-quality raw milk. The high cook temperatures, extended aging, and low moisture make them very safe cheeses, but more importantly, they depend upon raw milk's natural lipase and bacteria to provide enzymes to assist in creating flavor and texture during aging. If you don't have a source for superior-quality raw milk, go ahead and use pasteurized. I'll be giving you some tips on adding other bacteria and enzymes to give the pasteurized milk a leg up during aging.

CHOOSING CULTURE

The recipes in this chapter all call for thermophilic starter and ripening cultures. Historically, these traditional cheeses are almost always made from whey culture—the bacteria-rich whey left over from the previous day's cheesemaking. These whey cultures provide bacteria that one can never duplicate from a packet, another reason it's difficult to duplicate one of the classic grana cheeses. Let's go over some of the good culture choices that are available to the average cheesemaker.

The top three in Table 15–1 are blends that are ready to use without adding any other cultures. The first is a premixed blend sold in small packets for home cheesemakers. I have used it, and it works great for several varieties of aged, hard, and stretched curd cheeses. The next culture after the first group is plain *Streptococcus thermophilis* (ST), the next two are strains that can be purchased separately so you can make your own blends, and the final group is not needed for raw milk but can add helpful enzymes and ripening potential for cheeses made from pasteurized milk. Don't forget, that culture amounts and ratios may vary depending if milk is raw, it may develop acid more quickly in the vat; or if components are high, it may develop acid more slowly and therefore benefit from some additional acid developing bacteria.

CHOOSING COAGULANT

While you can use any of the three main varieties of coagulant—veal (calf) rennet, fermented chymosin, or microbial—they each create somewhat different results in the cheeses. Not only does each break down proteins at a different rate, but chymosin (from traditional calf rennet or from fermented chymosin) is mostly deactivated during heating, while microbial rennet, *Rhizomucor miehei*, is not quite as heat sensitive. This means that the amount of enzymes that remain and are active during the aging of the cheese will be different and so will the breakdown of protein and the flavor profile of the resulting cheese. As with everything to do with cheesemaking, try different approaches, measure carefully, take good notes, then document your observations. This is the only way you will become a truly accomplished cheesemaker!

TABLE 15-1. CULTURE CHOICES FOR EXTRA-HARD CHEESES

Name	Contains	Manufacturer	Small-Batch Dosage:* per 2 gal (8 L/17 lb)	Large-Batch Dosage: 26 gal (100 L/220 lb)	Characteristics and Application
Thermophilic	ST, LB, LH	Packaged by home cheese supply co.	1 packet	N/a	All-inclusive blend for several Italian and stretched curd–type cheeses
Thermo B	ST, LB	Abiasa	¼–½ tsp (1.5–3 g)	5–10 g	A blend usually used for stretched curd, mozzarella, but also Parmesan
TM 81 (Choozit)	ST, LB	Danisco	⅛ tsp (0.6 g)	2.5–5U	A blend usually used for stretched curd, mozzarella, but also Parmesean
Thermo C	ST, LH	Abiasa	¼–½ tsp (1.5–3 g)	5–10 g	A blend for hard aged cheeses
TA 60 Series (Choozit) (includes TA 61, 62)	ST (fast-acid producer)	Danisco	⅛–¼ tsp (0.3–0.6 g)	5–10U	Use to make your own blends by combining with LH 100 or LB 340
LH 100 (Choozit)	LBL, LH	Danisco	Dosage ranges from 10–50% of that of TA culture		Use with TA 60
LB 340 (Choozit)	LB	Danisco	Dosage ranges from 10–50% of that of TA culture		Use with TA 60 for Parmesan
LBC 80 (Choozit)	LBC	Danisco	1/16 tsp (0.1 g)	0.3 g	Used to assist flavor and aroma development, especially in pasteurized milk varieties
MA Series (Choozit) (includes MA 11, 14, 16, 19)	LLC, LL	Danisco	1/16 tsp (0.1 g)	1U	Mesophilic blend can help provide ripening for pasteurized milk varieties
Meso III	LLC, LL	Sacco	1/16 tsp (0.1 g)	1U	Mesophilic blend can help provide ripening for pasteurized milk varieties
Meso II	LLC	Sacco	1/16 tsp (0.1 g)	1U	Mesophilic blend can help provide ripening for pasteurized milk varieties

*Remember that culture is always best measured by weighing and dividing into unit doses based on the strength of the culture, instead of using volume or set weight.

USING LIPASE

When adding lipase to the piquant varieties such as Romano, remember that you will have at least three choices—calf, kid, and lamb lipase. They each have slightly different enzymes and therefore create different flavor profiles. Calf is the mildest and lamb the strongest. You can also adjust the amount added. For more on lipase, see chapter 1.

Curds cut to "wheat grain" size for grana-type cheeses

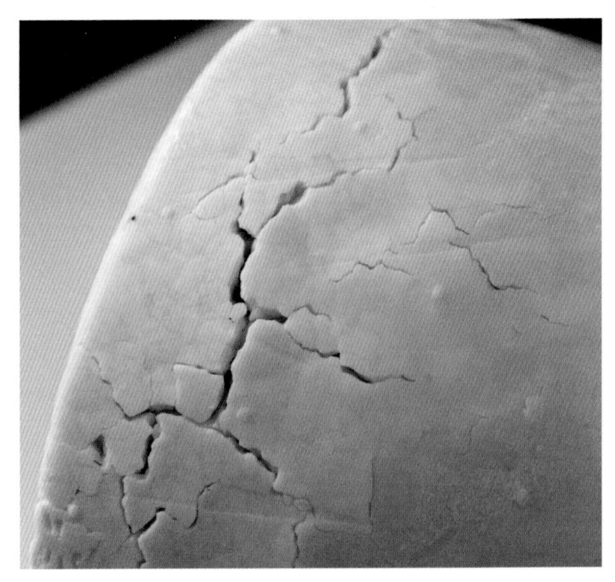

When curd is cooked too long in the vat, the wheel may come apart during brining.

CUTTING THE CURD

Curd is usually cut to the small size of wheat grains (about 0.1 inch or 3 mm). Often the cutting takes place slowly over time, starting with larger curds, then cutting them down to the goal size over many minutes. This allows for some initial whey drainage and firming of the curds. The curd size must be small before heating takes place, or whey will become trapped inside the curds when the high heat "cooks" the outside of the curd. Small curds will heat evenly. So pay attention to curd size as you are cooking, and immediately cut individual curds that are too large down to the goal size. If, at the end of cooking, there are any that are still too large, remove them if possible.

STIRRING AND COOKING

The temperature is increased steadily toward the goal over a fairly short time once the goal curd size has been reached. If you are working on a larger scale and do not have adequate heat added to the vat, it can be difficult to attain the final temperature goal. You can cheat a little: Remove a small portion of the whey when the temperature begins to stall in the vat, heat the whey on a stove to about 150°F (66 °C), then add it slowly back into the vat. Once the goal temperature is reached, it is usually held only briefly—long enough to ensure that the curds have reached the correct internal temperature.

Doing a texture test at this point is a bit different from what you would do with lower-temperature-cooked curds, as the high heat makes for a very plastic, malleable curd that is quite ready to knit back together. So don't expect the same feeling. Tear a few curds, or chew on them, and see if they feel consistent in texture throughout.

DRAINING AND PRESSING

Whey is drained from the vat or pot and curd is sometimes pressed in the vat or under the whey to consolidate the mass and reduce the possibility of tiny openings in the finished cheese. These mechanical-type openings are not desirable for extra-hard cheeses, and not only for aesthetic reasons: The little openings will trap whey during draining that will lead to more acid development in that area of the cheese after it is finished (postacidification). This will cause discoloration and poor texture.

As with most cheeses initial pressing should be lighter at first, then firmer so as to not close the rind of the cheese before any pockets of whey can be expelled. Once most of the whey has been removed, then you can increase the pressure to a higher level. A mechanical press might be necessary to accomplish the right texture, but the initial warmth of the curd, along with a form that drains well, can make pressing these cheeses with stacked weights quite possible.

Wheels are pressed until the correct pH is attained. Large wheels can take up to 24 hours, but this is quite dependent on room temperature and how much starter culture is still active in the curd. Remember the high-heat treatment will deactivate or slow much of the starter activity, which is one reason that the wheels are pressed for longer. Large wheels should be moved to the brine at a pH of about 5.5 and smaller forms at 5.4. During salting, the larger the size, the slower the uptake of salt and therefore the longer the pH will continue to develop.

SALTING

Grana-type cheeses can be brine- or dry-salted. They typically have a higher finished salt content than other hard cheeses, usually between 3 and 5 percent. Some are brined, then also rubbed with dry salt for several weeks. The large size of the wheels means that the salt concentration will initially be very high at the surface and will take some time to reach the center of the cheese, hence the longer salting period needed to help ensure that the center of the cheese will eventually have the right level of salt. The higher salt content also slows the breakdown of proteins and fats by limiting enzymatic activity, another reason that a long period of aging is necessary.

Brining times for grana types range from 7 to 12 hours per pound (14–24 hours per kg). It may seem as though this is too long, but because of the very dry nature of the curd, the salt uptake is slower. Don't forget to dry-salt the tops initially and turn the cheeses during brining.

AGING AND STORING

Plan on at least 12 months of aging for even a small wheel of these types of cheese—if you want to see the best results, that is! Larger wheels can age for 2 or more years. Aging is traditionally done at 64°F (18°C), quite a bit warmer than the usual aging temperature of 55°F (13°C). If you have a regular cellar or basement, these types might do quite well. For commercial production, I recommend aging at 55°F (13°C). If you must age them with your other cheeses (as I must), you have to settle for the lower temperature and understand that it will take even longer and won't have quite the same results as traditional versions. In addition, the humidity is usually kept lower, at 80 percent RH instead of the usual 85 to 95 percent. If you must age them at the higher relative humidity, expect a texture that is not quite as dry and be aware that you will be dealing with more mold growth on the rind during aging.

The recipes in this chapter offer you a great chance to compare the overarching sameness of these types alongside the myriad ways that such differences as milk fat, addition of lipase, changes in starter culture, changes in cooking temperature, and salt content can create unique, vastly different cheeses. As you make these recipes, try to compare them, and ask yourself what characteristic is changed due to small differences in the directions.

Romano-style cheese

CHERYL "CARLY" PAYNE, HOME CHEESEMAKER, OREGON

I have known Cheryl "Carly" Payne for about 7 years now, thanks to our shared love of Nigerian Dwarf goats and cheesemaking. Carly is not only a great friend, but she is one of the inspirations for writing this book: Her passion and desire to learn more are shared by so many for whom cheesemaking may not be a profession but is very much a part of their lives. She makes cheese for the most primal reason, the one without which cheese would not even exist: to preserve food for the future. The first time she was introduced to dairy goats, she says, it was a changing point in her life. She had not known that one animal could provide so much—from companionship to milk and cheese, and even meat. She was hooked.

Carly started making cheese while living in Alaska, where she was a professional dog groomer and champion dogsled racer. Living in Alaska as a goat farmer provided her with many tales that make any complaints about the weather in the Lower 48 a bit shameful—from blow-drying baby goat ears so they wouldn't become frostbitten (after being born at -20°F to having to wait until summer to be able to dig a deep enough grave to bury a lost charge. The harsh beauty of that environment also engrained in her a deep need to preserve food. She was even inspired by a story of Mongolian herders who dry their cheeses on the roofs of their yurts and modified the method to use a food dehydrator to preserve sticks of feta for portable, sustaining snacks.

She says that the fact that cheesemaking can be "as easy as it can be or as difficult as it can be" led her to focus her cheesemaking, for several years, on perfecting a Parmesan type made from the full-fat milk of her Nigerian Dwarf goats. The beautiful results of her work are featured in a photo earlier in this chapter.

Home cheesemaker Cheryl "Carly" Payne with one of her foundation milking does, Tiger Lily

CLASSIC ITALIAN-STYLE EXTRA-HARD CHEESE

Where would we be without the brilliance that is Parmigiano Reggiano, Italy's iconic grana cheese? From the fertile Emilia-Romagna region of northern Italy and the province of Reggio comes a cheese that is often called the king of all cheeses. The production of Parmigiano Reggiano is strictly controlled to ensure that the cheese remains superior—from restricted months of production to ensure the finest-quality food to a minimum aging time of 14 months. The entire process guarantees that when the name Parmigiano Reggiano is used the cheese will deliver what is promised: complex, amazing, melt-in-your-mouth, cheesy heaven.

Following close in quality is aged Grana Padano, made in surrounding provinces. Both Grana Padano and similarly made Asiago (from farther north) can be sold at different ages, meaning that the texture can vary from semihard to extra-hard, and the flavor will be more variable as well. All these factors also mean that both Grana Padano and Asiago are more affordable. All three of these Italian cheeses have PDO status. While many US companies make cheeses that bear the names Parmesan and Asiago, it is pretty tough to duplicate what *terroir* and a civilization have created.

INGREDIENTS

Milk: 4 gal (15 L/34 lb) partly skimmed milk (ideal fat content: 2–2.8%)
Culture: ½ tsp (3 g) Thermo B (or equivalent blend)
Calcium chloride (optional): maximum 1 tsp (5 ml) calcium chloride diluted in ½ cup (120 ml) cool, nonchlorinated water
Rennet: ¾ tsp (3.75 ml) single-strength rennet diluted just before use in ½ cup (120 ml) cool, nonchlorinated water
Salt: Heavy brine

STEPS

Prepare Equipment: Make sure all equipment is cleaned and sanitized and that your cheesemaking space is free from possible contaminants. Refer to chapter 6 for tips on proper equipment preparation.

Prepare Milk: Warm milk to 88 to 90°F (31–32°C).

Culture: Sprinkle cultures on top of milk. Let set for 2 to 5 minutes, then stir gently for 3 to 5 minutes.

Additions: Stir in calcium chloride solution. Wait 5 minutes.

Rennet and Coagulate: Stir in rennet solution with an up-and-down motion for 1 minute. Still milk. Hold temperature until clean break is achieved. The goal coagulation time is 20 to 30 minutes. If using flocculation as a guide, multiply the time to flocculation by 2 to 2.5 for the total time from adding rennet to cutting.

Cut: Cut curd mass into ¼-inch (6-mm) cubes slowly over 10 minutes. Maintain temperature at 88 to 90°F (31–32°C).

Stir: Stir gently for 10 to 15 minutes without changing temperature. Curds will shrink during this phase and will move easily in the vat. Very little acid development will take place at this temperature.

Stir and Heat: Continue to stir gently, and slowly increase the temperature to 108°F (42°C) over 30 minutes. Hold, and stir for 5 minutes. Continue to stir, and increase the temperature to 124 to 128°F (51–53°C) over 30 minutes. Let settle for 5 minutes. Goal whey pH is 6.5 to 6.4.

TIPS: It is difficult to meet these goal time and temperature parameters perfectly. If your make strays from them a bit, don't worry, but try to adjust so the total time and temperature goal is met. Curds will shrink to a very small size and be quite springy.

Drain and Press: Drain whey completely. Put curd in cloth-lined mold, and place in a mechanical press. Press with just enough pressure to feel the resistance of the curd and see some whitish, translucent whey draining from the form. After 15 minutes unmold, turn, and place back in lined mold. Press with more weight, watching that the draining whey is not too opaque. After 30 minutes, unmold and turn, place back in lined mold. Press with more weight. At this point very little whey will be draining. Increase weight as needed to continue to close the rind. Press for 6 to 12 hours. Goal pH of curd by the end of pressing is 5.5 to 5.4.

TIPS: A mechanical press is needed to knit the dry, small curds of this type of cheese back together. As with other high-pressure pressed cheeses, watch the whey and rind during turning to determine if enough pressure is being used.

Salt: Unmold, and immerse in heavy brine for 12 hours per pound (6 hours per kg) of the weight of a wheel. Turn in the brine at least one time during salting, and dry-salt the exposed tops.

Affinage: Remove from brine, and pat dry. Move to an aging environment at 55°F (13°C) and 80 percent RH for 8 to 18 months (larger wheels can age for much longer). Turn daily for the first month, then weekly. Rinds can be brushed and oiled, or waxed.

LARGE-BATCH GUIDELINES

Milk: 12 gal (46 L/100 lb) partly skimmed milk (ideal fat content: 2–2.8%)
Culture: 2–5 g Thermo B (or equivalent blend)
Calcium chloride (optional): 10–15 ml (about 1 tbsp)
Rennet: 8–11 ml single-strength rennet (4–5.5 ml double strength)
Salt: Heavy brine

EXTRA-HARD WASHED CURD CHEESE

Some say that Switerland's Sbrinz, a whole-milk, extra-hard cheese of similar complexity and flavor to Parmigiano Reggiano, is actually the oldest of the grana-type cheeses. Sbrinz varies from Parmesan types in several ways: For one thing it is made using whole milk, instead of partly skimmed, and some instructions call for washing the curd. The huge, 80-pound (36 kg) wheels are aged for 2 to 4 years, resulting in a mouth-watering complexity and granular, extra-hard texture. While California's Vella Cheese Dry Jack is made in a somewhat different style, the resulting cheese is quite on a par. As with the other members of the extra-hard family, if you try to duplicate them, patience will be one of the most important ingredients you can have.

INGREDIENTS

Milk: 4 gal (15 L/34 lb) whole milk (ideal fat content: 3.75–3.9%)
Culture: ½ tsp (3 g) Thermo B (or equivalent blend)
Lipase (optional): ¼ tsp (1.4 g) lipase dissolved in ¼ cup (60 ml) cool, nonchlorinated water
Calcium chloride (optional): maximum 1 tsp (5 ml) calcium chloride diluted in ¼ cup (60 ml) cool, nonchlorinated water
Rennet: ¾ tsp plus (4 ml) single-strength rennet diluted just before use in ¼ cup (60 ml) cool, nonchlorinated water
Salt: Heavy brine

STEPS

Prepare Equipment: Make sure all equipment is cleaned and sanitized and that your cheesemaking space is free from possible contaminants. Refer to chapter 6 for tips on proper equipment preparation.

Prepare Milk: Warm milk to 80°F (27°C).

Culture: Sprinkle cultures on top of milk. Let set for 2 to 5 minutes, then stir gently for 3 to 5 minutes.

Additions: Stir in lipase solution, then stir in calcium chloride solution. Wait 5 minutes.

Rennet and Coagulate: Stir in rennet solution with an up-and-down motion for 1 minute. Still milk. Hold temperature until clean break is achieved. The goal coagulation time is 35 to 40 minutes. If using flocculation as a guide, multiply the time to flocculation by 2 to 2.5 for the total time from adding rennet to cutting.

Cut: Cut curd mass into ¼-inch (6-mm) cubes, then to ⅛-inch (3-mm) pieces over 10 to 15 minutes.

Stir and Heat: Stir gently, and increase temperature to 129°F (54°C) slowly over 45 minutes.

Wash: Remove 1 to 2 quarts of whey (12–15% of total volume), stir, and add cold water to lower temperature by 3 to 4°F (2°C). Stir for 5 minutes. Let settle in vat for 20 to 30 minutes. Goal whey pH at the end of this phase is 6.4 to 6.5.

Drain and Press: Drain whey completely. Place curd in cloth-lined mold, and place in a mechanical press. Press with just enough pressure to feel the resistance of the curd and see some whitish, translucent whey draining from the form. After 15 minutes, unmold, turn, and place back in lined mold. Press with more weight, watching that the draining whey is not too opaque. After 30 minutes unmold and turn, place back in lined mold. Press with more weight. At this point very little whey will be draining. Increase weight as needed to continue to close the rind. Press for 6 to 12 hours. Goal pH of curd by the end of pressing is 5.5 to 5.4.

TIPS: A mechanical press is needed to knit the dry, small curds of this type of cheese back together. As with other high-pressure pressed cheeses, watch the whey and rind during turning to determine if enough pressure is being used.

Salt: Unmold, and immerse in heavy brine for 6 hours per pound (12 hours per kg) of the weight of a wheel. Turn in the brine at least one time during salting, and dry-salt the exposed tops.

Affinage: Remove from brine, and pat dry. Move to an aging environment at 55°F (13°C) and 80 percent RH for 8 to 18 months (larger wheels can age for much longer). Turn daily for the first month, then weekly. Rinds can be brushed and oiled, or waxed.

LARGE-BATCH GUIDELINES

Milk: 12 gal (46 L/100 lb) whole milk (ideal fat content: 3.75–3.9%)
Culture: 2–5 g Thermo B (or equivalent blend)
Calcium chloride (optional): 10–15 ml (about 1 tbsp)
Rennet: 11–13 ml single-strength rennet (5.5–6.5 ml double strength)
Salt: Heavy brine

A girolle is a tool used to cut rosettes from the top of a hard cheese.

EXTRA-HARD PIQUANT CHEESE

This spicy, piquant group is built on the strengths of the high-fat milk of sheep and traditional methods that introduce lipase (enzymes that break down fat) into the process. Traditional Italian Pecorino Romano, Siciliano, and Sardo PDO cheeses are hard to extra-hard, salty, and develop a distinctive aroma and flavor, thanks to the breakdown of milk fat. Caprino Romano is made from goat's milk and Vacchino Romano from cow's milk. If you decide to create your own Romano-type cheese, remember that you can vary its salt level and sharp, piquant flavor to suit your own tastes.

While Romano cheese originated in and around the Italian capital of Rome (hence the name Romano), only one regional producer, Fulvi, still produces Romano in that region. Most other producers are on the island of Sardinia (where Pecorino Sardo is also made). The Kurdish cheese Awshari is a piquant sheep's milk cheese of similar composition but made with far more basic techniques, such as pressing and kneading by hand and aging in salted animal skins.

A stack of aged, piquant-style hard cheeses

INGREDIENTS

Milk: 4 gal (15 L/34 lb) whole, high-fat milk (ideal fat content: 6.8%)
Culture: ½ tsp (3 g) Thermo B or C
Lipase: ¼–½ tsp (1.4–2.8 g) lipase (needed for traditional flavor)
Calcium chloride (optional): maximum 1 tsp (5 ml) calcium chloride diluted in ¼ cup (60 ml) cool, nonchlorinated water
Rennet: ½ tsp plus (3 ml) single-strength rennet diluted just before use in ½ cup (120 ml) cool, nonchlorinated water
Salt: Pure salt

STEPS

Prepare Equipment: Make sure all equipment is cleaned and sanitized and that your cheesemaking space is free from possible contaminants. Refer to chapter 6 for tips on proper equipment preparation.

Prepare Milk: Warm milk to 100 to 104°F (38–40°C).

Culture: Sprinkle cultures on top of milk. Let set for 2 to 5 minutes, then stir gently for 3 to 5 minutes.

Additins: Stir in lipase solution, then stir in calcium chloride solution. Wait 5 minutes.

Rennet and Coagulate: Stir in rennet solution with an up-and-down motion for 1 minute. Still milk. Hold temperature until clean break is achieved. The goal coagulation time is 15 to 20 minutes. If using flocculation as a guide, multiply the time to flocculation by 2 to 2.5 for the total time from adding rennet to cutting. The coagulation time is fast because of the higher temperature of the milk.

Cut: Cut curd mass into ⅜-inch (1-cm) cubes.

Stir: Maintain 100 to 104°F (38–40°C), and stir gently for 5 minutes. Let rest until curds are covered by a 4–5-inch (10–13-cm) layer of whey, usually about 10 minutes. Stir, and cut the curd down to ⅛-inch (3-mm) pieces.

> TIPS: Adjust the rest period between stirring and the second cutting, depending on how the curds want to mat. If they mat too much, the second cutting will shatter them. Either stir for longer than 5 minutes, or let rest for less than 10 minutes to decrease the likelihood of matting.

Stir and Heat: Stir gently, and increase temperature to 113 to 118°F (45–48°C) over 15 minutes. Hold temperature, and stir for 15 to 20 minutes until curds pass texture test (see "Checking Curd for Readiness" in chapter 2, page 52). Let settle in pot for 5 minutes.

Drain and Texture: Remove whey to level of curds. Gently hand-press the curds under the whey for about 5 minutes to form a consolidated mass. Drain off remaining whey. Goal whey pH by the end of draining is 6.3–6.4.

Press: Place curd mass in a cloth-lined form. Fold cloth over the top of curd, and set follower on top. Let drain without pressure for 15 minutes. Unmold, and turn cheese; cover, and let drain without pressure for 15 minutes longer. During the 15 minutes use a long meat skewer or knitting needle to pierce the mass several times to assist whey drainage. Unmold, turn, and begin adding weight to form the wheel and close the rind. Press for 30 minutes. Unmold, turn, and add more weight as needed to close the rind after 1 hour more of pressing. Continue to press to goal pH of 5.3 to 5.4, usually about 4–10 hours.

Salt: Unmold, and dip the wheels in heavy brine, then rub with a coating of salt. Pour brine over the cloths, and redress the wheel. Return to molds, and continue to press for 12 to 24 hours. Remove from forms, rub with dry salt, and press for 24 hours. Repeat one more time. Room temperature during the salting should be 55°F (13°C).

> TIPS: Romano is traditionally a very salty cheese with a final content of 5 to 7 percent (about twice as salty as similar extra-hard cheeses). If you don't like it, you can modify the salting to your tastes.

Affinage: Move to an aging environment at 55°F (13°C) and 85 percent RH for 8 to 12 months. Turn daily for the first month, then weekly. Rind can be rubbed with dry salt or medium brine as needed. Traditionally, Romano was rubbed with dark clay mixed with oil. As an alternative you can mix cocoa powder and oil and create a similar rind effect.

LARGE-BATCH GUIDELINES

Milk: 12 gal (46 L/100 lb) whole, high-fat milk (ideal fat content: 6.8%)
Culture: 2–5 g Thermo B or C
Lipase: 1 tsp (5.6 g)
Rennet: 7–9 ml single-strength rennet (3.5–4.5 ml double strength)
Salt: Goal content 5–7%

EXTRA-HARD CHEESE WITH EYES

Made in the Italian Alps, Montasio is a PDO cheese created from raw whole (sometimes partly skimmed) cow's milk, extra hard and with eyes. According to Steve Jenkins in his book *Cheese Primer,* Montasio originated as a monastery cheese, made from the milk of cows pastured on a terraced mountain of the same name. Montasio wheels are good sized at 11 to 19 pounds (5–9 kg) and aged for about a year. Small eyes occur evenly throughout the paste. The cheese can be made from whole or partly skimmed milk. Similarly, kefalotyri is a salty, extra-hard Greek or Balkan cheese made from goat's and sheep's milk. It too has numerous eyes throughout the white, dry paste. Kefalotyri, however, should be a little less salty and piquant than Romano. Eye formation is irregular and due to naturally occurring bacteria in the raw milk.

INGREDIENTS

Milk: 4 gal (15 L/34 lb) whole or partly skimmed milk
Culture: ½ tsp (3 g) Thermo C (or equivalent blend) and 1 skewer tip propionic acid bacteria (optional)

Oiling the rind of an aging Montasio-style cheese

Calcium chloride (optional): maximum 1 tsp (5 ml) calcium chloride diluted in ¼ cup (60 ml) cool, nonchlorinated water
Rennet: ¼ tsp plus (1.5 ml) single-strength rennet diluted just before use in ¼ cup (60 ml) cool, nonchlorinated water
Salt: Heavy brine

STEPS

Prepare Equipment: Make sure all equipment is cleaned and sanitized and that your cheesemaking space is free from possible contaminants. Refer to chapter 6 for tips on proper equipment preparation.

Prepare Milk: Warm milk to 95°F (35°C).

Culture: Sprinkle cultures on top of milk. Let set for 2 to 5 minutes, then stir gently for 3 to 5 minutes.

Additions: Stir in calcium chloride solution. Wait 5 minutes.

Rennet and Coagulate: Stir in rennet solution with an up-and-down motion for 1 minute. Still milk. Hold temperature until clean break is achieved. The goal coagulation time is 30 to 40 minutes. If using flocculation as a guide, multiply the time to flocculation by 2 to 2.5 for the total time from adding rennet to cutting.

Cut: Cut curd mass into ⅜-inch (1-cm) cubes, then continue to stir, and cut into ¼-inch (6-cm) pieces.

> TIPS: Traditional kefalotyri is cut by stirring with a wooden rake. You can try using a wire whisk with widely spaced wires (I cut two wires from a large whisk to give them better spacing).

Stir and Heat: Stir gently, and increase temperature slowly to 110 to 113°F (43–45°C) over 30 to 40 minutes. Let settle for 10 to 15 minutes. Curd texture should be even, slightly springy, and moister than the extra-hard cheeses earlier in this chapter.

Drain and Pitch: Drain whey to the level of the curds, then hand-press curds gently under the whey to consolidate into a single mass.

Press: Place mass into a cloth-lined mold. Knead and hand-press curd into mold. Add weight equal to that of the curd, and press for 15 minutes. Unmold, turn, and repress with weight equal to twice that of the curd for 30 minutes. Unmold, turn, and continue to press with as much weight as needed to close the rind after 1 more hour of pressing. Unmold, turn, and continue to press to a curd pH of 5.0 to 5.3, usually about 6 to 12 hours.

Salt: Unmold, and immerse wheels in heavy brine for 4 hours per pound (8 per kg) of the weight of a wheel.

Affinage: Pat dry, and move to an aging environment at 55 to 59°F (13–15°C) and 85 percent RH for 6 to 12 months. Turn daily for the first month, then weekly. Wash rinds with a light brine, or rub with dry salt for the first 6 weeks of aging.

> TIPS: Traditionally, multiple wheels are stacked on each other and rotated in the stack during aging.

LARGE-BATCH GUIDELINES

Milk: 12 gal (46 L/100 lb) whole or partly skimmed milk
Culture: 2–5 g Thermo C (or equivalent blend) and ½ g propionic acid bacteria (optional)
Calcium chloride (optinal): 10–15 ml (about 1 tbsp)
Rennet: 7–9 ml single-strength rennet (3.5–4.5 ml double strength)
Salt: Heavy brine (goal content 3.7%)

THINKING OUTSIDE THE VAT

SCENARIO 1: You want to make a cheese that is hard and dry but with lots of flavor. Some options include which of these?

A. Aging the wheel for a long period
B. Adding lipase to the milk
C. Using superior-quality raw milk
D. Including some thermophilic culture in the starter bacteria

SCENARIO 2: You want to make a mild cheese that is semisoft to semihard and ready to be consumed at 2 to 3 months of age. You can try which of the following?

A. Wax or vacuum-seal the cheese to keep it moist
B. Use a washed curd technique
C. Choose a culture with flavor- and aroma-producing bacteria
D. Cut the curd a bit larger

1. Answer: A, B, C, and D.
2. Answer: A, B, C, and D.

APPENDIX A: TROUBLESHOOTING GUIDE

TABLE A-1. PROBLEMS COMMON TO ALL CHEESE TYPES

Symptom	Likely Causes	Solutions
Bitterness	Low salt level, incorrect proteolysis pattern, inadequate aging, excessive use of calcium chloride	Measure salt or brine more carefully; analyze process for things leading to improper proteolysis, such as psychrotrophs in milk, outdated rennet, starter culture amount and type incorrect for milk and cheese type; aging time may be inadequate; reduce calcium chloride volume
Rancidity	High levels of free fatty acids from anything that causes lipolysis at a level beyond that desired	Limit the following: damaged milk fat globules, raw milk (with its intact lipase), psychrotophic bacteria that can produce lipases, elevated temperatures during aging, molds in the milk or introduced during the process
Early blowing: curd or cheese becomes frothy with gas within the first stages of cheesemaking	Growth of coliforms or yeasts (or both)	Good hygiene during milk collection, rapid chilling of milk, ensuring proper acid development (over ideal time span) by starter culture during cheesemaking

TABLE A-2. PROBLEMS WITH FRESH ACID- AND HEAT-COAGULATED CHEESES

Cheese Type	Symptom	Likely Causes	Solutions
Added or developed acid/high heat	Too tangy	Too much added acid	Try less next time or add a tiny amount of baking soda when curds are drained (but soft) to neutralize the acid
Added or developed acid/high heat	Too dry	Drained too long or milk too low in fat	Drain less time, use higher-fat milk, or stir in a bit of cream or butter when curds are drained but still soft
Added or developed acid/high heat	Pressed curd falls apart	Not enough time and/or pressure during draining, pressed curd is too warm	Increase time of pressing and final weight of pressing, cool curd before slicing
Developed acid, long set, long drain	At end of ripening, curd looks spongy or foamy	Contamination by coliforms' "early blowing"	Improve milk collection sanitation, chill milk more rapidly, and monitor pasteurization better
Added or developed acid/high heat	Curd is set, but pH is too high (not tangy enough)	Too much rennet, incorrect ripening temperature	Use less rennet solution next time, monitor ripening temperature
Added or developed acid/high heat	pH is too low (curd is too sour)	Too much culture or native bacteria in milk; ripening and/or draining temperature too high or too long	Decrease culture amount or pasteurize milk, monitor ripening temperature and time
Added or developed acid/high heat	Curd is chunky after draining	Curd at outside of draining bag drained quickly and trapped whey inside curd at center	Decrease room temperature during draining, or stir curd mass partway through draining

Cheese Type	Symptom	Likely Causes	Solutions
Curd Type	At cutting, curd looked ready but broke into little pieces when cut or stirred	pH too low or need to add calcium chloride	Monitor pH and temperature during draining, adjust temperature and amount of starter culture, add calcium chloride next time
Curd Type	At cutting, pH is right but is too firm, pH is not low enough, but curd texture is right	Too much rennet or not enough starter culture activity	Use less rennet or adjust starter culture volume and ripening temperature
Curd Type	At end of cooking time, curds are firm on the outside but mushy inside	Curds were heated too quickly in the beginning	Heat more slowly in the beginning, and stir more frequently over entire cooking
Curd Type	Finished curd is squeaky	Curd has not mellowed	Let set in fridge for a day before eating, squeak will go away
All	Soapy flavor and/or rancidity (see also Table A–1 for all cheese types)	Free fatty acids present due to breakdown of fat by lipolytic enzymes	Ensure milk is chilled rapidly and very clean to limit psychrotrophes and contaminating bacterial growth, mechanical abuse of milk prior to cheesemaking (pumping, agitation, etc.)
All	Chalky, gritty, or grainy texture	Incorrect finished pH	Monitor pH better
All	Oxidized, cardboard flavor	Exposure to oxidizing environment	Limit exposure to metals that are reactive and to sunlight and ozone generator (if in use)

TABLE A-3. PROBLEMS WITH WHITE MOLD SURFACE-RIPENED CHEESES

Symptom	Likely Causes	Solutions
No or too little white mold growth	Surface of cheese is too moist or too dry, lack of air exchange in environment, too much or too little *Geotrichum* or other yeast growth (cheese surface pH remains too low)	Dry the cheeses more carefully, improve air exchange in aging area, use yeasts in culture inoculation, spray on additional *P. candidum* mold spores
Texture too firm at end of aging	pH not low enough at end of draining, humidity too low during aging, not enough air exchange in aging area, high fat content of milk	Track pH during making, ensure proper humidity and air exchange in aging area, choose lower fat milk, wrap the cheeses earlier, choose correct cheese wrap paper
Texture too runny at end of aging, especially close to surface	Excessive *Geotrichum* growth, presence of other highly proteolytic microorganisms, too low pH (less than 4.6) at end of draining, too high a ripening temperature	Reduce or omit *Geotrichum* or choose different strain, heat treat or pasteurize milk to remove wild strains of yeasts and molds, monitor pH better, wrap cheeses and reduce ripening temperature once mold growth on surface is established
"Toad" skin (wrinkly, thickened rind)	Too much *Geotrichum*, too much salt, too high a ripening temperature, too moist at end of drying phase	Reduce or omit *Geotrichum*, measure salt amounts carefully, lower ripening temperature, ensure proper drying phase
Mucor (cat hair) mold growth	Too much moisture, pH too high, poor air exchange, contamination of environment	Choose mucor-resistant *Penicillium* strain, monitor pH carefully at end of draining, ensure proper drying phase, improve air exchange, ensure environment is free from unwanted molds
Bitterness	Breakdown of proteins to bitter peptides by *P. candidum* or other enzymes, such as from rennet	Wrap cheese when adequate mold growth appears, and slow ripening by moving to a lower temperature environment, use correct balance of *Geotrichum* to *Penicillium* or try different strains, use correct amount of starter culture
Ammonia odor	Production of ammonia by molds that are trapped close to cheese, simply overripe	Use breathable wrapping paper and ensure adequate air exchange, reduce storage temperature closer to refrigeration temperature (not freezing)

TABLE A-4. PROBLEMS WITH WASHED RIND SURFACE-RIPENED CHEESES

Symptom	Likely Causes	Solutions
No surface bacteria growth (no red color)	Surface pH too low, *B. linens* or other coryneform not present, too cool a ripening temperature, too low humidity	Add yeast culture to milk next batch, include yeast culture in washing solution, wash with coryneform solution more frequently, ensure proper aging temperature
Other molds growing on surface	Surface too hospitable for unwanted molds, contamination from environment	Increase washing frequency, use 3–5% or greater strength brine solution to discourage salt sensitive molds, ensure environment is free from unwanted molds
No softening by the end of aging	Surface bacteria lacking in activity, humidity too low	Ensure proper surface bacteria growth, increase humidity in aging environment

TABLE A-5. PROBLEMS WITH BLUE CHEESES

Symptom	Likely Causes	Solutions
No or little blue growth	Not enough oxygen, too high salt content, presence of *Geotrichum candidum* (only affects outside of cheese, as it needs more oxygen than blue does)	Add gas-producing starter to culture to help create air pockets, pierce more or sooner, ensure proper salt balance in cheese (1–3% in paste at beginning of ripening), remove contact with *Geotrichum* through other cheeses and aging environment (if not possible, choose to make a blue with many air pockets inside but a more closed rind to allow blue to get good growth before piercing occurs)
Brown spots on surface	Contaminating yeasts or molds	Use a light vinegar and brine solution to limit or remove growth on rind
Bitterness	Less than ideal proteolysis pattern during aging	Ensure proper aging conditions, consider using traditional rennet (suggested by many blue cheese makers)

TABLE A-6. PROBLEMS WITH STRETCHED CURD CHEESES

Symptom	Likely Causes	Solutions
Lack of stretch	pH too high, mineral content too high, temperature too low	Verify proper acidification, increase acidification development rate to ensure reduction of calcium content, increase temperature of curd during stretching

TABLE A-7. PROBLEMS WITH SEMIHARD TO EXTRA-HARD CHEESES

Cheese Type	Symptom	Likely Causes	Solutions
All: less likely in salted-curd cheeses such as cheddar	Late blowing (cheese expands with gas production later in aging)	Fermentation of lactate by *Clostridium* (usually *C. tyrobutyricum*) to butyrate and gas	Good hygiene during milk collection, avoiding the feeding of improperly fermented silage, proper salt levels in cheese, reduced ripening temperatures; some producers add product called lysozyme
Cheese with propionic eyes	No eye formation	Salt content over 1.2%, pH at draining below 6.3, residual lactose at salting leading to postacidification and crumbly texture that cannot trap gas	Monitor salting and pH at beginning of draining and just prior to salting
Cheese with eyes	Excessive eyes, slits, and cracks	Late fermentation	Reduce or limit use of *Lactobacillus helveticus*

APPENDIX B: RESOURCES FOR EDUCATION, SOURCING, AND ENJOYMENT

Rather than simply create a list of recommendations, I wanted to take the time to share a few more details about these resources, including my own review of those that I have any personal knowledge of or experience with. Keep in mind that things change, including my opinions, so I hope anything I share here won't come back to haunt me!

For some of these I haven't provided a web link as it is quite easy to search for them. If you aren't online, some of the resources will be very tough to access. Remember, you can use your public library as a resource for Internet searches and help finding things on the Web. For books it is tough to beat Amazon. Even most authors can't afford to sell their own books for as low as you can find them at Amazon.com. It is also a great resource for other products these days as well.

There was no way for me to adequately address resources for every country. While many of the books and products listed in this section are available across the globe, access to some will be limited. For this I apologize and hope that you will utilize some of the Internet forum resources to help you locate products that you can access in your part of the world. On that same note my book recommendations are pretty much limited to the English-speaking audience. My apologies to readers from other countries.

EDUCATION AND ENJOYMENT

There are a growing number of resources for cheesemakers wanting to continue to learn (and let's hope that is all of us!). It seems as though every few months a new cheesemaking book comes on the scene. Keep in mind that when a topic is popular, it is a publisher's job to try to capture their own little part (or preferably large part) of that market. If you like a publisher's books in general, chances are that they will produce a good book on cheese too. But some will not take the time and commit the resources to selecting the proper author, fact checking, and even providing appropriate illustrations.

The same is true of classes, seminars, and events. Some will be created simply to be a part of a thriving movement, and others will have real depth and worth. This is part of the price of popularity. So don't feel too overwhelmed with the wonderful plethora of options, and do your best to get a recommendation from a trustworthy source if you are going to be spending some money and valuable time to participate.

BOOKS AND PUBLICATIONS

FOR THOSE JUST GETTING STARTED

Home Cheesemaking by Ricki Carroll (Storey, 2002)
This was the first book on the scene, published originally in 1982 as *Cheesemaking Made Easy*. Since then it has sold at least 100,000 copies. There is a good reason this book remains so popular: It is great! It still tops my list for those just getting started. It is not meant to take you to a deep understanding of the process but instead is easy to read, simple to follow, and inspirational, with its large variety of recipes and motivating profiles of cheesemakers.

The Complete Idiot's Guide to Cheese Making by James R. Leverentz (Alpha Books, 2010)
Leverentz is president and founder of Leeners, which sells a lot of cheesemaking supplies. I know, some people are turned off by the Dummies and Idiots book series, but frankly, I think it is great to start from a point where you are humble enough to

know that you are not yet an expert. *The Complete Idiot's Guide* offers some additional information to the beginner and some great tips not found in *Home Cheesemaking*. It doesn't contain the number of recipes, but I think it should be included in any new cheesemaker's library.

Artisan Cheese Making at Home by Mary Karlin (Ten Speed Press, 2011)

This book is hot off the press as I write. It is a gorgeous book and will be quite inspirational for the avid hobbyist and foodie, with its glossy pages and quality hardcover. There are a few details missing that might lead to some issues, but overall, it is one I would recommend owning. Many of the recipes were provided by well-known cheesemakers and add a wonderful insight into the world of renowned cheesemakers.

FOR THE INTERMEDIATE AND ADVANCED

The Cheesemaker's Manual by Margaret P. Morris (Glengarry Cheesemaking, 2003)

This is really the only book I am aware of, other than the one you are holding in your hands, that provides information at the intermediate level on the making of the major categories of cheese (and includes recipes). The book is self-published, so a little more expensive than might be ideal, and the binding is a bit prone to releasing pages, but it is worth buying. Morris's company, Glengarry Cheesemaking and Dairy Supply, is a fantastic resource for cheesemakers of all levels and scales.

The Fabrication of Farmstead Goat Cheese by Jean-Claude Le Jaouen (The Cheesemakers Journal, 1987)

If you plan on making any mold-ripened cheeses—cow, goat, or sheep—you should eventually pick up this little book. Nothing else quite comes close to covering French bloomy rinded cheeses. You will learn a lot about their production, and the back includes a long list, along with details, about many of the classic French mold-ripened goat cheeses (you can make them with other milks, too, though).

American Farmstead Cheese by Paul Kindstedt, with contributions from others, including Peter Dixon (Chelsea Green, 2005)

When this book came out, it was the first of its kind—and still is the only one—to provide advanced cheese chemistry combined with business aspects. While not a cheesemaking book in the sense of teaching you recipes, it presents, in lay terms, the microbiology and chemistry of cheesemaking. It also discusses issues of food safety and business. It is a must-have, especially if you have any inkling of going pro.

The Joy of Cheesemaking by Jody M. Farnham and Marc Druart (Skyhorse Publishing, 2011)

This is a valuable book to add to your collection of cheesemaking books. (I do believe in owning many!) It contains some very helpful ways to measure cultures and do dilutions, as well as profiles of cheesemakers (yes, we're in there) and some great recipes. The cheesemaking portions were written by a cheesemaker and educator that I respect and admire greatly, Marc Druart (plus he has a super cute French accent).

On the Make, e-newsletter by *culture* magazine

I am excited about this new electronic publication. The industry has really been lacking a definitive, technical newsletter. I am hoping that this one, just out with its first, promising issues (I am lucky to be a contributor) is the answer to the need: www.culturecheesemag.com/.

ABOUT CHEESE IN GENERAL

Cheese Primer by Steve Jenkins (Workman, 1996)

This was my first cheese book, a gift from friends. What it lacks in glamour and color photos, it makes up for with insight and depth. Steve is an icon in the cheese world for good reason.

Cheese: Exploring Taste and Tradition by Patricia Michelson (Gibbs Smith, 2010)

A gorgeous book that will inspire you to taste, travel, and talk about cheese.

Mastering Cheese by Max McCalman and David Gibbons (Clarkson Potter, 2009)
This book is the brave and thorough work of one of cheese's first "celebrity" advocates. McCalman and Gibbons have two other books on cheese that are worth owning as well, but if you can only have one, get this one. The book is designed for the consumer of cheese, and I contend that cheesemakers are probably prime consumers as well as producers.

Cheese and Culture: A History of Cheese and its Place in Western Civilization by Paul Kindstedt (Chelsea Green, 2012)
You know cheese is important when a nearly three-hundred-page hardback book is published that is devoted solely to 9,000 years of cheese history. We have arrived!

culture magazine
If you love cheese, and I am assuming if you are reading this that you do, you must subscribe to this inspiring, gorgeous magazine. It is designed with food lovers in mind, but cheesemakers will gain from the stories of other cheesemakers from all over the world and be inspired by the beautiful photography: www.culturecheesemag.com/.

Cheese Connoisseur magazine.
Another lovely magazine devoted to the passion of eating cheese: www.cheeseconnoisseur.com.

BEING ENTERTAINED: CHEESEMAKER AND CHEESE ENTHUSIAST STORIES

Cheesemonger: A Life on the Wedge by Gordon Edgar (Chelsea Green, 2010)
Gordon Edgar is one of my favorite, no-nonsense cheesemongers. His book is a fun story about what brought him to cheese, and he shares some good insights about the cheese world along the way.

The Cheese Chronicles by Liz Thorpe (HarperCollins, 2009)
This is a nicely written story that gives some great insight into what goes on behind the scenes of one of the United States's better-known cheese counters.

Goat Song by Brad Kessler (Scribner, 2010)
This book poetically shares what most of us goat lovers feel. It is truly a lovely "read," as they say. And Brad and his wife are stellar cheesemakers!

The Year of the Goat: 40,000 Miles and the Quest for the Perfect Cheese by Margaret Hathaway (Lyons Press, 2009)
Here is another lovely book that shares a couple's journey to making cheese (and loving goats).

COOKING WITH AND SERVING CHEESE (MY FAVORITES, OF MANY)

The Cheesemonger's Kitchen by Chester Hastings (Chronicle Books, 2011)
This is not only a James Beard–nominated book (for its photography), but it is filled with recipes that are not "more of the same." If you love cheese and are looking for some new ways to use it in your cooking, this book is a must-have.

The New American Cheese by Laura Werlin (Stewart, Tabori & Chang, 2000)
As one of the pioneers in this current wave of cheese adoration here in the States, in this book Werlin profiles cheesemakers and their cheeses, then gives some luscious recipes in which to use them.

The All American Cheese and Wine Book by Laura Werlin (Stewart, Tabori & Chang, 2003)
In this book Werlin profiles both wine and cheesemakers, helps you learn how to pair wine and cheese, and shares some awesome recipes.

Cheese and Wine by Janet Fletcher (Chronicle Books, 2006)

Fletcher is another one of our favorite cheese people and cheese writers. One thing I love about this book is that she gives pronunciations of the cheese names! (For a country bumpkin such as me, that is a godsend.) This book focuses on pairings, plates, and cheese courses (She also wrote *The Cheese Course* [Chronicle Books, 2000]).

The Cheese Plate by Max McCalman and David Gibbons (Clarkson Potter, 2002)
This is a great pairing and cheese course from a cheese guy we like so much we named a goat after him (although with the adaptation of making it "Maxine" since it was a girl . . .).

In a Cheesemaker's Kitchen by Allison Hooper (Countryman Press, 2009)
Hooper, along with Bob Reese, started Vermont Butter & Cheese Creamery in 1984. Join Hooper in her virtual kitchen in this lovely cheese cookbook.

MOVIES AND VIDEOS

The Cheese Nun (PBS Home Video, 2006)
I waited far too long to purchase this video (or rent it)! Not only is it a fantastic story, but you will learn some pretty cool things about cheesemaking. Even since this BBC documentary was released, Sister Noella Marcellino has continued to aid cheesemakers with her research and consultations at events such as the American Cheese Society conference, where she also serves as a cheese judge.

Artois the Goat (Bogart Reininger Films, 2008)
This is the best cheese movie I have ever seen! Okay, I know it isn't a very crowded field, but I think it will hold top honors for me for some time. It isn't just about goats or goat cheese but instead is a quirky, romantic comedy where the true ingénue is cheese. Or rather, the making of cheese.

Cheese Slices (The LifeStyle Channel, 2007–)
A DVD series out of Australia and hosted by raw-milk cheese advocate and grand personality Will Studd, *Cheese Slices* is a globe-trekking tour of cheeses, both traditional and modern, from all parts of the planet. Don't count 100 percent on its content to tell you the accurate directions for the cheeses featured, but you will still take away some great information, visuals, and inspiration.

SCHOOLS AND CLASSES

This list has grown so much that I suggest to all readers to check out Vicki Dunaway's website, www.smalldairy.com/courses.html (which also includes information for countries other than the United States). Vicki is another cheese pioneer who until recently published *Creamline*, one of the original newsletters for artisan cheesemakers. Her website continues to offer up-to-date resources for cheesemakers. On the courses page that I have given you the link for, you will see another link to "just for fun" classes that include on-farm cheesemaking classes (such as we have here at our farm), tasting classes, and more.

ONLINE SOCIALIZING AND PEER SUPPORT

Cheeseforum.org
This global forum is awesome! Through it I met several of the cheesemakers that provided input for this book.

Gianaclis Caldwell
www.gianacliscaldwell.wordpress.com
This is my blog, where I update things for both of my books and include other articles and information that might be helpful.

Artisan Farmstead Living
http://artisanfarmsteadliving.blogspot.com

Curd Nerd
www. curd-nerd.com

I Make Cheese
www.imakecheese.com

Jos Voltu's blog
http://heinennellie.blogspot.com
Jos was a reader for this book and a recipe provider. His blog is filled with lots of inspiring cheesemaking stories. It is named after his parents, Heine and Nellie; the "n" in the middle is the Dutch conjunction "and."

New England Cheesemaking Supply Blog
http://cheesemakinghelp.blogspot.com/
If you need another reason to support this company, their website is accurate, up to date, and always educational.

CULTURE, INGREDIENTS, AND OTHER SUPPLIES

Dairy Connection, Wisconsin, USA
608-242-9030
www.dairyconnection.com
For small to mid-size artisans. One of the United States' main companies for cultures and supplies.

Fromagex, Canada
866-437-6624
www.fromagex.com
Artisan and industrial equipment and supplies. The complete line of Danisco and Chr. Hansen cultures. Global sales (except frozen bulk culture).

New England Cheese Supply, Massachusetts, USA
413-397-2012
www.cheesemaking.com
Supplies for the home and hobby cheesemaker. Kits, supplies, cultures (in small packets), recipes on website, tech support, and blog.

Glengarry Cheesemaking and Dairy Supply, Canada
888-816-0903
www.glengarrycheesemaking.on.ca
Small- and medium-size artisans in North America. Equipment, supplies, and cultures. Owner is Margaret Morris, who wrote *The Cheesemaker's Manual* (listed above).

SaltWorks, Washington, USA
800-353-7258
www.saltworks.us

OTHER EQUIPMENT—POTS, VATS, PRESSES, AND MORE

Central Restaurant Products, Indiana, USA
800-215-9293
www.centralrestaurant.com
Lots of stuff you might need: shelving, large stainless steel pots, food warmers, cutting boards, and so on.

Hamby Dairy Supply, Missouri, USA
800-306-8937
www.hambydairysupply.com

The Cheesemaker, Wisconsin, USA
414-745-5483
www.thecheesemaker.com
Steve Shapson provides classes, supplies, and a great assortment of versatile cheese presses. Steve is a passionate cheese maker with great customer service!

Another great source is any company that sells beer-brewing supplies, as they will often have not only some equipment but also other items such as cleaning agents.

CLEANING AND DISINFECTING SUPPLIES

If you live in a region with quite a few dairies, you may have a local source for cleaning chemicals and other supplies. Many regional companies have regular deliveries and service calls to producers. While it is not possible for me to attempt to list these here, I suggest contacting any dairy or cheese plant near you and finding out if they can refer you to a good source for cleaning chemicals. Another option is food service retailers, where bulk goods are sold to the public and to restaurants. These will have detergents and acid rinses, as well as other supplies you might need.

Hamby Dairy Supply, Missouri, USA
800-306-8937
www.hambydairysupply.com

Decon Labs, Pennsylvania, USA
800-332-6647
www.deconlabs.com
Enzymatic "Contrex EZ" powdered detergent is available on Amazon.com. I have not used this product, but the company says it works well on proteins and fats. So for those looking for small volumes of enzymatic cleaner, this might be your best bet.

Ecolab (worldwide)
800-392-3392
www.ecolab.com
Traditional and enzymatic cleaners, acid rinses, and more for the food industry. Usually available in larger pails and volumes. Will deliver orders as small as a single pail of product. Mixed reviews on responsiveness to small producers.

All QA Products, North Carolina, USA
800-845-8818
www.allqa.com
Sanitizer-strength testing supplies and more.

Zep, Georgia, USA
877-428-9937
www.zep.com
Cleaning products for many situations, large selection for food industry. Includes "GreenLink" products that are more ecologically friendly. Will work with even the smallest producer. When making first contact, ask for a representative for your area and then emphasize that you are processing dairy products.

LABORATORY SUPPLIES

Cole-Parmer (worldwide)
800-323-4340
www.coleparmer.com
They offer pH meters and more.

Nelson Jameson, Wisconsin, USA
800-826-8302
www.nelsonjameson.com
Lots of different supplies for food production, dairy, and cheesemaking. My choice for buying pH meters. Great customer support.

APPENDIX C: QUICK REFERENCE TABLES

I added this section in the hope that you will feel free to make photocopies of these tables, stick them in a protective sleeve, and have them easily accessible in your cheesemaking area. I know I have information such as this in various places in our cheese make room, but I wish I had done these tables before I started writing this book! It would have saved me a lot of time looking up redundant information.

BRINE MIXTURES

EXAMPLES:

1. For 1 gallon of **heavy brine**, mix 1 gallon of water with 2.52 pounds of salt plus 1 tablespoon (of 32 percent dilution) calcium chloride and 1 teaspoon white vinegar.
2. For 1 quart **light 5 percent washing brine**, mix 1 quart water with 0.11 pounds salt (about 3 tablespoons).
3. For 4 liters of **heavy brine**, mix 4 liters of water with 1.2 kg of salt plus 1 tablespoon calcium chloride (32 percent solution) and 1 teaspoon white vinegar.
4. For 1 liter **light 5 percent washing brine**, mix 1 liter of water with 0.06 kilograms salt.

TABLE C-1. COMMON BRINE SOLUTIONS: GALLONS*

Brine Use	% Salt by Weight of Brine**	° SAL at 60°F (16°C)	Lb. of Salt per Gal. Water***
Very light/ washing rinds	3%	11°	0.25
Light/ washing rinds	5%	19°	0.44
Medium light/ aging feta	8%	31°	0.74
Medium	16%	61°	1.60
Heavy brine/ initial salting	23%	88°	2.52

*Add 1 tbsp (15 ml) calcium chloride and 1 tsp (5 ml) vinegar (acetic acid) to each gallon (4 L) brine.
**Percentages and degrees are averaged; expect slight differences.
***1 gallon of water weighs 8.33 pounds.

TABLE C-2. COMMON BRINE SOLUTIONS: LITERS*

Brine Use	% Salt by Weight of Brine**	° SAL at 60°F (16°C)	Kg. of Salt per Liter Water***
Very light/ washing rinds	3%	11°	0.03
Light/ washing rinds	5%	19°	0.06
Medium light/ aging feta	8%	31°	0.09
Medium	16%	61°	0.19
Heavy brine/ initial salting	23%	88°	0.30

*Add 4 ml calcium chloride and 1 ml vinegar (acetic acid) to each liter of brine.
**Percentages and degrees are averaged; expect slight differences.
***1 L of water weighs 1 kg.

CALCIUM CHLORIDE MAXIMUM DOSAGE RATES

Note: You can usually dose calcium chloride (when it is used in a 32 percent solution) at the same rate as single-strength rennet, but some recipes that call for a slower or faster coagulation time than most other cheese types use less or more rennet than usual. In those cases the calcium chloride rate will be different.

TABLE C-3. CALCIUM CHLORIDE SOLUTION DOSAGE* RATES

Milk	Maximum Calcium Chloride: Milliliters**	Maximum Calcium Chloride: Teaspoons
1 gallon	1.25	¼
1 liter	0.30	1⁄16
10 pounds	1.8	¼ +
10 kilograms	3	¾

*Assumes standard dilution of calcium chloride solution at 32 to 33 percent
**The preferred, more accurate measurement

TABLE C-4. CALCIUM CHLORIDE CRYSTALS DOSAGE RATES (0.02%)

Milk	Maximum Calcium Chloride: Grams*	Maximum Calcium Chloride: Ounces
1 gallon	0.72	0.03
10 gallons	7.2	0.25
10 pounds	0.9	0.04
1 liter	0.2	0.009
10 liters	2.06	0.09
10 kilograms	2	0.09

*The preferred, more accurate measurement, rather than ounces. I assume, in this table, that if you are measuring milk by liter or kilogram you will not choose ounces to measure calcium chloride.

TABLE C-5. CULTURE ABBREVIATIONS AND NAMES

Mesophilic bacteria = orange
Thermophilic bacteria = red

Abbreviation	Current Name	Other, Common Name
LL	*Lactococcus lactis* ssp. *lactis*	*Streptococcus lactis*
LLC or LC	*Lactococcus lactis* ssp. *cremoris*	
LLD or LD	(Cit+) *Lactococcus lactis* ssp. *lactis* biovar. *diacetylactis*	*Lactococcus* ssp. biovar. *diacetylactis*
LMC or LM	*Leuconostoc mesenteroides* ssp. *cremoris*	*Leuconostoc cremoris*
ST	*Streptococcus thermophilus*	*Streptococcus salivarius* ssp. *thermophilus*
LB or LDB	*Lactobacillus delbrueckii* ssp. *bulgaricus*	
LBL	*Lactobacillus delbrueckii* ssp. *lactis*	
LH	*Lactobacillus helveticus*	
LBC	*Lactobacillus casei* ssp. *casei*	

CULTURE MEASUREMENTS

TABLE C-6. SAMPLE MEASUREMENTS FOR DIRECT-SET CULTURE*

Culture	Typical Usage	Teaspoon/unit Approximate*	Average weight	Units or grams per 26 gal (100 L/220 lb) **
Flora Danica	Fresh, soft ripened	⅛	0.2 g	5–10
MM and MA series	Fresh and hard	Scant ⅛	0.3 g	Soft 6–7 Hard 11–12
MA 4000	Semihard and hard	⅛	0.3 g	2.5–5
TA 50	Soft and semihard	⅛	0.3 g	Soft- 5–10 Semihard 1.25–5
TA 60	Stretched curd and hard	⅛	0.3 g	Hard, semi-hard 1.25–5 Italian grana 5–10
Thermo B and Thermo C	Stretched curd and hard	Scant ¼	1 g	10 g

*These measurements are estimates. Culture density and settling will cause variation.
**One unit will typically culture 50 lb (23 kg) of milk, but many milk types and cheese styles require different rates of use.

MEASUREMENT EQUIVALENTS

TABLE C-7. MILK AND WATER

	US Gallons	Pounds	Liters	Kilograms
1 US gallon milk*	1	8.6	3.8	3.9
1 US gallon water	1	8.3	3.8	3.8
1 liter milk	0.3	2.3	1	1.03
1 liter water	0.3	2.2	1	1

*Average weight for milk but will vary by fat content

TABLE C-8. CURD-SIZE EQUIVALENTS

Inches	Metric	Common Descriptor
⅛	3 mm	Rice, wheat grain
¼	6 mm	Rice, wheat grain, corn
⅜	1 cm	Corn, pea
½	1.3 cm	Hazelnut, thumbnail
¾–1 inch	2–3 cm	Hazelnut, walnut

TABLE C-9. DRY WEIGHTS: COARSE KOSHER SALT*

	Grams (rounded)	Pounds	Ounces	Cups	Tablespoons	Teaspoons
1 kilogram	1	2.20	35.30	–	–	–
1 pound	454	1.00	16.00	2+	–	–
1 ounce	28	0.06	1.00	–	2	–
1 cup	230	0.50	9.00	1	16	48
1 tablespoon	15	0.03	0.55	–	1	3
1 teaspoon	5	0.01	0.18	–	⅓	1

*Salt will vary by brand when comparing a volume measurement to a weight, and also depend on how compacted the salt is. The measurements above are less accurate the larger the volume! It is **always best to measure salt by a weight measurement** instead of a volume measurement.

TABLE C-10. FLUID MEASUREMENTS: RENNET, CALCIUM CHLORIDE

	Milliliters	Teaspoons	Tablespoons	Fluid Ounces	Cups
1 milliliter/cc	1	–	–	0.03	–
1 teaspoon	5	1	–	0.16	–
1 tablespoon	15	3	1	0.50	1/16
1 ounce	30	6	–	1.00	⅛
1 cup	240	48	16	8.00	1

TABLE C-11. COMPARISON OF "TINY" TEASPOONS AND OTHER EXTRA-SMALL MEASUREMENTS*

Measurement Name	Comparison to 1 Teaspoon	Conversion to Milliliters
Tad	Scant ¼ teaspoon	1.0 ml
Dash	Scant ⅛ teaspoon	0.5 ml
Pinch	1/16 teaspoon	0.25 ml
Smidge	1/32 teaspoon	0.12 ml
Drop	1/60 teaspoon	0.05 ml
Skewer tip	1/120 teaspoon	0.02 ml

*Teaspoon measurements are not necessarily standardized throughout the world. These comparisons are accurate for most US-sold measuring spoons.

RELATIVE HUMIDITY COMPARISON READINGS FOR USE WITH PSYCHROMETER

TABLE C-12. FAHRENHEIT SCALE FOR PSYCHROMETER READINGS

Dry Bulb Temp °F	Difference between Wet Bulb and Dry Bulb Temperatures (°F)				
	0	1	2	3	4
46	100%	93%	86%	78%	71%
50	100%	93%	87%	80%	74%
55	100%	94%	88%	82%	76%
57	100%	94%	89%	83%	78%

TABLE C-13. CELSIUS SCALE FOR PSYCHROMETER READINGS

Dry Bulb temp °C	Difference between Wet Bulb and Dry Bulb Temperatures (°C)				
	0	1	2	3	4
8	100%	87%	74%	62%	51%
10	100%	88%	76%	65%	54%
12	100%	88%	78%	67%	57%
14	100%	89%	79%	69%	60%

TEMPERATURE CONVERSION

(Rounded to tenths)

This is my favorite type of conversion chart. At first it might look complicated, but it is super easy once you get comfortable with the format. Take a look at each three-column set. Let's take an example. Say you want to know what 20°C converts to. Look down the center column and find 20; look to the right, and you will see that 20°C = 68°F. Try that. Now let's say you want to convert 20°F to Celsius. You can look down the same column at 20 and look to the left column. You will find that 20°F = −6.7°C. I hope that makes sense!

TABLE C-14. TEMPERATURE CONVERSION (CELSIUS TO FAHRENHEIT)

Celsius		Fahrenheit	Celsius		Fahrenheit	Celsius		Fahrenheit
−17.80	0	32.0	1.1	34	93.2	19.4	67	152.6
−17.20	1	33.8	1.7	35	95.0	20.0	68	154.4
−16.70	2	35.6	2.2	36	96.8	20.6	69	156.2
−16.10	3	37.4	2.8	37	98.6	21.1	70	158.0
−15.60	4	39.2	3.3	38	100.4	21.7	71	159.8
−15.00	5	41.0	3.9	39	102.2	22.2	72	161.6
−14.40	6	42.8	4.4	40	104.0	22.8	73	136.4
−13.90	7	44.6	5.0	41	105.8	23.3	74	165.2
−13.30	8	46.4	5.6	42	107.6	23.9	75	167.0
−12.80	9	48.2	6.1	43	109.4	24.4	76	168.8
−12.20	10	50.0	6.7	44	111.2	25.0	77	170.6
−11.70	11	51.8	7.2	45	113.0	25.6	78	172.4
−11.10	12	53.6	7.8	46	114.8	26.1	79	174.2
−10.60	13	55.4	8.3	47	116.6	26.7	80	176.0
−10.00	14	57.2	8.9	48	118.4	27.2	81	177.8
−9.50	15	59.0	9.4	49	120.2	27.8	82	179.6
−8.90	16	60.8	10.0	50	122.0	28.3	83	181.4
−8.30	17	62.6	10.6	51	123.8	28.9	84	183.2
−7.80	18	64.4	11.1	52	125.6	29.4	85	185.0
−7.20	19	66.2	11.7	53	127.4	30.0	86	186.8
−6.70	20	68.0	12.2	54	129.2	30.6	87	188.6
−6.10	21	69.8	12.8	55	131.0	31.1	88	190.4
−5.60	22	71.6	13.3	56	132.8	31.7	89	192.2
−5.00	23	73.4	13.9	57	134.6	32.2	90	194.0
−4.44	24	75.2	14.4	58	136.4	32.8	91	195.8
−3.89	25	77.0	15.0	59	138.2	33.3	92	197.6
−3.33	26	78.8	15.6	60	140.0	33.9	93	199.4
−2.78	27	80.6	16.1	61	141.8	34.4	94	201.2
−2.22	28	82.4	16.7	62	143.6	35	95	203.0
−1.67	29	84.2	17.2	63	145.4	35.6	96	204.8
−1.11	30	86.0	17.8	64	147.2	36.1	97	206.6
−0.56	31	87.8	18.3	65	149.0	36.7	98	208.4
0.00	32	89.6	18.9	66	150.8	37.2	99	210.2
0.56	33	91.4				37.8	100	212.0

APPENDIX D: SAMPLE MAKE SHEETS

There are many approaches to creating a make sheet or "cheese log." I have included two that we use here on our farm, one to make our feta style and one for a simple hard cheese. I am a fan of the table format that lists time, temperature, and pH goals at appropriate stages. But there are certainly many ways to design an equally workable set of directions. Please feel free to modify the following make sheets in any way you see fit!

The second table is for a simple tomme-style cheese (recipe is in chapter 14), such as our product Elk Mountain; you can see that with the addition of a few lines you could add steps for washing curd, stirring the curd, and even cheddaring. If you have access to a computer and spreadsheet software, simply type in these basic lines, then adapt them to any recipe. If you don't know goal pHs, you can develop them over time, depending on how well you track them and how you like the resulting cheese.

TABLE D-1. SAMPLE MAKE SHEET FOR FETA-STYLE CHEESES

Date	Cheesemaker			Wheels		
	Time Goal	**Actual Time**	**Temp Goal**	**Actual Temp**	**pH Goal**	**Actual pH**
Gallons/Pounds			<40°F (4.4°C)			
In vat					6.6–6.7	
2U Flora Danica/ 50 lb			80°F (27°C) then increase to 88°F (31°C) (add at 80°F [27°C] when using raw milk)			
Ripen	45–60 min		88°F (31°C)		0.10 drop maximum	
0.035 ml/lb double strength rennet	35–45 min		88°F (31°C)			
Cut curd ¾–1 inch (2–3 cm)			88°F (31°C)			
Rest	10–15 min					
Stir	20 min		88–90°F (31–32°C)			
Drain	5–10min		88–90°F (31–32°C)		6.4–6.5	
Hoop						
Flip	10 min		Room 72°F (22°C)			
Flip	15 min		Room 72°F (22°C)			
Dehoop	6–12 hours		Room 72°F (22°C)		4.5–4.7	
Brine (heavy)	72 Hours (8 hours per pound/16 hours per kg)		50–55°F (10–13°C)			
Storage Brine (8%)	Age up to 9 months		50–55°F (10–13°C)			

TABLE D-2. SAMPLE MAKE SHEET FOR SEMIHARD, LIGHTLY COOKED CHEESES

Date	Cheesemaker			Wheels		
	Time Goal	Actual Time	Temp Goal	Actual Temp	pH Goal	Actual pH
Gallons/Pounds						
In vat					6.6–6.7	
1U MA 4000 + 0.5U LM/50 lbs			80°F (27°C)			
0.035–0.045 ml/lb double-strength rennet	35–45 min		88°F (31°C)			
Cut curd			88°F (31°C)			
Rest	5 min					
Stir to 100°F	30 min		100°F (38°C)			
Hold temerature and stir to texture	10–20 min		100°F (38°C)		6.4–6.5	
Settle	5 min					
Whey off to curds, press	5–10 min					
Whey off, hoop, 8 pounds						
Flip, 8 pounds	10 min		Room 72°F (22°C)			
Flip, 10 pounds	15 min		Room 72°F (22°C)			
Flip, 10 pounds	30 min		Room 72°F (22°C)			
Dehoop	4–6 hours at goal pH		Room 72°F (22°C)		5.2–5.4	
Brine (heavy) 24 hours/7-pound wheel			50–55°F (10–13°C)			
Age	5–6 months		50–55°F (10–13°C) 85–90% RH			

GLOSSARY

Acetic acid. Food-grade acid. Vinegar is made up of mostly acetic acid.

Acidic. A substance with a pH below neutral 7. Flavors associated with acid are sour, tart, tangy.

Adjunct culture. Bacteria added to provide characteristics that will occur during aging. Synonyms often used are *ripening* or *secondary* culture, although these more correctly refer to microorganisms added to assist with ripening, such as mold and yeast cultures.

Advantageous. Something that offers or provides an advantage. Associated with advantageous microorganisms that enter the milk adventitiously (spontaneously without being added).

Adventitious. Something that occurs without planning and spontaneously. Associated with microorganisms that enter the milk from the environment and during milking.

Affinage. The French term, now widely used, for the care and tending of cheese while aging.

Affineur. The French term for someone whose profession and specialty is aging cheeses.

Alkaline. A substance with a pH above neutral 7. Flavors associated with alkaline are bitter. Synonym: Basic.

Alkaline phosphatase. A naturally occurring enzyme in milk that is destroyed by pasteurization. Not an enzyme that assists with cheesemaking, but its presence is used to test for proper pasteurization.

Amino acids. Singular units in a chain that makes up long protein strands. Subunit of peptides. In cheesemaking they are associated with protein breakdown and the creation of flavors and texture in cheese.

Annato. Natural orange coloring used in cheese. Made from the seeds of pods from a tropical and subtropical plant called the achiote tree.

Artisan. In cheesemaking the term can mean the production of cheese by hand or a small company that produces cheese by hand using milk from their own farm or other farms.

Ash. The mineral salts in a substance, obtained by incineration; the minerals are left as ash.

Ash, vegetable. Food-grade charcoal used to raise the pH on the surface of some cheeses or for a decorative effect.

Bacteriophage/phage. Viruses that attack bacteria and inject their own genetic material into the host. The process destroys the bacteria.

Bactofugation. A mechanical process sometimes used by large dairy producers that removes bacteria and somatic cells from warm milk using centrifugal force. Used in conjunction with pasteurization.

Bandaged rind. A cheese whose rind has been encased by a thin layer of fat-soaked cheesecloth and aged. Traditional treatment for cheddar. Synonym: clothbound.

Basic. See **Alkaline.**

Bloomy rind. A cheese whose surface is host to several types of yeasts and white (usually) molds. Also called "surface ripened."

Blowing. Applies to unwanted gas production within the curd or cheese that leads to the expansion of the mass. "Early blowing" occurs just after production and is usually caused by bacteria from the coliform family and yeasts. "Late blowing" occurs toward the end of aging, usually several months later, and is caused by bacteria from the clostridium family or propionic acid bacteria.

Botulism. Sometimes-fatal illness caused by the spore-forming (meaning resistant to heat treatment) bacteria *Clostridium botulinum*.

Brevibacterium linens. Commonly known as *B. linens*. The most commonly known and added surface-ripening bacteria used on washed rind (stinky) cheeses. As with other surface-ripening microorganisms, it can be present in the environment naturally. Part of the family of coryneform bacteria.

Brine. A saltwater solution in which cheeses are floated after being removed from the press to provide them with a salt content. Can also be used as a verb and mean the act of salting a cheese in brine.

Brine hydrometer. A tool or instrument designed to measure the salt concentration (salinity) of a solution. Synonym: salometer.

Buffer. A solution or substance that resists changes in pH. Buffer solutions, usually 7.00 and 4.01, are used to calibrate pH meters. The prime buffers in milk are the protein micelles and calcium phosphate.

Butter muslin. Finely woven cloth for draining dairy products with small particles. Cheesecloth with a thread count of 90 threads per square inch is often called butter muslin.

Butterfat. Term used to describe the fat, or lipids, in milk. Butterfat and cream are similar but not interchangeable.

Butyric. Fatty acid associated with a rancid flavor. If butyric fermenting bacteria, such as *Clostridium tyrobutyricum*, are present, they can produce gas and cause "late blowing."

Calcium. Mineral salt in milk. Exists alone and bonded with other minerals. Plays a key role in coagulation and finished texture of the cheese.

Calcium chloride. Food-grade additive used in milk to assist with coagulation and in brine to create a mineral balance (equilibrium) between the cheese to be brined and the brine solution.

Calcium lactate. Exists as crystals on the surface of some aged cheeses (usually rindless, vacuum sealed) when residual lactate combines with calcium in the cheese. Considered a defect.

Calibration. The act of aligning an instrument's readings with those of a known substance to ensure accuracy. For example, a pH meter is calibrated by soaking the electrode in solutions (buffers) with a known pH, then using a setting that tells the meter to align its readings with that of the buffer.

Capric. Short-chain fatty acid present in milk; in a greater degree in goat's milk. Associated with aroma and flavor development.

Caproic. See **Capric**

Caprylic. See **Capric.**

Carcinogen. A substance linked to the incidence of carcinoma (cancer) occurrence.

Carotene. A provitamin that can be converted by the body to vitamin A. Gives cow's milk its more yellow color when compared to carotene-free sheep, goat, and water buffalo milk.

Case hardening. Term used to describe the entrapment of whey within curds when the outer surface of the curd is dried and dehydrated too quickly while in the cooking phase in the vat.

Casein. Protein present in milk. Makes up 70 to 80 percent of total protein, depending on species and breed of animal. Main protein captured during cheesemaking. Consists of alpha (α), beta (β), and kappa (κ) caseins along with other variants.

Caves, cheese caves, holes in paste. "Caves" and "cheese caves" usually refer to a natural or constructed open space underground in which cheese can be aged. Also occasionally used (as is "holes in paste") to mean small opening within the paste, or interior, of a cheese; especially used by the producers of blue cheese.

Cheddar. Refers both to a cheese type and to the process of cheddaring, in which cooked curd is drained and allowed to mat together for an extended period in the vat. Mats of curd are usually turned and restacked several times during the cheddaring process.

Cheese iron. See **Trier.**

Cheesecloth. A fabric with an open weave that traps curd within but allows whey to drain. Can be made of natural fabrics (often linen, also called muslin) or synthetics. Comes in various weave consistencies (thread counts).

Chymosin. Enzyme produced in the ruminant's stomach (abomasum) and used to coagulate milk for cheesemaking. Is also produced through the fermentation of genetically engineered microorganisms.

Citrate. Exists naturally in milk as citric acid often bound to calcium and magnesium. Fermented by specific bacteria (called "citrate positive") to produce diaceytl (associated with a buttery flavor and aroma) and carbon dioxide for the formation of small eyes.

Citric acid. See **Citrate**.

Clabber. The curd produced when raw milk is allowed to ferment and coagulate at room temperature; also the act of coagulating milk.

Clean break. Describes the behavior of coagulated milk when a tool is used to lift the curd and observe how it separates, or breaks. A clean break is free from jagged tears, and the whey that drains from the break is clearish.

Clear coat. A synthetic, somewhat breathable coating that can be used prior to wax or in place of wax on a semihard to hard cheese. Can be clear or colored and may contain a mold inhibitor.

Clostridium. *Clostridium tyrobutyricum.*

Clothbound. See **Bandaged rind**.

Coagulant. Ingredient added to milk to cause it to thicken. Coagulants include animal sources such as chymosin and pepsin (rennet), microbial sources such as mucor meihi (vegetarian rennet), microbial sources such as fermented chymosin (also vegetarian), and plant sources such as cardoon thistle.

Coagulate. The act and process of adding coagulant. Synonymous with "curdle."

Coliforms. A large group of bacteria associated with soil fecal matter. Can ferment milk sugar and create gas—early blowing—in cheeses. One variation of *E. coli*, 0157:H7, is pathogenic, causing illness and even death.

Components. Refers to the constituents within milk such as fat, protein, lactose, and calcium.

Coryneform bacteria. A fairly large group of surface-ripening bacteria usually associated with washed rind (stinky) cheeses. Includes the most common surface-ripening bacteria, *Brevibacterium linens.*

Cream line. The visible line created when cream separates from milk and floats.

Cream wax. See **Clear coat.**

Creaming. The behavior of nonhomogenized milk that causes fat to separate and float or "cream."

Cream-top. Describes milk that has not been homogenized or fluid products made from such milk and that therefore experience a degree of separation of the cream even after production, such as in cream-top yogurt.

Cryoglobulin. Protein that exists in cow's milk that encourages clumping of fat globules and increases the rate of creaming. Cryoglobulin is denatured during homogenization.

Cryovac. Popular brand of vacuum-sealing equipment. Sometimes used as a verb to describe the act of vacuum sealing.

Culture. Bacteria and other microorganisms added to milk to produce changes, such as acid development. *Starter culture* refers to bacteria added whose primary job is the production of acid. *Ripening culture,* also called *adjunct* or *secondary,* is used to provide microorganisms that will facilitate changes during ripening or aging.

Curd dust. Fine particles of curd that are created when the coagulated network is shattered and small coagulated bits are present in the vat. Synonym: fines.

Curdle. See **Coagulate** and **Clabber**.

Deacidify. The process by which the acid level is lowered and the pH is increased.

Denatured. Term used to describe any change to a substance or molecule that causes it to no longer behave as it did before, or naturally.

Dew point. The point at which the air (at its current pressure and temperature) is fully saturated with water vapor and condensation occurs.

Diacetyl. By-product of the fermentation of citrate by specific bacteria, producing a buttery flavor in cheese.

Disaccharide. A sugar composed of two simple sugars. Milk sugar, lactose, is a disaccharide made up of one molecule of galactose and one of glucose.

D-lactate. Lactic acid produced through the fermentation of lactose. Its production also results in by-products of carbon dioxide (CO_2), ethanol, and small amounts of acetate. See also **L-lactate**.

Double créme. Term used to describe a surface-ripened cheese to which cream has been added to bring the finished total fat content up to at least 60 percent.

Dry salt. Term used to describe the act of salting a cheese by adding dry salt to the curd or to the surface of a pressed wheel.

Electrode. Part of an instrument, such as a pH meter, that has the ability to conduct electrical currents. Used in conjunction with a meter to measure the electrical potential of a substance, therefore determining and measuring the presence of hydrogen ions.

Electrolyte. A substance or solution that contains ions, allowing the substance to conduct electricity.

Enzyme. A substance that causes biochemical reactions to occur or occur more rapidly.

Exopolysaccharides (EPS). Natural compounds produced by some bacteria that have a slimy, viscous nature.

Eyes. Openings in cheese present either because of the lightness of pressing—leaving openings—or the formation of gas in the paste after pressing.

Farmstead. A cheesemaking operation using only the milk of its own animals to produce cheese.

Fatty acid. An acid associated molecularly with fats and oils.

Fermentation. The breakdown of a substance that produces acid and other by-products such as carbon dioxide. Usually associated with the breakdown of sugars (carbohydrates).

Fermented chymosin. Coagulant produced through the fermentation of microorganisms that have been genetically engineered to create chymosin.

Fines. See **Curd dust**.

Flocculation. The point at which proteins begin to cluster in milk during coagulation. Flocculation is visible when the coagulating milk is swirled or moved as small specs or *flocs*.

Follower. The portion of a cheese press or form that sits atop the cheese and evenly distributes the pressing weight. Named because it "follows" the cheese as it shrinks and is pressed.

Friable. A crumbly, brittle texture.

Galactose. One of the two simple sugars (monosaccharides) that compose lactose (milk sugar).

Globule. Small, spherical mass such as the milk fat globule.

Glucose. One of the two simple sugars (monosaccharides) that compose lactose (milk sugar).

Glycolysis. The breakdown, or degradation, of sugar.

Halotolerant. Microorganisms that are not adversely affected by the presence of salt.

Harp. Curved shape tool for cutting cheese curd while in the vat.

Heterofermentors. Bacteria capable of fermenting sugars into more than one resulting end product such as L- and D-lactate, CO_2, ethanol, and acetate.

Homofermentors. Bacteria capable of fermenting sugars into only one resulting end product: L- or D-lactate.

Homogenization. Mechanical treatment of milk in which the fat globules are reduced in size and cryoglobulin (a protein that encourages milk to cream) is denatured, resulting in fat that doesn't separate from the milk.

Humidity, absolute. The total amount of water vapor (measured by its mass) suspended in a volume of air.

Humidity, relative. The amount of absolute humidity relative to the maximum amount that same air could hold.

Hydrogen. A positively charged ion (+H) present in increasing amounts in acidic substances (below 7 pH).

Hydrometer. A tool or instrument for measuring the relative density of a solution. For example, a brine hydrometer measures the amount or percentage of salt dissolved in the solution.

Hydrophilic. Water loving. Term used to describe any substance that readily combines with water.

Hydrophobic. Water hating. Term used to describe any substance that repels water.

Hydroxyl. A negatively charged ion (–OH) present in increasing amounts in alkaline (above 7 pH) substances.

Hygrometer. A tool or instrument for measuring moisture and humidity.

Inoculate. The process of introducing microorganisms into a substance where they will grow. Synonymous in cheesemaking with adding culture or *culturing* the milk.

Ion. An atom or molecule with a positive or negative electrical charge. In cheesemaking calcium ions (in the milk) and hydrogen ions (which increase during acidification) are the most often referred to.

Isoelectric point. The point at which a substance loses its former electrical charge and becomes neutral.

Junket rennet. Coagulant made using enzymes from a ruminant's stomach (abomasum). Consists mostly of pepsin. Not a good choice for cheesemaking, as pepsin will break down proteins too rapidly after coagulation.

Kadova. Brand name of cheese molds made with a permanent mesh liner.

Kosher salt. Flaked salt. Comes in additive free or with anticaking agents included.

Lactase. Enzyme that breaks lactose into its two simple sugars, glucose and galactose. Some individuals do not produce lactase in their digestive tracts and are therefore lactose intolerant.

Lactate. (noun) The resulting product of the fermentation of lactose to lactic acid. Two different fermentation pathways produce either D-lactate or L-lactate. (verb) The act of an animal producing milk. When an animal is milking, it is referred to as a "lactation."

Lactic acid. Acid produced after the fermentation of lactose.

Lactic technology. Industry term for cheeses coagulated with acid and little or no rennet or other coagulant.

Lactose. Disaccharide sugar in milk comprising two simple sugars, glucose and galactose.

Lipase. Enzyme that breaks down fats (lipids). Present naturally in milk. Also present in the gastric system of animals. Commercially available (called pregastric lipase) and can be added to cheese milk to increase fat breakdown and flavor.

Lipolysis. The breakdown of fats (lipids) through enzymatic processes.

Lipophilic. Literally, "fat loving." Lipophilic substances can be dissolved and suspended within oil and fat. An example is fat-soluble vitamins.

Lipophobic. Literally, "fat hating." Lipophobic substances will not dissolve when combined with oil and fat.

Listeria. Psychrotrophic family of bacteria occurring naturally in the environment. One type, *Listeria monocytogenes*, is pathogenic and is known to cause severe illness (listeriosis) and even death in some cases.

L-lactate. Lactic acid produced by the fermentation of lactose. L-lactate production results in no other by-products. See also **D-lactate**.

Lysozyme. Commercially available antimicrobial effective against butyric-acid-fermenting bacteria. Used to prevent late blowing of cheeses.

Mastitis. An infection within the mammary system. Can be either mild and barely detectable (subclinical) or severe (acute). Mastitic milk is a poor choice for cheesemaking.

Mechanical eyes/openings. See **Eyes**.

Mellowing. Refers to the time after the salting of milled curds when the salt is allowed to dissolve and be partly absorbed by the curd.

Mesophile. Category of bacteria that prefer a warm temperature range of approximately 80 to 102°F (27–39°C).

Micelle. A cluster of molecules suspended within a liquid. In milk, protein micelles are a collection of caseins and calcium phosphate. Casein micelles are hydrophilic (water loving) and carry an overall negative electrical charge.

Microbial rennet. Enzymatic coagulant produced by microbes. Most common microbe used is *Rhizomucor miehei*.

Microfiltration. Mechanical method for removing bacteria, somatic cells, and spores from milk through the use of a semipermeable membrane.

Microflora. Technically refers to tiny plant life but is commonly used in cheesemaking and affinage to refer to any initially microscopic life that occurs, or is introduced, during the aging of cheese with a natural rind. Microflora are microorganisms.

Microorganism. Refers to all microscopic life, such as bacteria, viruses, and fungi (yeasts and molds).

Microperforated. Refers to cheese forms that are punctured with hundreds of tiny drain holes.

Mill. The act of cutting or breaking curds that have been compacted (and usually cheddared) into smaller pieces.

Mineralized. Refers to a cheese whose make process allows it to retain a higher proportion of minerals, especially calcium, giving it a more pliable texture. A *demineralized* cheese has lost a greater proportion of minerals and consequently has a more brittle, friable texture.

Mites. Microscopic pest often found on the rind of aged cheeses. Used intentionally to create the pitted rind of the French cheese Mimolette.

Mold. Either a form in which cheese is drained or a member of the fungi group, which includes yeasts and molds.

Molecule. A cluster of atoms bonded together (not to be used interchangeably with the terms *globule* or *micelle*).

Morge. Refers to a solution usually consisting of a light brine solution and surface-ripening bacteria and flora. Used to develop a sticky, washed rind cheese.

Mucor miehei. See ***Rhizomucor miehei.***

Natamycin. Natural antifungal produced by the bacteria *Streptomyces natalensis.* Included as an ingredient in some cheese rind coatings. Common brand used is Natamax.

Nonreactive. Indicates a material, in this case a pot or utensil, that will not erode or degrade in the presence of other substances, such as milk. Many metals, such as aluminum, will react with the acid produced by cheesemaking.

Nonstarter lactic acid bacteria (NSLAB). Refers to any adventitious bacteria that will produce lactic acid but at a rate too slow to assist with the initial "starting" of the cheese. They do contribute greatly to the flavor profile of raw-milk cheeses.

Paste/pate. Terms used to describe the interior of a cheese.

Pasteurization. Heat treatment designed to destroy or limit major foodborne pathogens. Legally defined and must conform to specific temperatures and times.

Pathogen. A disease- or illness-causing microorganism.

Pepsin. Proteolytic enzyme from a ruminant's stomach (abomasum) that can coagulate milk but also produces bitterness because of its high proteolytic activity.

Peptides. Short chains of amino acids. Multiple linked peptides make up a polypeptide.

pH. A measurement scale and system that designates acidic and alkaline solutions' positions in comparison to pure water, which is neutral.

Phage. See **Bacteriophage/phage.**

Phagicide. A treatment that kills bacteriophage.

Phosphate. An inorganic chemical naturally occurring in milk bonded to calcium.

Piquant. Peppery or hot, usually used to describe aged cheeses where lipase has played a role in the breakdown of fats, helping to create a spicy hotness.

Pitch. Term used to describe when curd is left to settle in the vat, without the whey being drained. The curd might be pressed under the whey with weights or left to compact on its own.

Plasmin. A proteolytic enzyme (protease) naturally present in milk in varying amounts. Varies by season, stage of lactation, and udder health. Negatively affects cheese yield and milk quality.

Plyban. Brand of synthetic, disposable cheesecloth in rectangular sheets especially designed for square cheese forms.

Poil de chat. Type of mold with a soft, furry texture. Translates as "cat's fur."

Polypeptides. Chains of amino acids. Subsections of a longer protein strand cut by enzymatic activity.

Protease. An enzyme that breaks down proteins. Sometimes called proteinase.

Protein. One or more strands of polypeptides, long chains of amino acids. Source of nutrition for living things. Milk protein—caseins and whey proteins—exists in a micellar form suspended in the liquid and within the liquid portion.

Proteolysis. The breakdown of proteins. Occurs in cheese gradually and in stages. Major contributor to flavor and texture during aging.

Provitamin. A substance that can be converted to a vitamin within the body. In milk beta-carotene is a provitamin for vitamin A.

Pseudomonas. A psychrotrophic spoilage bacterium.

Psychrometer. A specific type of hygrometer that measures relative humidity through the comparison of readings on two thermometers, one with a dry bulb and one with a wet bulb.

Psychrotroph. Cold-loving bacterium that thrives around 50°F (10°C) but can survive to much lower temperatures. Examples are *Listeria* and *Pseudomonas.*

Pure salt. Sodium chloride without any additives such as iodine (commonly added to table salt) or anticaking subtsances.

Renin. An enzyme produced by the renal glands (kidneys); not to be confused with renin, the previous name of the enzyme chymosin.

Rennet. More properly refers to animal-source coagulant but now commonly used to mean any enzymatic coagulant. See also **Coagulant**.

Rennet technology. Industry term for cheeses coagulated primarily by rennet or other coagulant as opposed to added or developed acid.

Rennin. Former name of the enzyme chymosin for which rennet was named.

Resolubilization. Term describing the behavior of protein in cheese when it reattaches water molecules during aging, leading to a softening of the texture.

Rhizomucor miehei. Microorganism that produces a protease capable of coagulating milk; the primary source for most microbial rennet produced today.

Rind rot. A situation in which a portion of the cheese rind degrades, usually due to too much moisture.

Ripening culture. Microorganisms added to the milk that will impart character and changes during ripening or aging.

Ruminant. An animal that has a compartmented upper digestive system (usually with four chambers), including a rumen. Ruminants regurgitate matter from the rumen and "chew a cud."

R-value. A substance's ability to resist temperature changes. Most often used to designate building materials and their ability to insulate.

Salinometer. See **Salometer**.

Salometer. A hygrometer that measures the salt content (salinity) of a solution.

Sanitize. To remove microorganisms from an object, usually through heat or chemicals used after cleaning. A sanitized object might or might not be considered sterilized—having no life forms—depending on the treatment used.

Scopulariopsis. A flakey tan to whitish mold, often found on cheese rinds. Full name is *Scopulariopsis brevicaulis*.

Secondary culture. A culture added to contribute character during aging and ripening.

Smear. Term used to describe the technique of creating a bacteria-surface-ripened cheese—a washed rind or "stinky" cheese. Sometimes used as a verb for the act of washing the rinds; also as a noun for the brine and bacteria mix used to wash the rinds.

Somatic cell. A cell originating in the body. Somatic cells occur naturally in milk (as they are shed from the interior of the udder during milking). High somatic cell counts, however, are indicative of an inflammatory process within the udder.

Spino. A traditional curd-cutting tool for cutting some Italian and Alpine cheeses. Shaped like a round whisk with fine blades.

Stabilized curd/paste. Surface white mold–ripened cheeses manufactured using techniques that limit the softening of the paste.

Standardization. Any number of processes or techniques by which the components of milk are adjusted to maintain a consistent level.

Starter culture. Bacterial cultures added at the start of the cheesemaking process to acidify the milk.

Sterilize. Techniques and processes by which all living matter is removed or destroyed on a surface or implement.

Stirred curd. Term describing cheeses made using a technique in which the drained cheese curd is stirred for a period of time in the warm vat.

Syneresis. Term meaning to lose fluid or drain.

Terroir. French term meaning "a sense of place or locality." Used to describe characteristics imparted to the milk and cheese from local conditions, food sources, aging environment, and so on.

Thermization. Process of heat-treating milk at temperatures under that of legal pasteurization; destroys some bacteria.

Thermophile. Heat-loving bacterium that grows well at temperatures of approximately 100 to 120°F (38–49°C). Most often used in Italian-style cheeses.

Titratable acidity. Measurement of the amount of acid within a substance using "titration"—adding small amounts of a base, sodium hydroxide, to a liquid to which a color reactor has been added that will change colors when the solution nears an alkaline state.

Trier. A tool for core-sampling a cheese. The weighted handle can be used to gently tap large wheels to detect air pockets (eye formation) and gas. Synonym: cheese iron.

Triglyceride. A glycerol (fat) with three fatty acids. Most of the fat in food is made up of triglycerides.

Triple crème. Term used to describe a surface-ripened cheese to which cream has been added to bring the finished total fat content up to at least 75 percent.

Tyrosine. An amino acid that can form crystals within long-aged cheeses such as Parmesan.

Ultrahigh temperature. Milk that has been heat-treated to 280°F (138°C) for 2 seconds. Does not need refrigeration. Not suitable for cheesemaking. Also known as UHT milk.

Ultrapasteurized. Milk heat-treated to 240°F (116°C) for 4 to 15 seconds. Not suitable for cheesemaking.

Washed curd. Technique in which whey is removed from the vat during the cook process and replaced with water or a light brine. Process decreases acid production during the time in the vat by removing a portion of lactose.

Washed rind. Refers both to the technique of washing the rind of a cheese to encourage certain microflora to develop and to the resulting category of cheese.

Water bath. Describes the use of a double-walled or two-part container in which the outer portion contains hot water that heats the contents of the interior container.

Whey. Liquid produced during cheesemaking; contains lactose, starter bacteria, whey proteins, and water-soluble vitamins.

Yield. The amount of cheese obtained in comparison to the amount of milk used. Can also refer to the amount of milk produced by a dairy animal.

NOTES AND REFERENCES

I have listed my sources in order of their usage level in this book. Any direct quotes are referenced within the text of the book.

BOOKS

Fox, Patrick F., Timothy P. Guinee, Timothy M. Cogan, and Paul L. H. McSweeney. *Fundamentals of Cheese Science*. Gaithersburg, Md.: Aspen Publishers, 2000.

Scott, R., R. K. Robinson, and R. A. Wilbey. *Cheesemaking Practice*. Gaithersburg, Md.: Aspen Publications, 1998.

McSweeney, P. L. H. (ed). *Cheese Problems Solved*. Boca Raton, Fla.: CRC Press, 2007.

Fox, Patrick F., Paul L. H. McSweeney, Timothy M. Cogan, and Timothy P. Guinee. *Cheese: Chemistry, Physics, and Microbiology*, Vol. 2. London: Elsevier Ltd, 2004.

Kosikowski, Frank V., and Vikram V. Mistry. *Cheese and Fermented Milk Foods*, Vols. 1 and 2. Westport, Conn.: F. V. Kosikowski, LLC, 1997.

Biss, Kathy. *Practical Cheesemaking*. Wiltshire, UK: Crowood Press, 2002.

Le Jaouen, Jean-Claude. *The Fabrication of Farmstead Goat Cheese*. Massachusetts: Cheesemakers' Journal, 1990.

Moore, John T. *Chemistry for Dummies*. Hoboken, N.J.: Wiley Publishing, 2011.

Kindstedt, Paul. *American Farmstead Cheese*. White River Junction, Vt.: Chelsea Green Publishing, 2005.

Morris, Margaret P. *The Cheesemaker's Manual*. Lancaster, Ontario, Canada: Glengarry Cheesemaking, 2003.

Jenness, Robert, Elmer H. Marth, Noble P. Wong, and Mark Keeney. *Fundamentals of Dairy Chemistry*. New York: Springer, 1988.

Russell, N. J., and G. W. Gould. *Food Preservatives*. New York: Springer, 2003.

INTERNET: ONLINE PAPERS AND WEBSITES

"Cultures." Madison, WI: *Dairy Connection*. http://www.dairyconnection.com/cultures.jsp (May 30, 2010).

Eknæs, Margrete, Øystein Havrevoll, Harald Volden, and Knut Hove. "Fat Content, Fatty Acid Profile and Off-Flavours in Goats Milk—Effects of Feed Concentrates with Different Fat Sources during the Grazing Season," *Animal Feed Science and Technology* 152, 1–2 (2009): 112–122. http://www.mendeley.com/research/animal-feed-science-technology-fat-content-fatty-acid-profile-offflavours-goats-milk-effects-feed-concentrates-different-fat-sources-during-grazing-season/ (January 30, 2010).

Vines, Dwight T., and Annel K. Greene. "Fundamentals of Milk Composition," *Clemson Extension Dairy Science* DSEL 102 (December 2002): http://www.clemson.edu/psapublishing/PAGES/ADVS/DSEL102.pdf (January 30, 2010).

Sefèíková, M., J. Sefèík, and V. Báles. "Kinetics of Casein Micelle Destabilization," Department of Chemical and Biochemical Engineering, Faculty of Chemical Technology, Slovak University of Technology (May 19, 2000), http://www.chempap.org/papers/546aa345.pdf (January 2010).

Hill, Arthur H. "Dairy Chemistry and Physics," University of Guelph, Ontario, Canada, http://www.foodsci.uoguelph.ca/dairyedu/chem.html (January 2010).

Haenlein, George F. W. "Composition of Goat Milk and Factors Affecting It," *Goat Connection*, (October 28, 2002) http://goatconnection.com/articles/publish/article_70.shtml (January 30, 2010).

Wiking, Lars, Jan Stagsted, Lennart Björck, and Jacob H. Nielsen. "Milk Fat Globule Size Is Affected by Fat Production in Dairy Cows," *International Dairy Journal*, 14, 10 (October 2004): http://www.sciencedirect.com/science/article/pii/S0958694604000780 (January 2010).

Fankhauser, David B. "Rennet Preparation," (December 2008) http://biology.clc.uc.edu/fankhauser/Cheese/Rennet/rennet_preparation.html (January 2010).

Cornell University. "Heat Treatments and Pasteurization," (February 2007): http://milkfacts.info/Milk%20Processing/Heat%20Treatments%20and%20Pasteurization.htm (February 2010).

Zeng, S. S., K. Soryal, B. Fekadu, B. Bah, and T. Popham. "Predictive Formulae for Goat Cheese Yield Based on Milk Composition," *Small Ruminant Research* 69 (2007): 180–186. http://ddr.nal.usda.gov/bitstream/10113/6912/1/IND43895826.pdf (February 2010).

Coker, Christina, Craig Honoré, Keith Johnston, and Lawrie Creamer. "Manufacture and Use of Cheese Products," Food Science Section and Cheese and Milkfat Technology Section: New Zealand Dairy Research Institute, http://nzic.org.nz/ChemProcesses/dairy/3D.pdf (February 2010).

O'Connor, Charles. *Traditional Cheesemaking Manual*. Addis Ababa, Ethiopia: International Livestock Centre for Africa (ILCA), 1993. http://www.fao.org/sd/erp/toolkit/BOOKS/Cheese.pdf (February 2010).

Hill, Arthur. "Process and Quality Control Procedures," Ontario, Canada: University of Guelph, http://www.foodsci.uoguelph.ca/cheese/sectionb.htm (March 2010).

Horstmeyer, Steve. "Relative Humidity . . . Relative to What?" (2008), http://www.shorstmeyer.com/wxfaqs/humidity/humidity.html (March 2010).

Karabıyıklı, Şeniz, and Mehmet Karapınar. "Identification of Lactic Acid Bacteria in the Fermentation of Kopanisti Cheese," *The Association of Food Technology*, 33, 6 (2006): http://www.gidadernegi.org/EN/Genel/BelgeGoster.aspx?17A16AE30572D313AAF6AA849816B2EFE1D77730CC5A4452 (May 2010).

INDEX

ABOUT THE AUTHOR

Gianaclis Caldwell has been teaching all levels of cheesemaking for years, as well as speaking and teaching about the business of farmstead cheese, both at her family's licensed cheese dairy, Pholia Farm, and other venues, including the American Dairy Goat Association annual convention, the American Cheese Society Conference, and the *Mother Earth News* fairs.

Gianaclis's aged, raw-milk cheeses have been recognized and applauded by America's foremost authorities on cheese. Pholia Farm cheeses have been included in many major books on artisan cheese, the latest being Max McCalman's *Mastering Cheese*, in which her Elk Mountain cheese is included in a short list of "rock stars of the 21st century." Her Hillis Peak cheese was the centerfold cheese in the Winter 2010 issue of *culture* magazine. She was one of the spotlighted cheesemakers in *Cheesemaking* by *Hobby Farms* magazine for their Popular Kitchen Series.

She is also the author of *The Farmstead Creamery Advisor* (Chelsea Green, 2010), a thorough guide to building and running a small, on-farm cheese business.

Amelia Caldwell

ABOUT THE FOREWORD AUTHOR

Ricki Carroll, aka the "Cheese Queen," is the author of *Home Cheese Making* (Storey, 2002) and the cofounder and owner of New England Cheesemaking Supply Company.

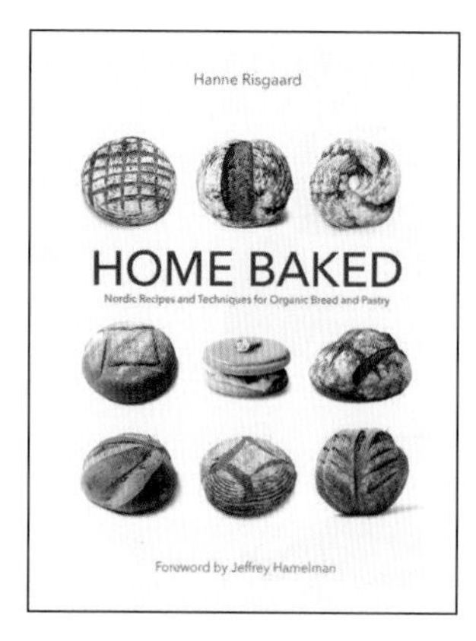

HOME BAKED
Nordic Recipes and Techniques for Organic Bread and Pastry
HANNE RISGAARD
ISBN 9781603584302
Hardcover • $39.95

CHEESEMONGER
A Life on the Wedge
GORDON EDGAR
ISBN 9781603582377
Paperback • $17.95

WILD FLAVORS
One Chef's Transformative Year Cooking from Eva's Farm
DIDI EMMONS
ISBN 9781603582858
Hardcover • $34.95

TASTE, MEMORY
Forgotten Foods, Lost Flavors, and Why They Matter
DAVID BUCHANAN
9781603584401
Paperback • $17.95